Advances in

VIRUS RESEARCH

VOLUME 65

ADVISORY BOARD

Advances in
VIRUS RESEARCH

Edited by

KARL MARAMOROSCH

Department of Entomology
Rutgers University
New Brunswick, New Jersey

AARON J. SHATKIN

Center for Advanced Biotechnology
and Medicine
Piscataway, New Jersey

VOLUME 65

ELSEVIER

AMSTERDAM • BOSTON • HEIDELBERG • LONDON
NEW YORK • OXFORD • PARIS • SAN DIEGO
SAN FRANCISCO • SINGAPORE • SYDNEY • TOKYO
Academic Press is an imprint of Elsevier

Elsevier Academic Press
525 B Street, Suite 1900, San Diego, California 92101-4495, USA
84 Theobald's Road, London WC1X 8RR, UK

This book is printed on acid-free paper. ∞

For all information on all Elsevier Academic Press publications
visit our Web site at www.academicpress.com

ISBN-13: 978-0-12-039867-6
ISBN-10: 0-12-039867-2

PRINTED IN THE UNITED STATES OF AMERICA
05   06   07   08   09   9   8   7   6   5   4   3   2   1

CONTENTS

Kakugo Virus from Brains of Aggressive Worker Honeybees

Tomoko Fujiyuki, Hideaki Takeuchi, Masato Ono, Seii Ohka,
Tetsuhiko Sasaki, Akio Nomoto, and Takeo Kubo

RNA Viruses Redirect Host Factors to Better Amplify Their Genome

Anna M. Boguszewska-Chachulska and Anne-Lise Haenni

Tomato Spotted Wilt Virus Particle Assembly and the Prospects of Fluorescence Microscopy to Study Protein–Protein Interactions Involved

MARJOLEIN SNIPPE, ROB GOLDBACH, AND RICHARD KORMELINK

Influenza Virus Virulence and Its Molecular Determinants

DIANA L. NOAH AND ROBERT M. KRUG

Alteration and Analyses of Viral Entry with Library-Derived Peptides

KEITH BUPP AND MONICA J. ROTH

A Decade of Advances in Iridovirus Research

Trevor Williams, Valérie Barbosa-Solomieu, and V. Gregory Chinchar

ADVANCES IN VIRUS RESEARCH, VOL 65

KAKUGO VIRUS FROM BRAINS OF AGGRESSIVE WORKER HONEYBEES

Tomoko Fujiyuki,* Hideaki Takeuchi,* Masato Ono,†
Seii Ohka,‡ Tetsuhiko Sasaki,§ Akio Nomoto,‡ and Takeo Kubo*

*Department of Biological Sciences, Graduate School of Science
The University of Tokyo, Bunkyo-Ku, Tokyo, 113-0033, Japan
†Center of Excellence Integrative Human Science Program, Research Institute and Graduate
School of Agriculture, Tamagawa University, Machida, Tokyo, 194-8610, Japan
‡Department of Microbiology, Graduate School of Medicine, The University of Tokyo
Bunkyo-Ku, Tokyo, 113-0033, Japan
§Honeybee Science Research Center, Tamagawa University
Machida, Tokyo, 194-8610, Japan

I. INTRODUCTION

Aggressive behavior is observed among various species with diverse outcomes, such as establishing territory, dominance, discipline, obtaining prey avoiding predation, etc. (Wilson, 1975). When animals

Copyright 2005, Elsevier Inc.
All rights reserved.
0065-3527/05 $35.00
DOI: 10.1016/S0065-3527(05)65001-4

encounter other animals, they immediately decide whether to attack the enemy or escape from it. To judge the situation appropriately, the animal must assess information from the environment as well as the movement of the other animal. Therefore, aggression is a very complicated output behavior that occurs after the processing of much information. Aggressive behaviors of social animals, which live in kin groups, are more complex because they sometimes imply altruism. One example is the honeybee's aggressive behavior toward natural enemies. Honeybees are eusocial insects; female adults are differentiated into reproductive females and sterile females. The behaviors of sterile females are ecologically altruistic. Altruistic behavior is defined as behavior that provides advantages for others at the expense of one's own fitness (Wilson, 1975). For example, it is altruistic that workers who cannot bear their own daughters take care of the daughters produced by the queen. Among these labors, aggression is a typical altruistic behavior because the aggressors die as a result of the aggression to protect the colony, as mentioned in the next chapter. Altruistic aggression is also observed in other animal species (Wilson, 1975). Most of the mechanisms underlying these behaviors remain unknown, but their clarification will contribute to a better understanding of animal behaviors and sociality. We have been studying the molecular basis of aggressive behaviors in the honeybee.

The advantages of studying aggressive behaviors in the honeybee are as follows: (1) It is easy to distinguish individuals based on their behaviors, because there is a division of labor in the sterile workers based on their age after eclosion. Therefore, the workers that exhibit aggressive behaviors are expected to have specific brain functions to express the programmed aggressive behaviors. (2) The brain structure of the honeybee has been extensively analyzed compared to other social insects (Ehmer and Gronenberg, 2004) and many experimental methods, such as molecular biology, electrophysiology, neurohistology, etc., can be applied to study the honeybee brain (Giurfa, 2003; Kunieda and Kubo, 2004; Menzel, 2001). (3) Honeybee aggression is often aimed toward human beings, resulting in stinging incidents. The Africanized honeybee in particular, a European honeybee and African honeybee hybrid, is highly aggressive and is called the killer bee because its sting can kill humans (Guzman-Novoa and Page, 1994b). These bees broadly inhabit North and South America, and stinging is a serious public health issue there. Therefore, analysis of the neural and molecular mechanisms underlying honeybee aggression is expected to contribute to public health, agriculture, and apiculture.

For these reasons, we screened the genes preferentially expressed in the brains of aggressive workers. These experiments led to the discovery of a novel insect picorna-like virus, termed Kakugo virus, in the brains of worker honeybees that exhibited aggressive behaviors against hornets (Fujiyuki et al., 2004). It is possible that viral infection in the brain has a role in regulating the aggressive behaviors of the honeybee.

In this article, we describe the findings on Kakugo virus, and discuss the possible relationship between viral infection and aggressive honeybee behavior. We also discuss other animal viruses that are thought to be related to the aggressive behaviors of the host.

II. Sociality of the European Honeybees (*Apis mellifera* L.)

A. Caste Differentiation and the Division of Labor of the Workers

Honeybees are eusocial insects, defined as follows: female, and in some cases also male (e.g., termite), differentiate into reproductive and sterile adults (reproductive and worker castes, respectively), and the workers cooperate in caring for the young. Furthermore, there is a lifespan overlap of at least two generations that maintain the colony (Wilson, 1975). The honeybee is considered to have developed the most complicated insect sociality.

The European honeybee (*Apis mellifera* L.) colony usually contains a few dozen males (drones) from spring to autumn and thousands of females (Winston, 1987). Females fall into one of two castes: queens (reproductive caste) and workers (sterile caste). There is usually only one queen in a colony with reproductive capacity. During their 30–40-day lifespan from spring to autumn, workers are devoted to the labors required to maintain colony viability, rather than their own reproduction. In addition, workers divide labors according to age after eclosion (age-polyethism). Young workers (less than 12 days old) are engaged in feeding the brood and the queen by secreting royal jelly from the hypopharyngeal glands in the head. Middle-aged workers perform guard functions at the hive entrance and attack their natural enemies, such as hornets (Breed et al., 1990, 1992). Older workers forage for nectar and pollen. The queen is engaged in reproduction throughout the course of her life and functions as the "ovary" of the colony. The queen and the workers die when separated from the colony.

Thus, the colony behaves as a single individual and is thus called a "superorganism." This feature is characteristic to eusocial animals like the honeybee and is useful for searching for gene involvement in the behaviors described below.

B. Aggressive Behavior of the Workers as Altruistic Behavior

Worker honeybees (guard bees) attack their enemies using their stingers. In Japan, the most prominent natural enemy to the honeybee is the giant hornet (*Vespa mandarinia japonica*). When workers attack their enemies using their stingers, the stingers are often detached from the worker's abdomen due to its unique hook-shaped structure (Winston, 1987). The stinger is connected to a venom sac, and the sting apparatus and venom sac are connected to the ganglion and the muscle (Dade, 1962). When the workers sting the enemy, the connection between the sting apparatus and the ganglion is broken and the muscle attached to the sting apparatus begins convulsing to inject the stored venom into the enemy through the stinger. Additionally, the venom sac releases alarm pheromones to induce further attack by other workers. The loss of the stinger, however, is a mortal wound for the bee and is suicidal for the workers.

The Asian honeybee (*Apis cerana*), a honeybee subspecies, exhibits a different style of aggression against the giant hornet (Ono *et al.*, 1995). Many workers attack the hornet by surrounding it. The hornet is killed by the high temperature generated by the mass of bees. This defense is not fatal to the attacking Asian honeybees as defense is to the European honeybees. Therefore, we used the European honeybees to analyze self-sacrificing aggressive behaviors.

The aggressive behavior of honeybees must be regulated appropriately in the colony, because if all of the workers are engaged in suicidal defense behavior, the colony can no longer survive. Therefore, the ratio of attacking workers in the colony should be maintained at appropriate levels. An analysis of the molecules involved in the aggressive defense behaviors of the workers can provide clues as to how the aggressive behaviors are regulated in the colony.

C. Analysis of the Molecular Basis Underlying the Social Behavior of the Honeybee

The age-polyethism of the workers is intriguing, and behavioral and physiological alterations that occur along with age-polyethism have been studied extensively. First, the physiological changes accompanying the transition from nurse bee to forager bee have been examined. Nurse bees feed larvae with royal jelly protein secreted from the hypopharyngeal gland. When they become forager bees, the same gland changes to secrete carbohydrate-metabolizing enzymes, such as alpha-glucosidase, glucose oxidase, and amylase, which are required to process nectar into honey (Kubo et al., 1996; Ohashi et al., 1996, 1999, 2000).

The brain structure changes in conjunction with the behavioral changes during age-polyethism (Sigg et al., 1997). There are also task-dependent alterations of gene expression in the brains (Whitfield et al., 2003). Gene expression profiles characteristic to aggressive workers, however, have never been reported.

In contrast, some loci related to the worker aggressive behaviors have been identified using quantitative trait loci analyses. The aggressiveness of the honeybee differs depending on the strain (Guzman-Novoa and Page, 1994a). The genetic elements related to aggressiveness have been analyzed, especially in the Africanized honeybee (also called the killer bee), because of the dangers posed by this behavior to humans (Arechavaleta-Velasco et al., 2003; Hunt et al., 1998, 1999; Lobo et al., 2003). It is unknown, however, whether the genes of the loci are involved in regulating aggressiveness in the brains. The corresponding genes have also not been identified. Therefore, we focused on genes expressed exclusively in the brains of the aggressive workers as candidate genes involved in the regulation of aggressive behaviors.

Mushroom bodies (MB) are considered important regions for the integration of sensory information and learning and memory in insect brains. The expression of genes for proteins involved in calcium signaling is enriched in the MB, suggesting that synaptic plasticity based on calcium signaling is enhanced in the MB of the honeybee brain (Kamikouchi et al., 1998, 2000; Takeuchi et al., 2002). In addition, the expression of genes encoding a novel transcription factor (Mblk-1), a tachykinin-related neurosecretory peptide, and a noncoding nuclear RNA (Ks-1) is upregulated in the MBs of the honeybee (Kunieda et al., 2003; Park et al., 2002, 2003; Sawata et al., 2002; Takeuchi et al., 2001, 2004). These findings suggest that the honeybee has highly advanced brain functions unique to the species. Screening of the genes expressed specifically in the MBs of aggressive workers may lead to an understanding of the molecular basis of aggression.

III. IDENTIFICATION OF A NOVEL INSECT PICORNA-LIKE VIRUS, KAKUGO VIRUS,
FROM THE BRAINS OF AGGRESSIVE WORKER HONEYBEES

In this section, we describe the identification and characterization of
Kakugo virus.

A. Kakugo *RNA, a Novel RNA Identified in the*
Aggressive Worker Brain

To screen the candidate genes expressed uniquely in the brains of
aggressive workers, we first established a system to collect the workers
based on their aggressiveness. For this, we set the colony in a botanical
garden and used the giant hornet, *Vespa mandarinia japonica*, a natural enemy of the honeybee, as a decoy to induce aggression. Some of the
guard workers attacked the giant hornet (attackers; Fig. 1). The rest of
the guard bees merely observed it and did not attack. Some of the
workers inside the hive fled the hornet (escapers). Nonaggressive
guard bees did not flee the hornet. These types of aggressive behaviors
against the enemies can also be regarded as one of the labors divided
among workers, since not all of the bees exhibited aggressive behavior.
Because they show an apparently different response to the same

FIG 1. Attacking behavior of the honeybees. The guard bees engulf a forager giant
hornet, *Vespa mandarinia japonica*. (See Color Insert.)

target, we expected that brain functions might differ between attackers and escapers.

We compared the gene expression pattern in the brains of attackers and escapers using a differential display (DD) method. The DD method is commonly used to compare gene expression patterns between different samples (Liang and Pardee, 1992). For this, mRNAs from the MBs of the attackers and escapers were collected and reverse-transcribed (RT) and polymerase chain reactions (PCR) were performed using fluorescent oligo dT primer and random 10mer primers. The PCR products were compared by sequence gel electrophoresis. By screening approximately 8700 bands using 174 sets of primer pairs, 458 differing bands were detected between the brains of attackers and escapers. The DD method, however, can detect many false-positive bands because the PCR is performed under conditions of low specificity. Therefore, it is also necessary to confirm differences in gene expression levels. Fourteen positive bands were subcloned and analyzed using semi-quantitative RT-PCR with gene-specific primers. As a result, two were confirmed as attacker-specific genes. The ratio of genes confirmed to be differentially expressed among candidates identified by DD was generally around 20–30%. Therefore, our results are consistent with those of previous reports. One of the two genes is named *Kakugo*. In Japanese, the word "Kakugo" means to be prepared to act, even if it means dying. Other candidate bands are now being analyzed using a cDNA microarray (Fujiyuki *et al.*, unpublished; Paul *et al.*, unpublished; Takeuchi *et al.*, 2002).

B. Sequence Analysis of Kakugo RNA and Its Identification as a Viral Genome

In the DD method, short DNA fragments (300–900 bp in length) can be analyzed. The fragment of the *Kakugo* cDNA originally identified by the DD method was 519 bp and seemed to correspond to a part of an open reading frame (ORF). We then performed the cDNA cloning of *Kakugo* RNA. A full-length cDNA of 10152 bp was identified and sequenced using a Rapid Amplification cDNA Ends (RACE) method. It contained one ORF consisting of 2893 amino-acid residues, preceded by a 5' untranslated region (UTR) of 1156 bp. Unexpectedly, the amino-acid sequence encoded by the ORF shared significant homology with polyproteins encoded by the genomic RNA of insect picorna-like viruses (Figs. 2, 3). The putative *Kakugo* protein has at least two virion protein (VP) domains and motifs for helicase, protease, and RNA dependent RNA polymerase in order from the N terminus (Fig. 2).

Kakugo protein	VP	Hel	Pro	RdRp
1 Sacbrood virus		36(%)	29(%)	37(%)
Perina nuda picorna-like virus		24	22	28
Infectious flacherie virus		23	22	26
2 Hepatitis A virus		29	28	28
Foot-and-mouth disease virus		30	18	26
Poliovirus		25	23	24
Encephalomyocarditis virus		23	23	28
3 Acute bee paralysis virus		23	25	29
Rhopalosiphum padi virus		22	23	33
Drosophila C virus		20	23	30
Plautia stali intestine virus		19	21	29

FIG 2. Schematic representation of the polyprotein structure of the Kakugo virus and homology with that of other picorna(-like) viruses. Putative amino-acid sequence of *Kakugo* polyprotein compared with other picorna-like viruses: (1) Iflaviruses, (2) mammalian picornaviruses, (3) Cripaviruses. The amino-acid sequences of the helicase (hel), protease (pro), and RNA dependent RNA polymerase (RdRp) domains were compared. Sequence identities with *Kakugo* protein in each domain are indicated below the corresponding domains.

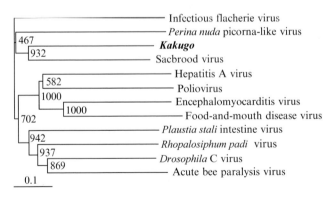

FIG 3. Phylogenetic analysis of genomic RNAs of Kakugo virus and other picorna-like viruses. Trees were constructed with the highly conserved fragment of RdRp amino-acid sequences encompassing motifs 1 to 8 in RdRps of various picorna-like viruses using the neighbor-joining method. The statistical significance of branch order was estimated by performing 1000 replications of bootstrap resampling of the original aligned amino-acid sequences.

The amino-acid residues required for enzymatic activities of helicase, protease, and RNA dependent RNA polymerase are also conserved in *Kakugo* protein. These findings suggest that *Kakugo* RNA has characteristics of the genomic RNA of a novel insect picorna-like virus.

It is possible that *Kakugo* represents a honeybee gene with a similar sequence to that of the viral genome transcribed from the honeybee genome. *Kakugo* was, however, not detectable in the honeybee genome by Southern blot analysis. Furthermore, the *Kakugo* sequence was also not found in the honeybee genome database. These results strongly suggest that *Kakugo* is the genomic RNA of a novel insect picorna-like virus, and that the brains of the aggressive workers are infected by this virus.

C. Infectivity and Virion Construction of Kakugo Virus

If *Kakugo* RNA represents the RNA genome of a novel RNA virus, the virion structure and infectivity of Kakugo virus can be measured. To investigate whether *Kakugo* constitutes a virion, sucrose density gradient centrifugation was performed using the lysate of Kakugo virus infected bees. As a result, a part of the *Kakugo* RNA was fractionated to the same fraction as that of the poliovirus virion, with a sedimentation coefficient of approximately 160S. These results suggest that *Kakugo* constitutes a poliovirus-like virion. This sedimentation coefficient was also similar to that of other insect picorna-like viruses that infect honeybees (Bailey and Ball, 1991).

Next, to examine whether the *Kakugo* RNA constitutes an infectious virus, we injected the lysate prepared from aggressive workers that contained *Kakugo* RNA into the heads of nonaggressive bees. We then quantified *Kakugo* RNA at different time points after inoculation. The amount of *Kakugo* RNA was markedly increased at three days after injection, although there was no significant increase in the amount of *Kakugo* RNA after the injection of the lysate prepared from the heads of noninfected foragers. These results indicate that *Kakugo* RNA is the genomic RNA of an infectious virus, Kakugo virus.

D. Kakugo Virus Is not Detected in the Brains of Nurse Bees or Forager Bees

To examine the relationship between Kakugo virus infection and honeybee aggressive behaviors, it is important to examine what percentage of workers is infected with Kakugo virus. As described previously, the role of the workers shifts from nursing, guarding, and foraging according to age after eclosion. *Kakugo* RNA was detected only in the brains of the attackers, but not in those of the nurse bees and foragers, suggesting that the attackers are specifically infected with the Kakugo virus. We also examined the body parts of the attackers infected with Kakugo virus and found that *Kakugo* RNA was present in the brains. These results clearly indicate that there is a close relationship between Kakugo virus presence in the brain and aggressive behavior.

This finding is very interesting because it is the first to suggest that a viral infection in the brain can be directly related to the aggressive behavior of the host. The question of whether Kakugo virus infection alone induces aggressive behavior in honeybees is a very important topic for future research.

According to the age-related division of labor, attackers become foragers, but Kakugo virus was not detected in the brains of foragers. This appears to be a contradiction.

There are at least two possible explanations:

1. Kakugo virus is always present, albeit sometimes latent, in the honeybee colony and selectively amplifies in the brain of the aggressive guard bees. When the infected and aggressive guard bees become older (foragers), the physiology of the foragers may no longer be suitable for the existence of the Kakugo virus, which, thus, might disappear from the brains of the foragers. The amplified Kakugo virus might heighten the aggressiveness of the guard bees.

2. Alternatively, some workers become infected with Kakugo virus by accident. If this is the case, infected workers will become aggressive like guard bees and will die before they become foragers because of the possible pathogenicity of the Kakugo virus, as well as the increased risk associated with the attacking behavior. These issues could be examined by investigating the relationship between honeybee aging and the ratio of infected bees.

E. Homology with Deformed Wing Virus

We observed that the sequence of *Kakugo* RNA has a very high homology with the genome of deformed wing virus (DWV), which is supposed to be responsible for deformed wing syndrome (Bailey and Ball, 1991). Two groups from Italy and the United States have registered the genome in GenBank. Kakugo virus and DWV are 98% identical in nucleotide and amino-acid sequence. The differences between them are a nucleotide substitution, deletion, insertion, and the length of the 5' UTR, suggesting that they are closely related viruses. The DWV causes wing malformation if it infects the pupae, and increased mortality when it infects adults. Bees with deformed wings, however, were not observed among the attacker bees in our experiments. Thus, it is possible that we detected a distinct characteristic of infection by the same virus or two very closely related viruses.

IV. COMPARISON OF FEATURES OF KAKUGO VIRUS WITH OTHER INSECT PICORNA-LIKE VIRUSES THAT INFECT HONEYBEES

Until recently, a few picorna-like viruses were known to infect honeybees. In this chapter, we briefly describe the characteristics of these viruses, especially sacbrood virus and DWV, and compare their features with those of Kakugo virus.

A. General Features

1. Characteristics and Detailed Classification of Insect Picorna-like Viruses that Infect Honeybees

Seventeen other viruses are known to infect the honeybee (Allen and Ball, 1996). Of these, the complete genome sequence of three viruses has been published: acute bee paralysis virus (ABPV) (Govan *et al.*, 2000), black queen cell virus (BQCV) (Leat *et al.*, 2000), and sacbrood virus (SBV) (Ghosh *et al.*, 1999) (Table I). All three are insect picorna-like viruses. Their virions are icosahedral, approximately 30 nm diameter, and contain an RNA genome. These viruses all have certain pathogenicity to the honeybee. The pathogenicity of Kakugo virus, however, is not known. To date, only a portion of these viruses has been analyzed at the sequence level. Although the DWV genome has been proposed, it was just registered in the database and is not yet published.

The aforementioned insect picorna-like viruses are divided into two subgroups: infectious flacherie-like viruses (Iflaviruses) and cricket paralysis viruses (Cripaviruses). The difference between the two groups is in the genome structure (Fig. 4), although the sequences corresponding to nonstructural proteins are similar to each other. The genome of Iflaviruses is monocistronic. Virion proteins and nonstructural proteins are coded in the N-terminus and C-terminus,

TABLE I

List of the Picorna-like Viruses that Infect Honeybees

Virus	Pathogenicity	Tissue tropism	Infection route	Related behavior
Infectious flacherie-like viruses				
Sacbrood virus	Death of larvae	Adult brain, adult hypopharyngeal gland	Oral	Precocious aging
Kakugo virus	Unknown	Adult brain	Unknown	Aggression?
Deformed wing virus*	Deformed wing	Larvae	Mite	Unknown
Cricket paralysis-like viruses				
Acute bee paralysis virus	Paralysis	Salivary gland	Oral	Unknown
Black queen cell virus	Death of queen	Queen larvae/pupae	Oral	Unknown

* The sequence data is unpublished.

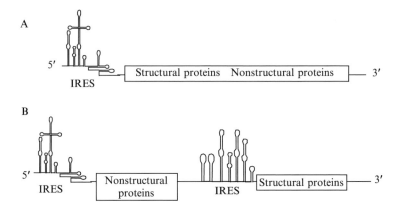

Fig 4. Genome structure of picornaviruses and Iflaviruses (A) and Cripaviruses (B). Note that the genomic RNAs of Iflaviruses are monocistronic with single IRES at the 5' UTR whereas those of Cripaviruses contain two ORFs, each beginning with an IRES. Also, note that the order of the structural and nonstructural proteins encoded by these viral RNAs is opposite.

respectively. In the 5' UTR, there might be an internal ribosome entry site (IRES), similar to mammalian picornaviruses, for the initiation of the polyprotein translation. The 5'UTR is sufficiently long to contain an IRES. On the other hand, the genomes of Cripaviruses are dicistronic. Two coding regions for nonstructural proteins and structural proteins exist in the 5' and 3' regions, respectively, separated by a UTR. They contain two IRES in the 5' UTR and an internal region between the two coding regions (Sasaki and Nakashima, 1999; Wilson *et al.*, 2000). The genomic structure of Kakugo virus is coincident with the Iflaviruses as *Kakugo* is monocistronic. Kakugo virus is also phylogenetically related to Iflaviruses (Fig. 3).

2. Tissue Tropism

The insect picorna-like viruses show tissue tropism. The mechanism by which these viruses selectively infect certain tissues, however, remains unknown. The mechanisms underlying tissue tropism have been analyzed in picornaviruses, especially in the poliovirus (PV). The major determinant of tissue tropism of enteroviruses, including the human poliovirus (hPV), has originally been suggested to be the viral receptor (Holland, 1961). The receptor is a protein with a immunoglobulin-like domain (Evans and Almond, 1998). Transgenic mice carrying the hPV receptor (hPVR) gene can be infected with hPV (Koike *et al.*, 1991), indicating that the virus receptor determines the species specificity. On the other hand, various cell types express hPVR, but only restricted cell types support viral replication in hPVR-transgenic mice (Ren and Racaniello, 1992a). These findings indicate that additional factors contribute to tissue tropism. One such factor is the transport system to the cells expressing the receptor. PV is transported via two possible pathways: through the blood-brain barrier accompanied by viremia and from peripheral motor neurons retrograde to the brain (Gromeier and Wimmer, 1998; Ohka *et al.*, 1998, 2004; Ren and Racaniello, 1992b). Such specific pathways can contribute to the tissue tropism. Recently, the most important determinant for PV tissue tropism has been identified as the mechanism of polyprotein translation from viral genome based on IRES activity. IRES activity is cell-type selective, which can result in tissue tropism (Kawamura *et al.*, 1989; Yanagiya *et al.*, 2003). In the case of Kakugo virus, it might infect and amplify selectively in the brains of the attacker bees. There might be molecular mechanisms that selectively function in attacker brains, such as a certain receptor molecule, or a regulatory system for the IRES activity. Kakugo virus might also have a selective delivery system unique to the brains of attacker bees.

B. Sacbrood Virus

Among the picornaviruses that infect the honeybee, SBV belongs to the group of Iflaviruses. The name of SBV is derived from the effects of the infection at the larval stage, which causes malformation of the brood-like sac and the death of the infected larvae. SBV infection at the adult stage, however, does not cause any morphologic abnormality. SBV is very interesting as its infection is reported to change the behavior of the workers. In adult honeybees, SBV most likely infects the hypopharyngeal gland and the brain (Bailey, 1969; Bailey and Fernando, 1972). The mechanisms by which tissue tropism occurs remain unknown. It is suggested that the longevity of the infected adults is shorter and workers become foragers earlier than normal (Bailey and Fernando, 1972). Also, fewer infected bees return to the colony than healthy bees. This symptom is interesting from an ecological viewpoint. In normal colonies, SBV infection rarely spreads to the whole colony. This might be explained, in part, by the behavioral alterations in the workers infected with SBV, as behavioral changes might contribute to the reduction of the amount of the virus in the colony. In contrast, SBV infection in the hypopharyngeal gland might be advantageous to the virus. As described previously, the hypopharyngeal gland is functionally altered according to the age-polyethism of the workers. The glands of nurse bees secrete royal jelly and milk, which are fed to larvae. On the other hand, the glands of forager bees produce glucose oxidase and glucosidase, which have roles in honey production from flower nectar (Ohashi *et al.*, 1996, 1999). SBV might orally infect the other individuals when the infected worker feeds royal jelly to the larvae (Bailey and Ball, 1991). Thus, the behavioral alterations caused by SBV infection might be advantageous to the honeybees, whereas the tropism to the hypopharyngeal gland might be advantageous to the virus.

It will be interesting to examine the mechanisms that make worker bees change their behavior. SBV infection in the brain, however, has not yet been analyzed in detail.

Another example by which infection by a microbial organism changes the behavior of the honeybees is infection by *Nosema apis*, a protozoan. The *Nosema* infection causes the same symptoms as SBV —after infection, the workers begin foraging earlier (Wang and Moeller, 1970). Their efficiency in nectar and pollen collection decreases (Anderson and Giacon, 1992). Notably, a colony infested with *Nosema apis* is sometimes also infected with SBV (Anderson, 1990). Thus, the behavioral alterations caused by *Nosema apis* infection might be partly a result of SBV infection. There are more cases in which animal behavior is changed by protozoan infection than that of viral infection (Moore, 2002). The mechanisms underlying behavioral changes in animals infected with protozoan also remain unknown.

C. Deformed Wing Virus

DWV was first identified in Japan and has been detected in European honeybees in many countries (Bailey and Ball, 1991). The DWV infection at the pupal stage causes the deformed wing syndrome after eclosion and decreased longevity of adults (Bailey and Ball, 1991). Infection at the adult stage has little impact on longevity (Martin, 2001).

It is suggested that DWV is transmitted to the honeybee by the mite *Varroa destructor*, which infests honeybee colonies (Bowen-Walker *et al.*, 1999). If the colony is infested with the mite, workers have malformed wings (Bailey and Ball, 1991). The correlation between mite infestation and DWV infection is high, raising the possibility that the malformation of the wing is caused by DWV, which infects the Varroa mite.

D. Discussion

The similarities between Kakugo virus, SBV, and DWV are particularly interesting. For example, if the similarities between Kakugo virus and DWV extend to transmission by Varroa mite infestation, the possible relation between Kakugo virus infection and honeybee aggressive behaviors might be further supported by the following observations.

Varroa mite infestation induces workers to start foraging earlier than those in normal colonies (Janmaat and Winston, 2000). This precocious behavioral alteration is positively correlated with the intensity of the aggressiveness of the colony (Giray *et al.*, 2000). These possible links would be consistent with our hypothesis that Kakugo virus infection stimulates aggressive behavior. This awaits confirmation, however, by determining whether the Varroa mite also transmits the Kakugo virus. This hypothesis is also supported by findings on the Africanized bee, which is more aggressive than normal European honeybees. Both Varroa mite infestation and DWV are present in Africanized honeybee colonies (Calderon *et al.*, 2003). Furthermore, the onset of foraging behavior begins earlier in Africanized honeybees than in European honeybees (Winston and Katz, 1982). It is possible that mite infestation, and thus Kakugo virus and/or DWV infection, make the honeybees aggressive.

Other viruses that infect the honeybee, ABPV, SBV, and BQCV, which are all picorna-like viruses, also infect the Varroa mite (Bailey and Ball, 1991). If Kakugo virus is also transmitted to the honeybee by the mite, the next subject of investigation should be to determine whether the attacker bees are infected specifically with Kakugo virus or universally with all of these picorna-like viruses. As Ball points out, many of the viruses that infect the honeybee are usually latent and begin to amplify only under specific conditions (Ball, 1996). If this is the case, infection by some of these picorna-like viruses might also be related to the aggressive behaviors of the honeybees.

It will be important to identify the receptor molecule, infection route, and translation mechanism in order to understand the reason Kakugo virus is detected specifically in the brains of the attacker bees.

The honeybee colony is usually kept clean by the workers. However, various microorganisms, such as viruses, mites, and protozoa, get into the colony and affect the host honeybee. Social behaviors can defend the host from the microorganisms (Loehle, 1995). Interaction with these microorganisms might have contributed to the development of social and aggressive behaviors specific to the honeybee (Schmid-Hempel, 1998). We hope that our analysis of the Kakugo virus can be used as a model to understand why the honeybee has evolved specific aggressive behavior.

V. EXAMPLES OF VIRAL INFECTIONS RELATED TO BEHAVIORAL CHANGES IN HOST ANIMALS

In this section, we describe viral infections related to behavioral changes in the host animal. We also discuss how viral infection affects behavior, and the ecological significance of the altered behavior.

A. *Intrinsic Behavioral Alterations Caused by Viral Infection*

Some animal viruses are related to the host's intrinsic behaviors (Table II). For example, Dengue virus (DENV) is a causative agent of the severe clinical disease dengue fever. It is transmitted to humans from the vector mosquito. Mosquitoes (*Aedes aegypti*) infected with DENV feed for longer periods of time than uninfected mosquitoes (Platt *et al.*, 1997). This behavioral alteration is advantageous to the

TABLE II

VIRUSES THAT CAUSE BEHAVIORAL ALTERATIONS IN THE HOST ANIMALS

Virus	Host	Behavioral alterations	Changes in host brain
Viruses that cause intrinsic behavior changes			
Kakugo virus	Honeybee	Aggression?	Unknown
Dengue virus	Mosquito	Prolonged feeding	Unknown
Nuclear polyhedrosis virrus	Lepidopteran	Climbing	Unknown
Polydnavirus	Parasitoid wasp	Unknown (endosymbiosis)	Unknown
Viruses that cause pathological behavior changes			
Rabies virus	Mammals	Rabies	Apoptosis in hippocampus
Borna disease virus	Mammals	Aggressive behavior, learning deficit, etc.	Apoptosis and increase of cytokine in hippocampus and cerebellum, alteration in monoamine level
Herpes simplex virus	Mammals	Aggressive behavior, etc.	Increase of IL-1 and prostaglandin E2, alteration of cytokine production
Tick-borne encephalitis virus	Mammals	Aggressive behavior, etc.	Increase of testosterone
Simian immunodeficiency virus	Monkey	Hypersensitivity to stimuli	Neurodegeneration synaptic alteration, changes in neuron size

virus because prolonged feeding can enhance DENV transmission. DENV infects the salivary glands and brains of mosquitoes (Linthicum *et al.*, 1996). There is no evidence, however, as to how DENV infection increases the feeding time. The putative receptor molecule for DENV in mosquitoes was identified, and the molecular distribution was correlated with DENV tissue tropism (Mendoza *et al.*, 2002). Analysis of the function and localization of the molecule and the viral infected neurons in the brain will provide some clues as to how DENV infection affects feeding behavior.

Nuclear polyhedrosis virus (NPV) infections in lepidopteran hosts also induce behavioral alterations (Moore, 2002). NPV infected larvae climb and migrate to higher positions on the plant and on to the external parts of the leaves (Goulson, 1997; Vasconcelos *et al.*, 1996). Exposure to rainfall increases the infectivity to the plant, suggesting that the behavioral changes might be advantageous to the virus in that it can disperse to a broader range of the plant and infect other larvae. The altered behavior will also contribute to transmission to predators and scavengers of the lepidopteran hosts (Lee and Fuxa, 2000).

Another example is polydnavirus (PDV) as the endosymbiont for parasitoid wasps. The relationship between polydnavirus and the parasitoid wasp is well-studied as an example of virus-host symbiosis. The wasp injects the PDV into their insect host larvae, which enables the parasitoids to evade or suppress their hosts' immunity (Federici and Bigot, 2003). This relationship is favorable both for the virus and the host parasitoid wasps; it is interesting to note when and how this relation was established. The origin of the relationship is estimated to be approximately 73 million years ago (Whitfield, 2002).

B. Behavioral and Neuronal Abnormalities Caused by the Pathogenic Viruses

How can these viruses affect the behaviors of their host animal? The neuronal and molecular mechanisms of the intrinsic behavioral alteration accompanied by viral infection are poorly understood. There are many findings, however, regarding the alterations in brain function that occur following viral infections that induce abnormal behaviors. Studying the effects of pathogenic viral infection can lead to hints about the mechanisms of intrinsic behavioral changes related to viral infections.

One of the most well-known viruses that alters animal behavior is the rabies virus (RV). It is a causative agent for rabies, which is an acute, progressive, and incurable viral encephalitis (Rupprecht *et al.*, 2002). The virus is most often transmitted via a bite from a rabid animal. The history of RV studies is very old because RV also infects humans and many people die of RV infection (Hemachudha *et al.*, 2002). After the inoculation of RV to the mouse brain, apoptosis is observed in the hippocampus and cerebral cortex (Jackson and Rossiter, 1997). The apoptosis is caused by an RV glycoprotein (Faber *et al.*, 2002). Why the infection causes behavioral alterations in the host animals is a question that remains to be answered, in spite of RV's extensive analysis.

Borna disease virus (BDV) is relatively well-studied in relation to the manner of infection and the resulting behavioral change in the host from molecular, biological, and neurological perspectives. BDV is one of the mononegaviridae (Briese *et al.*, 1994), and the infection causes behavioral abnormalities, such as abnormal play behavior, learning deficits, and so on, in rats (Gonzalez-Dunia *et al.*, 1997; Hornig *et al.*, 2001). BDV-infected rats have an attenuated freezing response to auditory stimuli (Pletnikov *et al.*, 1999). The brains of neonatal rats infected with BDV have increased apoptosis and cytokine mRNA levels in hippocampus and cerebellum (Hornig *et al.*, 1999) (similar symptoms are also observed in RV infection [Baloul and Lafon, 2003]). Neonatal BDV infection produces regional alterations in serotonin and norepinephrine levels, which are selectively upregulated in the hippocampus and cerebellum, respectively (Pletnikov *et al.*, 2002). Serotonin has been genetically implicated in aggressive behavior, though it is yet unclear whether it is the causal factor for aggressive behavior, due to controversial findings (Korte *et al.*, 1996; Saudou *et al.*, 1994). Therefore, there is possibly a common molecular basis underlying both the regulation of intrinsic aggressive behavior and the physiological alteration by viral infection. Some studies in mice indicate that behavioral and neurological abnormalities due to BDV infection originate from one of the BDV proteins (Kamitani *et al.*, 2003). This raises the possibility that one viral protein can affect the regulatory mechanisms of animal behavior and the modification by viruses enhances the behavior.

Herpes simplex virus (HSV) 1 infects the host's brain tissues and causes fever and behavioral changes, such as increased aggressive behavior. The infection is accompanied by enhanced gene expression interleukin-1 and prostaglandin E2 (Ben-Hur *et al.*, 2001). Alterations in cytokine production affect animal behaviors (Zalcman *et al.*, 1998).

Tick-borne encephalitis virus (TBEV) also induces aggression in infected male mice accompanied by an increase of basal testosterone (Moshkin *et al.*, 2002). Therefore, even when a virus does not directly infect neurons, it can induce behavioral alterations mediated by host endocrine physiology, which is eventually reflected by brain function.

Simian immunodeficiency virus (SIV) infection causes monkeys to be abnormally sensitive to auditory and visual stimuli (Prospero-Garcia *et al.*, 1996). This retroviral infection causes neurodegeneration and synaptic alteration (Burudi and Fox, 2001), leading to changes in neuron size (Montgomery *et al.*, 1998). Moreover, synaptic transport of retroviral protein causes neurotoxicity in the brain (Bruce-Keller *et al.*, 2003). These alterations in the neural circuit might contribute to the behavioral changes caused by viral infections.

Similar cases are also observed for some human viruses. One disease symptomized by aggression in humans is schizophrenia (Steinert et al., 1999). Schizophrenically induced aggression is an important clinical problem. Therefore, the relationship between schizophrenia and viral infection has been extensively examined. BDV infection is associated with human neuropsychiatric disorders (Nakamura et al., 2000). It is possible that viral infection is involved in human mental disorders and behavioral alterations. In humans, however, we can only reveal the relationship, but not causality, between viral infection and human behaviors. Appropriate models of virus-induced neuropsychiatric disorders have not yet been developed.

C. Discussion

Although there are various cases where viral infection alters honeybee social behaviors and viral infection affects host aggressiveness, Kakugo virus is unique in that a direct relationship to aggressive behavior of the host animal is suggested. In the examples of other viruses, the behavioral changes caused by the viral infection seem to be rather pathogenic. In contrast, Kakugo virus infection seems well suited to honeybee sociality.

It will be very important to determine whether Kakugo viral infection really affects honeybee aggressive behaviors. It is also important to analyze the route of infection, mechanisms of brain tropism, and the effect of the infection on brain neurons.

What is the importance of the relationship between Kakugo virus and the honeybee? For the virus, one advantage is the increased chance to infect a broad range of new hosts, such as natural enemies against the honeybee, similar to NPV. For the honeybee, one advantage is that they have evolved to commit suicide to repel attackers from their colony, which will be important if Kakugo virus is really as pathogenic to the bee as DWV. One of the costs of sociality is increased transmission of infectious diseases (Hart, 1990). Thus, the evolution of the behaviors of the social animals might have been due to positive selection to protect against pathogens, though we should be cautious when attempting to ascribe an increase in the fitness of the honeybee to behavioral alterations caused by viral infection (Poulin, 1995). It is likely that more viruses than are currently known are involved in the regulatory system of animal behaviors.

In most virus/host relationships, the viral strategy seems to take priority to that of the host animal, and the advantages of the relationship are more profitable for viruses. Viral infection occurs in all living entities, and arms races between viruses and hosts have repeatedly continued over time. Thus, it is not surprising that viruses have evolved to alter host behaviors to benefit themselves. Conversely, when the behavioral changes are also profitable for the host animals, the behaviors are maintained in the host animals.

VI. Conclusions and Future Prospects

It is possible that Kakugo viral infections regulate honeybee aggressive behaviors. If this is true, the study of this new virus might lead to the further development of several new scientific fields. We expect that the analysis of how Kakugo virus can affect the honeybee's aggressive behavior at the molecular and neurologic level will contribute not only to virology but also to behavioral neurology. For example, some of the genes whose expression is induced by BDV infection are common to the genes implicated in aggressive behavior of the host. Similarly, it is possible that some of the genes induced by Kakugo viral infection are involved in intrinsic aggressive behaviors of the honeybee. Therefore, the identification and functional analysis of novel genes induced in the honeybee brains after Kakugo viral infection might clarify the novel mechanisms involved in host aggressive behaviors.

We also expect that this study will contribute to the ecologic and evolutional fields. Behavioral characteristics of the animals, including aggressiveness, sociality, and altruism, differ strikingly from species to species, suggesting that the animals experienced different selective pressures during their evolution. It is likely that some species have been altered and/or modified due to the interactions with viruses and/or microorganisms that infect them. We expect that the relationship between Kakugo virus and the honeybee might be a good model to address these questions.

Acknowledgments

Many thanks to Dr. Hans J. Gross (Würzburg University) for giving us the opportunity to write this review. This work was supported by Grants-in-Aid from the Bio-oriented Technology Research Advancement Institution (BRAIN), the Ministry of Education, Science, Sports, and Culture of Japan, the Terumo Life Science Foundation, and The Naito Foundation. T.F. is the recipient of a Grant-in-Aid for JSPS Fellows.

REFERENCES

Allen, M., and Ball, B. (1996). The incidence and world distribution of honeybee viruses. *Bee World* **77**:141–162.

Anderson, D. L. (1990). Pests and pathogens of the honeybee (*Apis mellifera* L) in Fiji. *Journal of Apicultural Research* **29**:53–59.

Anderson, D. L., and Giacon, H. (1992). Reduced pollen collection by honeybee (Hymenoptera: *Apidae*) colonies infected with *Nosema apis* and sacbrood virus. *J. Econ. Entomol.* **85**:47–51.

Arechavaleta-Velasco, M. E., Hunt, G. J., and Emore, C. (2003). Quantitative trait loci that influence the expression of guarding and stinging behaviors of individual honeybees. *Behav. Genet.* **33**:357–364.

Bailey, L. (1969). The multiplication and spread of sacbrood virus of bees. *Ann. Appl. Biol.* **63**:483–491.

Bailey, L., and Ball, B. V. (1991). "Honeybee pathology." Academic Press Inc., San Diego, CA.

Bailey, L., and Fernando, E. F. W. (1972). Effects of sacbrood virus on adult honey-bees. *Ann. Appl. Biol.* **72**:27–35.

Ball, B. (1996). Honeybee viruses: A cause for concern? *Bee World* **77**:117–119.

Baloul, L., and Lafon, M. (2003). Apoptosis and rabies virus neuroinvasion. *Biochimie* **85**:777–788.

Ben-Hur, T., Cialic, R., Itzik, A., Barak, O., Yirmiya, R., and Weidenfeld, J. (2001). A novel permissive role for glucocorticoids in induction of febrile and behavioral signs of experimental herpes simplex virus encephalitis. *Neuroscience* **108**:119–127.

Bowen-Walker, P. L., Martin, S. J., and Gunn, A. (1999). The transmission of deformed wing virus between honeybees (*Apis mellifera* L.) by the ectoparasitic mite *Varroa jacobsoni* Oud. *J. Invertebr. Pathol.* **73**:101–106.

Breed, M. D., Robinson, G. E., and Page, J. R. E. (1990). Division of labor during honeybee colony defense. *Behav. Ecol. Sociobiol.* **27**:395–401.

Breed, M. D., Smith, T. A., and Torres, A. (1992). Role of guard honeybees (Hymenoptera: *Apidae*) in nestmate discrimination and replacement of removed guards. *Ann. Entomol. Soc.* **85**(5):633–637.

Briese, T., Schneemann, A., Lewis, A. J., Park, Y. S., Kim, S., Ludwig, H., and Lipkin, W. I. (1994). Genomic organization of Borna disease virus. *Proc. Natl. Acad. Sci. USA* **91**:4362–4366.

Bruce-Keller, A. J., Chauhan, A., Dimayuga, F. O., Gee, J., Keller, J. N., and Nath, A. (2003). Synaptic transport of human immunodeficiency virus- Tat protein causes neurotoxicity and gliosis in rat brain. *J. Neurosci.* **23**:8417–8422.

Burudi, E. M. E., and Fox, H. S. (2001). Simian immunodeficiency virus model of HIV-induced central nervous system dysfunction. *Adv. Virus Res.* **56**:435–468.

Calderon, R. A., van Veen, J., Arce, H. G., and Esquivel, M. E. (2003). Presence of deformed wing virus and Kashmir bee virus in Africanized honeybee colonies in Costa Rica infested with *Varroa destructor. Bee World* **84**:112–116.

Dade, H. A. (1962). "Anatomy and dissection of the honeybee." International Bee Research Association, Cardiff, UK.

Ehmer, B., and Gronenberg, W. (2004). Mushroom body volumes and visual interneurons in ants: Comparison between sexes and castes. *J. Comp. Neurol.* **469**:198–213.

Evans, D. J., and Almond, J. W. (1998). Cell receptors for picornaviruses as determinants of cell tropism and pathogenesis. *Trends Microbiol.* **6**:198–202.

Faber, M., Pulmanausahakul, R., Hodawadekar, S. S., Spitsin, S., McGettigan, J. P., Schnell, M. J., and Dietzschold, B. (2002). Overexpression of the rabies virus glycoprotein results in enhancement of apoptosis and antiviral immune response. *J. Virol.* **76:**3374–3381.

Federici, B. A., and Bigot, Y. (2003). Origin and evolution of polydnaviruses by symbiogenesis of insect DNA viruses in endoparasitic wasps. *J. Insect Physiol.* **49:**419–432.

Fujiyuki, T., Takeuchi, H., Ono, M., Ohka, S., Sasaki, T., Nomoto, A., and Kubo, T. (2004). Novel insect picorna-like virus identified in the brains of aggressive worker honeybees. *J. Virol.* **78**(3)**:**1093–1100.

Ghosh, R. C., Ball, B. V., Willcocks, M. M., and Carter, M. J. (1999). The nucleotide sequence of sacbrood virus of the honeybee: An insect picorna-like virus. *J. Gen. Virol.* **80**(Pt 6)**:**1541–1549.

Giray, T., Guzman-Novoa, E., Aron, C. W., Zelinsky, B., Fahrbach, S. E., and Robinson, G. E. (2000). Genetic variation in worker temporal polyethism and colony defensiveness in the hone bee, *Apis mellifera*. *Behavioral Ecology* **11:**44–55.

Giurfa, M. (2003). Cognitive neuroethology: Dissecting nonelemental learning in a honeybee brain. *Curr. Opin. Neurobiol.* **13:**726–735.

Gonzalez-Dunia, D., Sauder, C., and De La Torre, J. C. (1997). Borna disease virus and the brain. *Brain Res. Bull.* **44:**647–664.

Goulson, D. (1997). *Wipfelkrankheit*: Modification of host behaviorr during baculoviral infection. *Oecologia* **109:**219–228.

Govan, V. A., Leat, N., Allsopp, M., and Davison, S. (2000). Analysis of the complete genome sequence of acute bee paralysis virus shows that it belongs to the novel group of insect-infecting RNA viruses. *Virology* **277**(2)**:**457–463.

Gromeier, M., and Wimmer, E. (1998). Mechanism of injury-provoked poliomyelitis. *J. Virol.* **72:**5056–5060.

Guzman-Novoa, E., and Page, J. R. E. (1994a). Genetic dominance and worker interactions affect honeybee colony defense. *Behav. Ecol.* **5**(1)**:**91–97.

Guzman-Novoa, E., and Page, R. E. (1994b). The impact of Africanized bees on Mexican beekeeping. *American Bee Journal* **134:**101–106.

Hart, B. L. (1990). Behavioral adaptations to pathogens and parasites: Five strategies. *Neurosci. Biobehav. Rev.* **14:**273–294.

Hemachudha, T., Laothamatas, J., and Rupprecht, C. E. (2002). Human rabies: A disease of complex neuropathogenetic mechanisms and diagnostic challenges. *Lancet Neurology* **1:**101–109.

Holland, J. J. (1961). Receptor affinities as major determinants of enterovirus tissue tropisms in humans. *Virology* **15:**312–326.

Hornig, M., Briese, T., and Lipkin, W. I. (2001). Bornavirus tropism and targeted pathogenesis: Virus-host interactions in a neurodevelopmental model. *Adv. Virus Res.* **56:**557–582.

Hornig, M., Weissenbock, H., Horscroft, N., and Lipkin, W. I. (1999). An infection-based model of neurodevelopmental damage. *Proc. Natl. Acad. Sci. USA* **96:**12102–12107.

Hunt, G. J., Collins, A. M., Rivera, R., Page, R. E., Jr., and Guzman-Novoa, E. (1999). Quantitative trait loci influencing honeybee alarm pheromone levels. *J. Hered.* **90**(5)**:**585–589.

Hunt, G. J., Guzman-Novoa, E., Fondrk, M. K., and Page, J. R. E. (1998). Quantitative trait loci for honeybee stinging behavior and body size. *Genetics* **148:**1203–1213.

Jackson, A. C., and Rossiter, J. P. (1997). Apoptosis plays an important role in experimental rabies virus infection. *J. Virol.* **71:**5603–5607.

Janmaat, A. F., and Winston, M. L. (2000). The influence of pollen storage area and *Varroa jacobsoni* Oudemans parasitism on temporal caste structure in honeybees (*Apis mellifera* L.). *Insectes Sociaux* **47**:177–182.

Kamikouchi, A., Takeuchi, H., Sawata, M., Natori, S., and Kubo, T. (2000). Concentrated expression of Ca2+/calmodulin-dependent protein kinase II and protein kinase C in the mushroom bodies of the brain of the honeybee *Apis mellifera* L. *J. Comp. Neurol.* **417**(4):501–510.

Kamikouchi, A., Takeuchi, H., Sawata, M., Ohashi, K., Natori, S., and Kubo, T. (1998). Preferential expression of the gene for a putative inositol 1,4,5- trisphosphate receptor homologue in the mushroom bodies of the brain of the worker honeybee *Apis mellifera* L. *Biochem. Biophys. Res. Commun.* **242**(1):181–186.

Kamitani, W., Ono, E., Yoshino, S., Kobayashi, T., Taharaguchi, S., Lee, B. J., Yamashita, M., Kobayashi, T., Okamoto, M., Taniyama, H., Tomonaga, K., and Ikuta, K. (2003). Glial expression of Borna disease virus phosphoprotein induces behavioral and neuro- logical abnormalities in transgenic mice. *Proc. Natl. Acad. Sci. USA* **100**:8969–8974.

Kawamura, N., Kohara, M., Abe, S., Komatsu, T., Tago, K., Arita, M., and Nomoto, A. (1989). Determinants in the 5' noncoding region of poliovirus Sabin 1 RNA that influence the attenuation phenotype. *J. Virol.* **63**:1302–1309.

Koike, S., Taya, C., Kurata, T., Abe, S., Ise, I., Yonekawa, H., and Nomoto, A. (1991). Transgenic mice susceptible to poliovirus. *Proc. Natl. Acad. Sci. USA* **88**:951–955.

Korte, S. M., Meijer, O. C., De Kloet, E. R., Buwalda, B., Keijser, J., Sluyter, F., Van Oortmerssen, G., and Bohus, B. (1996). Enhanced 5-HT$_{1A}$ receptor expression in forebrain regions of aggressive house mice. *Brain Res.* **736**:338–343.

Kubo, T., Sasaki, M., Nakamura, J., Sasagawa, H., Ohashi, K., Takeuchi, H., and Natori, S. (1996). Change in the expression of hypopharyngeal-gland proteins of the worker honeybees (*Apis mellifera* L.) with age and/or role. *J. Biochem. (Tokyo)* **119**(2):291–295.

Kunieda, T., and Kubo, T. (2004). *In vivo* gene transfer into the adult honeybee brain by using electropotation. *Biochem. Biophys. Res. Commun.* **318**:25–31.

Kunieda, T., Park, J. M., Takeuchi, H., and Kubo, T. (2003). Identification and character- ization of Mlr1,2: Two mouse homologues of Mblk-1, a transcription factor from the honeybee brain(1). *FEBS Lett.* **535**:61–65.

Leat, N., Ball, B., Govan, V., and Davidson, S. (2000). Analysis of the complete genome sequence of black queen-cell virus, a picorna-like virus of honeybees. *J. Gen. Virol.* **81**:2111–2119.

Lee, Y., and Fuxa, J. R. (2000). Transport of wild-type and recombinant nucleopolyhe- droviruses by scavenging and predatory arthropods. *Microb. Ecol.* **39**:301–313.

Liang, P., and Pardee, A. B. (1992). Differential display of eukaryotic messenger RNA by means of the polymerase chain reaction. *Science* **257**(5072):967–971.

Linthicum, K. J., Platt, K., Myint, K. S., Lerdthusnee, K., Innis, B. L., and Vaughn, D. W. (1996). Dengue 3 virus distribution in th mosquito *Aedes aegypti*: An immunocyto- chemical study. *Med. Vet. Entomol.* **10**:87–92.

Lobo, N. F., Ton, L. Q., Hill, C. A., Emore, C., Romero-Severson, J., Hunt, G. J., and Collinf, F. H. (2003). Genomic analysis in the *sting-2* quantitative trait locus for defensive behavior in the honeybee, *Apis mellifera. Genome Res.* **13**:2588–2593.

Loehle, C. (1995). Social barriers to pathogen transmission in wild animal populations. *Ecology* **76**:326–335.

Martin, S. J. (2001). The role of Varroa and viral pathogens in the collapse of honeybee colonies: A modelling approach. *Journal of Applied Ecology* **38**:1082–1093.

Mendoza, M. Y., Salas-Benito, J. S., Lanz-Mendoza, H., Hernandez-Martinez, S., and Angel, R. M. D. (2002). A putative receptor for dengue virus in mosquito tissues:

Localization of a 45-KDa glycoprotein. *American Journal of Tropical Medecine and Hygiene* **67**:76–84.

Menzel, R. (2001). Searching for the memory trace in a mini-brain, the honeybee. *Learn. Mem.* **8**:53–62.

Montgomery, M. M., Wood, A., Stott, E. J., Sharp, C., and Luthert, P. J. (1998). Changes in neuron size in cynomolgus macaques infected with various immunodeficiency viruses and poliovirus. *Neuropathol. Appl. Neurobiol.* **24**:468–475.

Moore, J. (2002). "Parasites an the behavior of animals." Oxford University press, New York, NY.

Moshkin, M., Gerlinskaya, L., Morozova, O., Bakhvalova, V., and Evsikov, V. (2002). Behavior, chemosignals and endocrine functions in male mice infected with tick-borne encephalitis virus. *Psychoneuroendocrinology* **27**:603–608.

Nakamura, Y., Takahashi, H., Shoya, Y., Nakaya, T., Watanabe, M., Tomonaga, K., Iwahashi, K., Ameno, K., Momiyama, N., Taniyama, H., Sata, T., Kurata, T., Torre, J. C., and Ikuta, K. (2000). Isolation of Borna disease virus from human brain tissue. *J. Virol.* **74**:4601–4611.

Ohashi, K., Natori, S., and Kubo, T. (1999). Expression of amylase and glucose oxidase in the hypopharyngeal gland with an age-dependent role change of the worker honeybee (*Apis mellifera* L.). *Eur. J. Biochem.* **265**:127–133.

Ohashi, K., Sasaki, M., Sasagawa, H., Nakamura, J., Natori, S., and Kubo, T. (2000). Functional flexibility of the honeybee hypopharyngeal gland in a dequeened colony. *Zoolog. Sci.* **17**:1089–1094.

Ohashi, K., Sawata, M., Takeuchi, H., Natori, S., and Kubo, T. (1996). Molecular cloning of cDNA and analysis of expression of the gene for α-glucosidase from the Hypopharyngeal gland of the Honeybee *Apis mellifera* L. *Biochem. Biophys. Res. Commun.* **221**:380–385.

Ohka, S., Matsuda, N., Tohyama, K., Oda, T., Morikawa, M., Kuge, S., and Nomoto, A. (2004). Receptor (CD155) - dependent endocytosis of poliovirus and retrograde axonal transport of the endosome. *J. Virol.* **78**:7186–7198.

Ohka, S., Yang, W. X., Terada, E., Iwasaki, K., and Nomoto, A. (1998). Retrograde transport of intact poliovirus through th axon via th fast transport system. *Virology* **250**:67–75.

Ono, M., Igarashi, T., Ohno, E., and Sasaki, M. (1995). Unusual thermal defense by a honeybee against mass attack by hornets. *Nature* **377**(6547):334–336.

Park, J. M., Kunieda, T., and Kubo, T. (2003). The Activity of Mblk-1, a Mushroom Body-selective Transcription Factor from the Honeybee, Is Modulated by the Ras/MAPK Pathway. *The Journal of Biological Chemistry* **270**:18689–18694.

Park, J. M., Kunieda, T., Takeuchi, H., and Kubo, T. (2002). DNA-binding properties of Mblk-1, a putative transcription factor from the honeybee. *Biochem. Biophys. Res. Commun.* **291**:23–28.

Platt, K. B., Linthicum, K. J., Myint, K. S. A., Innis, B. L., Lerathusnee, K., and Vaughn, D. W. (1997). Impact of dengue virus infection on feeding behavior of *Aedes Aegypti*. *American Journal of Tropical Medicine and Hygiene* **57**(2):119–125.

Pletnikov, M. V., Rubin, S. A., Schwartz, G. J., Moran, T. H., Sobotka, T. J., and Carbone, K. M. (1999). Persistent neonatal borna disease virus (BDV) infection of the brain causes chronic emotional abnormalities in adult rats. *Physiol. Behav.* **66**:823–831.

Pletnikov, M. V., Rubin, S. A., Vogel, M. W., Moran, T. H., and Carbone, K. M. (2002). Effects of genetic background on neonatal Borna disease virus infection-induced neurodevelopmental damage II. Neurochemical alterations and responses to pharmacological treatments. *Brain Res.* **944**:108–123.

Poulin, R. (1995). "Adaptive" changes in the behavior of parasitized animals: A critical review. *Int. J. Parasitol.* **25**:1371–1383.

Prospero-Garcia, O., Gold, L. H., Fox, H. S., Koob, G. F., Bloom, F. E., and Henriksen, S. J. (1996). Microglia-passaged simian immunodeficiency virus induces neurophysiological abnormalities in monkeys. *Proc. Natl. Acad. Sci. USA* **93**:14158–14163.

Ren, R., and Racaniello, V. R. (1992a). Human poliovirus receptor gene expression and poliovirus tissue tropism in transgenic mice. *J. Virol.* **66**:296–304.

Ren, R., and Racaniello, V. R. (1992b). Poliovirus spreads from muscle to the central nervous system by neural pathways. *J. Infect. Dis.* **166**:747–752.

Rupprecht, C. E., Hanlon, C. A., and Hemachudha, T. (2002). Rabies re-examined. *Lancet infectious disease* **2**:327–343.

Sasaki, J., and Nakashima, N. (1999). Translation initiation at the CUU codon is mediated by the internal ribosomal entry site of an insect picorna-like virus *in vitro*. *J. Virol.* **73**:1219–1226.

Saudou, F., Amara, D. A., Dierich, A., Le Meur, M., Ramboz, S., Segu, L., Buhot, M. C., and Hen, R. (1994). Enhanced aggressive behavior in mice lacking 5-HT1B receptor. *Science* **265**:1875–1878.

Sawata, M., Yoshino, D., Takeuchi, H., Kamikouchi, A. K. O., and Kubo, T. (2002). Identification and punctate nuclear localization of a novel noncoding RNA, Ks-1, from the honeybee brain. *RNA* **8**:772–785.

Schmid-Hempel, P. (1998). "Parasites in social insects." Princeton University Press, Chichester, West Sussex.

Sigg, D., Thompson, C. M., and Mercer, A. R. (1997). Activity-dependent changes to the brain and behavior of the honeybee, *Apis mellifera* (L.). *J. Neurosci.* **17**:7148–7156.

Steinert, T., Wiebe, C., and Gebhardt, R. P. (1999). Aggressive behavior against self and others among first-admission patients with schizophrenia. *Psychiatr. Serv.* **50**:85–90.

Takeuchi, H., Fujiyuki, T., Shirai, K., Matsuo, Y., Kamikouchi, A., Fujinawa, Y., Kato, A., Tsujimoto, A., and Kubo, T. (2002). Identification of genes expressed preferentially in the honeybee mushroom bodies by combination of differential display and cDNA microarray. *FEBS Lett.* **513**(2–3):230–234.

Takeuchi, H., Kage, E., Sawata, M., Kamikouchi, A., Ohashi, K., Ohara, M., Fujiyuki, T., Kunieda, T., Sekimizu, K., Natori, S., and Kubo, T. (2001). Identification of a novel gene, *Mblk-1*, that encodes a putative transcription factor expressed preferentially in the large-type Kenyon cells of the honeybee brain. *Insect Mol. Biol.* **10**(5):487–494.

Takeuchi, H., Yasuda, A., Yasuda-Kamatani, Y., Sawata, M., Matsuo, Y., Kato, A., Tsujimoto, A., Nakajima, T., and Kubo, T. (2004). Prepro-tachykinin gene expression in the brain of the honeybee *Apis mellifera*. *Cell Tissue Res.* **316**:281–293.

Vasconcelos, S. D., Cory, J. S. R., W. K., Sait, S. M., and Haials, R. S. (1996). Modified behavior in baculovirus-infected lepidopteran larvae and its impact on the spatial distribution of inoculum. *Biological control* **7**:299–306.

Wang, D. I., and Moeller, F. E. (1970). The division o labor an queen attendance behavior of Nosema-infected worker honeybees. *J. Econ. Entomol.* **63**:1539–1541.

Whitfield, C. W., Cziko, A. M., and Robinson, G. E. (2003). Gene expression profiles in the brain predict behavior in individual honeybees. *Science* **302**:296–299.

Whitfield, J. B. (2002). Estimating the age of the polydnavirus/braconid wasp symbiosis. *Proc. Natl. Acad. Sci. USA* **99**:7508–7513.

Wilson, E. O. (1975). "Sociobiology: The new synthesis." The Belknap Press of Harvard University Press, Cambridge, MA.

Wilson, J. E., Powell, M. J., Hoover, S. E., and Sarnow, P. (2000). Naturally occurring dicistronic cricket paralysis virus RNA is regulated by two internal ribosome entry sites. *Mol. Cell. Biol.* **20:**4990–4999.

Winston, M. L. (1987). "The biology of the honeybee." Harvard University Press, Cambridge, MA.

Winston, M. L., and Katz, S. J. (1982). Foraging differences between cross-fostered honeybee workers (*Apis mellifera*) of European and Africanized races. *Behavioral Ecology and sociobiology* **10:**125–159.

Yanagiya, A., Ohka, S., Hashida, N., Okamura, M., Taya, C., Kamoshita, N., Iwasaki, K., Sasaki, Y., Yonekawa, H., and Nomoto, A. (2003). Tissue-specific replicating capacity of a chimeric poliovirus that carries the internal ribosomal entry site of hepatitis C virus in a new mouse model transgenic for the human poliovirus receptor. *J. Virol.* **77:**10479–10487.

Zalcman, S., Murray, L., Dyck, D. G., Greenberg, A. H., and Nance, D. (1998). Interleukin-2 and -6 induce behavioral-activating effects in mice. *Brain Res.* **811:**111–121.

RNA VIRUSES REDIRECT HOST FACTORS TO BETTER AMPLIFY THEIR GENOME

Anna M. Boguszewska-Chachulska* and
Anne-Lise Haenni*,†,‡

*Institute of Biochemistry and Biophysics, Polish Academy of Sciences
ul. Pawinskiego 5a, 02–106 Warsaw, Poland
†Institut Jacques Monod, CNRS, Universités Paris VI et VII, 2 Place Jussieu-Tour
43,75251 Paris Cedex 05, France
‡Laboratorio de Inmunovirologia, Faculty of Medicine, University of Antioquia
A.A. 1226, Medellin, Colombia

I. Introduction

As obligatory intracellular parasites, viruses depend on their host for all steps of their reproduction, starting from uptake by the cell to the final encapsidation of progeny virions and exit from the cell, as well as their spread within the host. Thus, host factors can be expected to participate at most—if not all—levels of virus reproduction.

The present review aims at providing an updated view of the host factors presently believed to participate in replication/transcription of RNA viruses. However, it does not deal with viruses that resort to reverse transcription, nor does it discuss the many protein-modifying enzymes such as phosphorylating enzymes whose indirect participation in viral genome amplification is undeniable. The reader may wish to turn to various other reviews dealing with certain aspects of genome

29

DOI: 10.1016/S0065-3527(05)65002-6

amplification of RNA viruses for complementary information (Ahlquist *et al.*, 2003; Buck, 1996; De and Banerjee, 1997; Lai, 1998).

Determining what host factors are required for viral protein synthesis has been rather straightforward since viruses rely entirely on the host protein synthesizing machinery to produce their own proteins (Ehrenfeld, 1996), although in certain cases they recruit additional cell proteins not normally involved in translation regulation. On the other hand, searching for host factors involved in transcription and replication of RNA viruses has proven to be much trickier. Originally, the prevailing seemingly simple view of virus genome replication was modeled on the replication of RNA phages such as $Q\beta$ (Klovins and van Duin, 1999; Schuppli *et al.*, 2000). Yet it soon became apparent that the situation is far more complex for eukaryotic RNA viruses than for RNA phages.

Multiplication of the genome of most RNA viruses takes place in the cytoplasm. However, influenza virus (*Orthomyxoviridae*) (Wang *et al.*, 1997) and Borna disease virus (*Paramyxoviridae*) (Pyper *et al.*, 1998) replicate in the nucleus. An interesting consequence of this nuclear localization is that transcription in these viruses can be accompanied by splicing, a process requiring the host's splicing machinery (Lamb and Horvath, 1991; Tomonaga *et al.*, 2002). In addition, an increasing number of RNA viruses require the nucleus in certain steps of their life cycle (Hiscox, 2003). Arenaviruses cannot complete their replication cycle in cells enucleated soon after infection (Borden *et al.*, 1998). On the other hand, certain (+) as well as (−) strand RNA viruses grow in enucleate cells, virus yield being related to the length of the virus growth cycle (Follett *et al.*, 1975). Several viruses target some of their proteins to the nucleus (Hiscox, 2003; Urcuqui-Inchima *et al.*, 2001), whereas others such as hepatitis C virus (HCV; *Flaviviridae*) code for proteins that contain nuclear localization signals but do not seem to enter the nucleus (Song *et al.*, 2000). Finally, infection by members of the *Picornaviridae*, such as poliovirus and coxsackievirus, and by vesicular stomatitis virus (VSV; *Rhabdoviridae*) (Belov *et al.*, 2000; Gustin and Sarnow, 2001) triggers relocation of certain nuclear proteins to the cytoplasm, suggesting that these proteins probably participate in the life cycle of the virus (Hiscox, 2003).

For viruses that produce subgenomic (sg) RNAs, a distinction must be made between two mechanisms, replication of the entire genome and transcription, which usually entails amplification of only certain regions of the genome to yield the sg mRNAs. Since sgRNAs are generally 3' coterminal, an internally located open reading frame (ORF) in the genomic (g) RNA can become 5' proximal in the sgRNA,

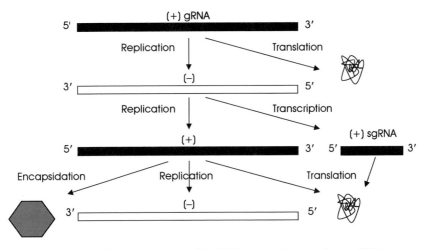

FIG 1. Replication/transcription of (+) RNA viruses that produce sgRNAs.

a more favorable situation for the eukaryotic translation apparatus. To date, most of the information available concerning the role of host factors in viral RNA amplification deals with replication, while information concerning host factors and transcription is less abundant except for (−) strand RNA viruses and for viruses of the *Coronaviridae, Arteriviridae,* and *Closteroviridae* families (Miller and Koev, 2000).

All RNA viruses contain within their genome the information for the synthesis of an RNA-dependent RNA polymerase that shall be referred to as the RdRp or polymerase. The viral polymerase and other viral proteins as well as host factors participate in RNA transcription and replication. Yet, how the switch occurs between transcription and replication remains largely enigmatic. During replication, the RNA strand complementary to the genome strand is produced in far lower amounts than the newly synthesized genome strands. How this imbalance is regulated and what factors may be involved in this bias is also far from clear. Nevertheless, some results have appeared suggesting that certain host factors participate in synthesis of one but not of the other strand.

The sequence of events that leads to virus amplification depends on whether the viral RNA entering the cell is of (+) or of (−) polarity.

The incoming genome of many (+) strand RNA viruses (Fig. 1) serves as mRNA for virus protein synthesis including the RdRp, and is then replicated yielding the complementary (−) RNA strand. In many virus groups, the (−) strand serves as a template to transcribe the sgRNAs that will serve as mRNAs (Miller and Koev, 2000). The (−) strand is also replicated, producing nascent full-length (+) strands; these are used as mRNA for protein synthesis, as a template for further (−) strand synthesis, and ultimately as a genome to be encapsidated, yielding progeny virus particles.

Exceptions to this general scheme exist. Red clover necrotic mosaic tombusvirus (Sit *et al.*, 1998) and citrus tristeza closterovirus (Gowda *et al.*, 2001) generate (−) strand sgRNAs that might function as templates for sg mRNA synthesis. Production of sgRNAs in viruses of the *Coronaviridae* and *Arteriviridae* families (both of the order *Nidovirales*) involves a mechanism similar to splicing known as discontinuous sgRNA synthesis: the transcripts are both 3′ and 5′ coterminal (Fig. 2). In this strategy, synthesis of a sgRNA proceeds on a template. The initiated sgRNA, together with the polymerase, then switches from its position to another complementary region called transcription-regulating sequence (TRS) on the template to which it can base-pair and resume synthesis. The step at which the nascent RNA chain is transferred from one region of the template to another remains controversial—it could occur during (+) or (−) strand synthesis. The recent demonstration of the presence of (−) sg strands favors the latter possibility (Pasternak *et al.*, 2001; Sawicki *et al.*, 2001). TRSs present in the 3′ end of the leader and at the 5′ end of the sgRNA body regions in the gRNA could allow the nascent RNA chain to be transferred to the leader TRS by TRS-TRS (sense-antisense) base-pairing, or conversely could allow the RNA chain in the leader to be transferred to the sgRNA.

The scheme for (−) strand RNA viruses is different from that of (+) strand RNA viruses because the incoming RNA cannot serve as a template for protein synthesis. Consequently, all viruses with a (−) strand RNA genome encapsidate their polymerase complex, including host factors. The genome tightly bound to the nucleocapsid protein as a ribouncleoprotein (RNP) with its associated polymerase (De and Banerjee, 1997) is either replicated into complementary (+) strands, or transcribed to produce sg mRNAs. The mechanism of transcription depends on whether the virus contains a nonsegmented or a segmented genome.

In nonsegmented (i.e., monopartite) (−) RNA viruses (order: *Mononegavirales*), such as the *Paramyxoviridae* and *Rhabdoviridae*, the 3′ and the 5′ termini of the (−) sense gRNA are known as the 3′ leader promoter and 5′ trailer region, respectively (Fig. 3). The 3′ region of the complementary (+) sense RNA or antigenome, known as the trailer, directs synthesis of the gRNA. In transcription, synthesis of

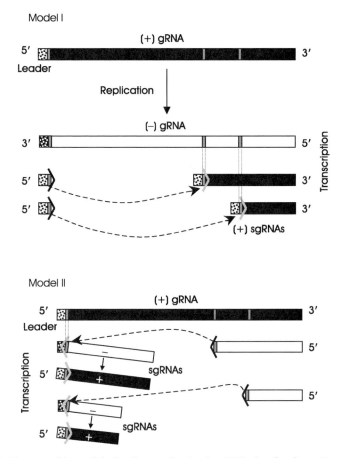

Fig 2. Two possible models for the synthesis of sgRNAs by the discontinuous transcription strategy used by the *Nidovirales* (*Coronaviridae* and *Arteriviridae*). Model I, during (+) strand synthesis (leader-primed transcription); Model II, during (−) strand synthesis (recombination during (−) RNA strand synthesis). Thin vertical grey bars, transcription-regulating sequences (TRSs) which determine the stop and start points during discontinuous transcription. The Leader is composed of the 5′-positioned nucleotides (stippled box) and the TRS in the (+) RNA strand. (Adapted with permission from Pasternak *et al.*, 2001.)

complementary (+) sgRNAs begins at the 3′ end (3′ leader) of the incoming (−) strand genome producing a short leader RNA, followed by monocistronic capped and poly(A)-tailed mRNAs by sequential transcription. The 5′ trailer region of the genome and the intercistronic regions separating the ORFs are not transcribed (Kolakofsky *et al.*, 2004).

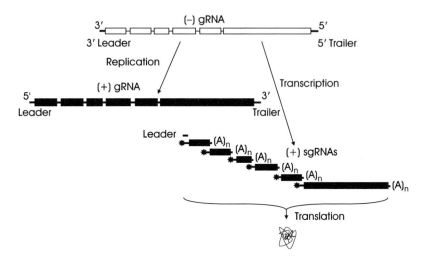

F<small>IG</small> 3. Strategy of replication/transcription used by nonsegmented (−) RNA viruses. Asterisk, cap structure; 3′ Leader, 3′ leader promoter; 5′ Trailer, 5′ trailer region.

Transcription of segmented (−) strand RNA viruses such as the *Orthomyxoviridae, Arenaviridae, Bunyaviridae,* and Tenuiviruses requires a primer to initiate synthesis of the mRNAs. This is achieved by cap-snatching in which the replicase complex, or a protein thereof, binds to the 5′ region of cell mRNAs, cleaves off the cap together with generally 7–15 nucleotides from the 5′ end of the cell mRNA, and uses this fragment as a primer to initiate synthesis of the viral mRNAs (Bouloy *et al.*, 1978; Nguyen and Haenni, 2003). Hence, the viral mRNAs are capped and have cell-derived nucleotides at their 5′ end; they also contain a poly(A) tail at their 3′ end. Replication, on the other hand, is not primer-dependent, and the RNAs complementary to the genome are devoid of cap and of poly(A) tail.

II. How to Study Host Factors Specifically Involved in Viral RNA Transcription/Replication

Host factors participate in viral genome amplification based on their specific binding to or copurification with replicase subunits, their binding to specific regions of the viral RNA, their influence on replication/transcription, the occurrence of host mutations that influence virus multiplication, and the reduced effect on replication/transcription observed in the presence of small interfering RNAs (siRNAs) directed against specific candidate host proteins.

The unambiguous identification of host proteins that persistently copurify with viral polymerases during multiple purification steps has been fraught with difficulties, in spite of the fact that certain cell proteins bind with high affinity to the purified polymerase, or that cell proteins are often encapsidated together with the polymerase as in the case of the (−) strand RNA viruses. Similar difficulties have been encountered in searching for specific host factors that interact with the viral genome. These difficulties are exacerbated by the fact that it is often difficult to distinguish between factors that directly affect RNA replication/transcription from those that have an indirect effect on viral RNA synthesis, such as by interacting with host factors that in turn are directly involved in RNA amplification. Gel-shift mobility assays, UV-crosslinking, coimmunoprecipitation, and the yeast two- or three-hybrid systems have served to search for interactions between host factors and the viral polymerase and/or the viral RNA. Yet the relevance of such interactions remains equivocal in the absence of functional RNA amplification assays. In many cases, particular caution must therefore surround published data, and further experiments are needed to determine whether copurification and/or interaction with viral proteins or RNA is fortuitous or has a functional significance. Technologies such as the systematic search for host genes became involved in virus RNA amplification using an ordered genome-wide set of deletion mutants (Kushner et al., 2003) and RNA interference (Ahlquist, 2002; Isken et al., 2003; Lan et al., 2003; Li et al., 2002a), are beginning to provide promising results that nevertheless still require further verification.

One of the major hurdles faced when attempting to identify host factors specifically involved in viral RNA replication/transcription is how to discriminate these factors from those involved in translation. As has become apparent, several of the host factors shown to affect viral RNA synthesis are factors known to be involved in protein synthesis such as, for example, translation factors. In addition, some of the factors identified to date appear to influence viral RNA amplification as well as viral protein synthesis, and translation and replication are frequently tightly associated. This situation adds a further level of complication to the identification of host factors involved in a specific viral function. It will hopefully become possible to overcome this problem by examining the effect of a given host factor on the replicase and its activity, and in parallel its effect on translation *in vitro* and *in vivo*. Such experiments have been undertaken

(Choi *et al.*, 2002) and have already provided interesting information concerning transcription in a coronavirus.

On the other hand, to identify host factors that specifically favor synthesis of RNA strands complementary to the genome but not synthesis of further genome strands, one can take advantage of the observation that a few additional nontemplated nucleotides positioned at the 5' end of a viral genomic RNA can still allow synthesis of the complementary strand, but can hinder subsequent synthesis of new genome strands. Experiments based on this approach have been used to distinguish certain aspects of (−) from (+) strand synthesis of poliovirus (Murray and Barton, 2003).

In spite of all the pitfalls outlined previously, several specific host factors actively participating in viral RNA transcription/replication have been identified and the regions of host protein/replicase or host protein/viral RNA interaction determined. The present review centers exclusively on those factors that appear functionally important for viral amplification.

III. Interaction with the Replicase Complex

Table I presents a list of the viruses for which a specific host factor associates with the polymerase, affecting viral genome amplification. It also indicates the usually accepted cell function of the factor and the viral polymerase or polymerase subunit to which the host factor binds.

The replication complex of RNA phages, such as $Q\beta$ is composed of four subunits: the virus-encoded RNA polymerase, and three host factors, the protein elongation factors Tu (EF-Tu) and Ts (EF-Ts), and the ribosomal protein S1 (Table I). Another host factor, designated HF, is also required for (−) strand synthesis from the (+) strand RNA template. In the generally accepted scenario, the polymerase provides the polymerizing activity, whereas S1 allows the enzyme complex to bind to two (S and M) internal sites in the viral genome (Table II; Fig. 4). The long-standing observation that the polymerase subunit binds to internal sites on the viral genome remained a paradox for several decades. Indeed, if the enzyme is to faithfully replicate the template, it must somehow also lie close to the 3' end of the template. It has now been established that this is achieved at least in part by elaborate secondary structures in the viral RNA (Fig. 4) that bring the 3' end of the viral RNA in close proximity to the internal sites where the polymerase is positioned (Klovins and van Duin, 1999), facilitating this interaction.

TABLE I
Host Factors Associating with Viral RNA Replicase Complexes

Virus	Family/Genus (genome polarity)	Host factor	Function in host	Virus target	References
Qβ	*Leviviridae* (+)	EF-Tu, EF-Ts, S1	Translation	Replicase	Blumenthal and Carmichael (1979)
		HF	Translation regulation		Klovins and van Duin (1999); Schuppli *et al.* (2000)
BMV	*Bromoviridae* (+)	eIF3	Translation	2a	Quadt *et al.* (1993)
TMV	Tobamovirus (+)	eIF3	Translation	126 kDa, 183 kDa	Osman and Buck (1997)
TEV	*Potyviridae* (+)	eIF4E	Translation	NIa	Schaad *et al.* (2000)
RV	*Togaviridae* (+)	Rb	Cell growth	NSP90	Atreya *et al.* (1998)
HCV	*Flaviviridae* (+)	α-actinin	Transport	NS5B	Lan *et al.* (2003)
		hPLIC1	Protein degradation		Gao *et al.* (2003)
		p68	RNA metabolism		Goh *et al.* (2004)
VSV	*Rhabdoviridae* (−)	EF1α, β, γ	Translation	L	Das *et al.* (1998)
MV	*Paramyxoviridae* (−)	tubulin	Transport	L	Moyer *et al.* (1990)
CDV	*Paramyxoviridae* (−)	hsp72	Protein metabolism	NC	Oglesbee *et al.* (1996)
IV	*Orthomyxoviridae* (−)	NPI-5	Splicing	NP	Momose *et al.* (2001)

BMV, brome mosaic virus; TMV, tobacco mosaic virus; TEV, tobacco etch virus; RV, rubella virus; HCV, hepatitis C virus; VSV, vesicular stomatitis virus; MV, measles virus; CDV, canine distemper virus; IV, influenza virus; EF, elongation factor; HF, host factor; eIF, eukaryotic initiation factor; Rb, retinoblastoma; hPLIC1, human homolog 1 of protein linking integrin-associated protein and cytoskeleton; hsp, heat shock protein; NPI, nucleoprotein interactor. Protein membrane components interacting with virus replicases are presented in Table IV.

TABLE II
Host Factors Associating with Viral RNAs

Virus	Family (genome polarity)	Viral genome target region (strand polarity)	Host factor	Function in host	References
Qβ	*Leviviridae* (+)	Internal (+)	S1	Translation	Blumenthal and Carmichael (1979);
		Internal (+), 3'(+)	HF	Translation regulation	Klovins and van Duin (1999); Schuppli *et al.* (2000)
PV	*Picornaviridae* (+)	5'(+)	PCBP	mRNA stability	Gamarnik and Andino (1997); Parsley *et al.* (1997); Walter *et al.* (2002)
		3'(+)	PABP	mRNA stability	Herold and Andino (2001)
			Nucleolin	rRNA processing	Waggoner and Sarnow 1998
MHV	*Coronaviridae* (+)	3'(−), IG (−)	hnRNP A1	Splicing	Li *et al.* (1997); Shen and Masters (2001); Shi *et al.* (2000); Shi *et al.* (2003)
		5'(+), 5'(−)	PTB	Splicing	Choi *et al.* (2002); Huang and Lai (1999);
		3'(+)	Mitochondrial aconitase	Citric acid cycle	Nanda and Leibowitz (2001)
WNV	*Flaviviridae* (+)	3'(−)	TIA-1, TIAR	Translation, apoptosis	Li *et al.* (2002b)
BVDV	*Flaviviridae* (+)	5' UTR, 3' UTR	NF90/NFAR-1, NF45, RHA	Transcription/ translation regulation	Isken *et al.* (2003)
HPIV-3	*Paramyxoviridae* (−)	Leader RNA (−), Leader RNA (+)	GAPDH	Glyolysis	De *et al.* (1996)

PV, poliovirus; MHV, mouse hepatitis virus; WNV, West Nile virus; BVDV, bovine viral diarrhea virus; HPIV-3, human parainfluenza virus-3; IG (−), intergenic region in (−) RNA; UTR, untranslated region; Leader RNA (−), 3' end of (−) RNA; Leader RNA (+), 5' end of (+) RNA; HF, host factor; PCBP, poly(C)-binding protein; PABP, poly(A)-binding protein; hnRNP A1, heterogeneous nuclear ribonucleoprotein A1; PTB, polypyrimidine tract-binding protein; TIA-1, T-cell-activated intracellular antigen; TIAR, TIA-1-related; RHA, RNA helicase A; GAPDH, glyceraldehyde-3-phosphate dehydrogenase.

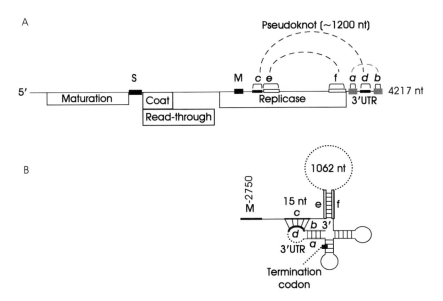

FIG 4. (A) Genetic map and genome organization of Qβ RNA. Highlighted are the S and M sites on the genome to which the enzyme complex binds, as well as various RNA–RNA interaction sites within the 3′ half of the genome. (B) Enlarged schematic representation of RNA–RNA interactions. A pseudoknot is formed between regions c and d. Not to scale. (Adapted with permission from Klovins and van Duin, 1999.)

The eukaryotic initiation factor 3 (eIF3) stimulates binding of the mRNA to the 40S ribosomal subunit and stabilizes the interaction between the 40S and initiator Met-tRNA. It has been shown that the 2a polymerase-like protein involved in brome mosaic bromovirus (BMV) amplification interacts with a 41-kDa protein, one of the 10 subunits of eIF3 or of a closely related protein (Quadt et al., 1993) thereby greatly stimulating (−) strand RNA synthesis in vitro. Similarly, the membrane-bound replication complex of tobacco mosaic tobamovirus (TMV) contains a subunit of eIF3, the 56 kDa component which is the RNA-binding subunit of the wheat germ eIF3. Hence, the host components of the two viral polymerases correspond to different subunits of eIF3. Antibodies against the 56 kDa subunit inhibit TMV RNA synthesis in vitro (Osman and Buck, 1997).

NIa forms part of the replication complex of Potyviruses, such as tobacco etch potyvirus (TEV; Schaad et al., 2000), by providing the viral genome-linked protein (VPg) and proteolytic functions. It interacts with the protein initiation factor eIF4E, a factor that binds to the cap structure and facilitates recruitment of the mRNA as well as other factors on the 40S ribosomal subunit. To investigate this interaction further, full-length TEV transcripts were produced harboring different mutated versions of the eIF4E coding sequence. After transfection of protoplasts with these *in vitro*-transcribed constructs, cells containing the mutated eIF4E transcripts accumulated far lower amounts of virus than those harboring the wild-type eIF4E transcript.

The tumor suppressor retinoblastoma protein (Rb) is a nuclear phosphoprotein that binds to many proteins and plays a fundamental role in suppression of various neoplasms. Its capacity to suppress cell proliferation is caused mainly by binding and inhibiting the transcription factor E2F (Taya, 1997). The NSP90 protein, the putative RdRp of rubella virus (RV; *Togaviridae*), binds to the Rb *in vitro* and *in vivo*. This interaction requires the presence of at least some Rb in the cytoplasm where amplification of RV is believed to be confined (Forng and Atreya, 1999). Rb facilitates RV replication, since in null-mutant (Rb−/−) mouse embryonic fibroblasts, the level of RV replication is lower than in wild-type (Rb+/+) cells (Atreya *et al.*, 1998). NSP90 contains the Rb-binding motif LXCXE. In mutant viruses containing a C to R mutation in this motif, replication is reduced to less than 0.5% of the wild-type virus, and deletion of the motif is lethal (Forng and Atreya, 1999), clearly demonstrating the requirement of the Rb protein for efficient RV replication.

Interaction of the HCV RNA polymerase (NS5B) with α-actinin, a member of a superfamily of actin cross-linking proteins, was demonstrated using various *in vitro* techniques. It was confirmed by reducing the expression of α-actinin by RNA interference in the HCV replicon system, resulting in decrease of HCV RNA levels (Lan *et al.*, 2003). Likewise, a ubiquitin-like protein hPLIC1 (the human homolog 1 of the protein linking integrin-associated protein and cytoskeleton) also interacts with NS5B, and its overexpression in the replicon system reduces the levels of NS5B and of replicon RNA probably by promoting NS5B degradation via a ubiquitin-dependent pathway (Gao *et al.*, 2003). Using the two-hybrid system to screen a human cDNA library with NS5B as bait, it was demonstrated that the cellular RNA helicase p68 is another protein that interacts with the HCV polymerase (Goh *et al.*, 2004). The C-terminal part of NS5B is responsible for this interaction. The p68 helicase is redistributed from the nucleus to the cytoplasm in NS5B expressing cells.

The use of two different cell systems to study the effect of p68 on replication led to opposite results when the level of p68 was reduced using an siRNA: in an HCV replicon-bearing cell line, no effect on HCV replication was observed, while in cells transiently transfected with a full-length HCV construct, synthesis of the negative strand was inhibited. Overexpression of NS5B and its C-terminally truncated mutant in the transiently transfected cell line also inhibited negative strand synthesis, probably due to sequestration of p68. The helicase activity of p68 does not seem to be required for replication, suggesting that it acts as a transcription factor rather than as a helicase (Goh *et al.*, 2004).

The RdRp or L protein of VSV has been overexpressed in recombinant baculovirus-infected cells. The purified, transcriptionally active fraction contains the subunits α, β, and γ of EF-1. The α subunit binds to the RdRp with the highest affinity yielding a partially active enzyme, and addition of the β and γ subunits significantly enhances enzyme activity. All three subunits are packaged with the L protein in the virions (Das *et al.*, 1998).

Tubulin binds to the RdRp (L protein) of measles virus (MV; *Paramyxoviridae*) and may be part of the polymerase complex of this virus, as suggested by its coimmunoprecipitation with the RdRp from extracts of MV-infected cells (Moyer *et al.*, 1990). It is required for efficient viral RNA replication and transcription, since anti-tubulin antibodies inhibit viral RNA synthesis and addition of purified tubulin stimulates MV RNA synthesis *in vitro*.

Experiments performed with canine distemper virus (CDV; *Paramyxoviridae*) (De and Banerjee, 1997; Lai, 1998) strongly suggest that heat shock proteins may also be required for optimal transcription when closely associated with the nucleocapsids (NCs) of this virus. Indeed, when the CDV NCs from non-heat-shocked and heat-shocked cells are compared, an increase in—and/or accumulation of—viral transcripts is observed in the heat-shocked cells. The heat-shock protein hsp72 seems to directly enhance viral RNA transcription *in vitro* (Oglesbee *et al.*, 1996).

Three proteins, PB1, PB2, and PA compose the polymerase of influenza A virus. PB2 cleaves 5' oligonucleotides containing the cap structure from cellular mRNAs in the cap-snatching scenario. PB1 is the catalytic subunit for viral RNA synthesis and uses the capped oligonucleotides as primers for transcription. The function of the phosphoprotein PA remains unclear. In addition, the nucleoprotein NP is required for synthesis of full-length RNA (Momose *et al.*, 2001), and as such it forms part of the replicase complex. Various proteins have been shown to interact with the NP. Among them is the nucleoprotein interactor 5 (NPI-5, also known as RAF-2p48, BAT1 or UAP56), a cellular splicing factor (Momose *et al.*, 2001). It interacts with free

NP, but not with NP-RNA complexes, and stimulates synthesis of the viral RNA. Its interaction with NP is disrupted by the addition of large amounts of viral RNA that bind to NP, suggesting that NPI-5 may facilitate loading of NP on the RNA, thus acting as a chaperone. In influenza virus-infected cells, NPI-5 disappears from spliceosomes, where it is concentrated in uninfected cells. Its relocalization is similar to that of another splicing-related host protein, the NS1-binding protein (NS1-BP), which depends on the presence and distribution of the viral nonstructural protein NS1 (Wolff *et al.*, 1998). The same distribution pattern of the three proteins (NPI-5, NS1-BP, and NS1) in infected cells suggests that NPI-5 may indirectly contribute to shutoff of host protein synthesis in these cells, due to inhibition of splicing and 3′-terminal processing of pre-mRNAs by the NS1 protein (Momose *et al.*, 2001; Wolff *et al.*, 1998).

IV. Binding of Host Proteins to the Viral RNA Genome

The search for host factors involved in viral RNA amplification has also focused on cellular proteins that specifically bind to the 3′ or 5′ untranslated regions (UTRs) of either the (−) or (+) RNA strand, as well as to intergenic regions in the *Nidovirales* and *Mononegavirales*.

Several host factors that bind to specific regions of viral RNAs have been identified. All those whose implication in viral RNA replication has also been clearly demonstrated are presented in Table II.

The poly(C)-binding protein (designated PCBP, hnRNP E or αCP) isoforms 1 and 2 regulate the stability and expression of several cellular mRNAs (Herold and Andino, 2001; Walter *et al.*, 2002). They are involved in poliovirus RNA translation and replication (Gamarnik and Andino, 1998). When bound to the internal ribosome entry site (IRES) of poliovirus RNA, PCBP probably enhances translation (Blyn *et al.*, 1997), but it also regulates RNA synthesis since its binding to the disrupted stem-loop IV of the IRES negatively affects replication (Gamarnik and Andino, 2000). It also binds to the 5′ cloverleaf structure of the poliovirus RNA upstream of the IRES (Gamarnik and Andino, 1997, 1998; Parsley *et al.*, 1997) to which the poliovirus 3CD polymerase also binds, and where it is essential for initiation of (−) RNA synthesis. Studies have aimed at investigating the functions of the different PCBP isoforms as well as identifying the domains of these factors involved in poliovirus translation and replication using PCBP-depleted cell extracts. Using this approach, it was shown that PCBP2 rescues both RNA replication and translation initiation, whereas

PCBP1 rescues only replication (Walter *et al.*, 2002). PCBP thus seems to be involved in the switch from viral RNA translation to replication that may result from a change in the relative homo- and hetero-dimer concentrations of both isoforms.

The ternary ribonucleoprotein (RNP) complex formed between the 5′ cloverleaf structure, 3CD, and PCBP can interact *in vitro* with the poly(A)-binding protein PABP bound to the poly(A) tail, thus circularizing the RNA (Herold and Andino, 2001). Such a circularized RNP complex consisting of an RNA–protein–protein–RNA bridge is required to initiate (−) strand RNA synthesis.

Nucleolin is an abundant nucleolar protein. It is highly modified by phosphorylation and methylation, and at times also by ADP-ribosylation. One of the multiple functions of this RNA-binding protein is to promote processing of ribosomal RNA in the presence of U3 snoRNP (Hiscox, 2002). Upon poliovirus infection, it relocalizes to the cytoplasm and interacts strongly with the poliovirus 3′ UTR (Waggoner and Sarnow, 1998). Cell-free extracts depleted of nucleolin produce fewer infectious particles than control extracts at early times after infection. This is overcome at later times for reasons that remain undefined. Ongoing viral gene expression is necessary for relocalization of nucleolin to the cytoplasm and is not the result of inhibition of host protein synthesis, since addition of cycloheximide to uninfected cells does not trigger relocalization of nucleolin.

An interesting example of seemingly contradictory results obtained using different approaches to identify host factors required for viral genome amplification, is the case of the possible role of the heterogeneous nuclear ribonucleoprotein (hnRNP) A1 in coronavirus sg mRNA synthesis. hnRNP A1 binds to the pyrimidine tract-binding protein (PTB) and is believed to be part of a splicing complex involved in alternative RNA splicing (Lai, 1998; Miller and Koev, 2000), although new data indicate that it may act as a splicing silencer as demonstrated for human immunodeficiency virus-1 (*Retroviridae*) (Tange *et al.*, 2001). On one hand, by UV-cross-linking and immunoprecipitation, it was reported that hnRNP A1 binds to the common TRSs in the mouse hepatitis virus (MHV; *Coronaviridae*) RNA of the complementary (−) leader and (−) intergenic regions, which conform to the consensus PTB-binding motif. Hence, hnRNP A1 was proposed to bring together the two elements of the viral RNA involved in transcription (Huang and Lai, 2001; Li *et al.*, 1997). Moreover, MHV RNA transcription and replication were stimulated by overexpression of hnRNP A1, and inhibited by dominant-negative mutants of hnRNP A1 (Shi *et al.*, 2000). On the other hand, however, the hnRNP A1-binding motif found in MHV transcription-regulating sequences is not conserved in other *Nidovirales*, and MHV replication and transcription were not affected when a cell line not expressing hnRNP A1 was used (Shen and Masters, 2001). The protein is thus dispensable for MHV replication and transcription. Furthermore, recent results indicate that hnRNP A1-related proteins can replace hnRNP A1 for MHV RNA synthesis in hnRNP A1-depleted cells (Shi *et al.*, 2003). Therefore, similar functions may be exerted by closely related proteins, making it difficult to assign conclusive roles to certain cellular proteins in viral genome amplification.

PTB is a member of the hnRNP family of proteins and shuttles between the nucleus and the cytoplasm. It binds to the polypyrimidine tract of the adenovirus major-late pre-mRNA, and regulates alternative splicing of certain pre-mRNAs. Its role in stimulating internal initiation of translation of the genome of certain RNA viruses containing an IRES has attracted considerable attention (Anwar *et al.*, 2000; Hunt and Jackson, 1999; Kaminski and Jackson, 1998). In the cases where it favors internal initiation, this appears to be achieved by maintaining the IRES in the appropriate configuration for efficient translation (Kaminski and Jackson, 1998).

In MHV RNA, PTB binds to a stretch of four UCUAA repeats in the 5′ (+) leader and to the 5′ UTR of the complementary (−) strand. Binding to these sites is required for transcription (Huang and Lai, 1999). To further elucidate the role of PTB in MHV RNA synthesis and discriminate between an effect on transcription and translation, defective interfering (DI) RNAs harboring different reporter genes were transfected into stable cell lines overexpressing full-length or truncated PTB (Choi *et al.*, 2002). When these overexpressing cell lines were infected with a DI RNA—coding for chloramphenicol acetyl transferase (CAT) preceded by the intergenic region—that had to be transcribed during MHV infection to undergo translation, CAT activity was much lower than in control cell lines, suggesting that PTB overexpression had a dominant-negative effect on translation and/or transcription. To determine whether PTB was affecting translation or transcription, cell lines overexpressing PTB were transfected with an *in vitro* transcribed DI RNA containing part of the MHV ORF1a fused to the luciferase gene and flanked by the MHV 5′ and 3′ UTRs. The same level of luciferase activity was obtained in cells overexpressing PTB as in control cells. Likewise, synthesis of a reporter protein in a reticulocyte lysate previously depleted of PTB was the same as in a control undepleted lysate. These last two series of experiments demonstrate that PTB is directly involved in RNA synthesis, not in translation. Moreover, PTB interacts *in vivo* and *in vitro* with the viral

N protein, an element of the replication complex, supporting the idea that PTB is part of the viral transcription/replication machinery (Choi *et al.*, 2002).

PTB interacts specifically with the IRESs of several picornaviruses enhancing translation. However, as opposed to its proposed function in directly enhancing MHV RNA transcription, PTB probably indirectly participates in poliovirus RNA synthesis. It has been suggested that early in infection, PTB enhances poliovirus translation. However, because the poliovirus translation products accumulate, and because PTB is cleaved by the viral protease $3C^{pro}$ and/or $3CD^{pro}$, during late stages of infection the concentration of PTB would no longer be sufficient to favor translation, leading to a switch from translation to replication (Back *et al.*, 2002).

Mitochondrial aconitase is involved in the citric acid cycle; it is a posttranscriptional regulator and possesses iron-regulatory functions (Nanda and Leibowitz, 2001). It is one of several proteins that bind specifically to the MHV 3' (+) RNA region harboring the poly(A) tail. Early in infection such binding increases production of infectious virus particles as determined by Western blots of the infected murine cell extracts, in particular when the cell cultures are supplemented with iron. Binding of mitochondrial aconitase to the 3' UTR of the viral genome might increase the stability of the mRNA, thereby favoring translation and the production of virus particles at early times after infection.

The multifunctional TIA-1 (T-cell-activated intracellular antigen) and TIAR (TIA-1-related) proteins belong to a family of RNA-binding proteins and are involved in translation regulation and apoptosis; they have also been identified in mammalian stress granules, where they promote recruitment of untranslated mRNAs during environmental stress (Dember *et al.*, 1996; Kedersha *et al.*, 1999; Li *et al.*, 2002b). One of the host proteins that bind to the West Nile virus (WNV; *Flaviviridae*) 3' hairpin in the (−) RNA strand has recently been identified as TIAR (Li *et al.*, 2002b). TIA-1 also binds to the same region in the viral RNA. Moreover, virus growth is inefficient in TIAR knockout cells, and when such cells are complemented with a vector expressing TIAR, WNV growth is partially restored. Thus, TIAR and possibly also TIA-1 are functionally important for WNV replication. This has not been shown for Sendai virus (SeV; *Paramyxoviridae*), although its trailer RNA binds TIAR, and the protein is involved in virus-induced apoptosis (Iseni *et al.*, 2002).

A group of cellular proteins belonging to a family with diverse functions and containing a common double-strand (ds) RNA, binding motif (dsRBM) was identified when searching for proteins interacting with the 3' UTR of the bovine viral diarrhea virus (BVDV; *Flaviviridae*) genome. Interaction of the three proteins NF90/NFAR-1, NF45, and RNA helicase A (RHA), known as the "NFAR" group to both the 3' and 5' UTRs, was unambiguously demonstrated (Isken *et al.*, 2003). Mutations of the proteins' interaction sites in the 3' UTR decreased protein binding and RNA replication (measured as progeny (+) RNA synthesis). Similarly, transfection of cell lines supporting BVDV replication with siRNAs directed against RHA led to 60–85% reduction of viral replication. The data suggest that the NFAR proteins may mediate circularization of the viral RNA leading to a switch from translation to replication. Interestingly, this group of proteins is also believed to be part of the antiviral response of the cell that seems to be subverted by the virus for the purpose of its propagation. The direct involvement of the NFAR proteins in viral replication and not in translation will require confirmation by further experiments.

Glyceraldehyde-3-phosphate dehydrogenase (GAPDH) is a key enzyme of glycolysis. It appears as multiple isoforms consistent with its multifunctional nature (Takagi *et al.*, 1996). It interacts with various proteins and with RNAs including the leader RNA (−) and leader RNA (+) of human parainfluenza virus type 3 (HPIV-3; *Paramyxoviridae*) (De *et al.*, 1996) and has helix-destabilizing activity. Specific phosphorylated forms of the protein associate with the viral RNA and inhibit transcription *in vitro* (Choudhary *et al.*, 2000). Thus, GAPDH is a negative regulator of viral gene expression in this case. Specific, highly phosphorylated forms of GAPDH are encapsidated in the HPIV-3 virions (Choudhary *et al.*, 2000).

The La autoantigen is a ubiquitous phosphoprotein and an RNA-binding protein. Although it interacts with the UTRs of several RNA viruses (De Nova-Ocampo *et al.*, 2002; Duncan and Nakhasi, 1997; Gutiérrez-Escolano *et al.*, 2000, 2003; Lai, 1998; Wolin and Cedervall, 2002), to date, it has not been shown to have a clear functional effect on viral genome amplification and therefore is not discussed further here.

V. Interaction with Cytoskeletal Proteins

Tubulin and actin activate RNA transcription of members of the *Rhabdoviridae* and *Paramyxoviridae* families (Table III). These cytoskeletal proteins might stabilize the various components of the transcription/replication complexes, or could transport or orient the viral RNAs and the complexes to specific cellular compartments in the appropriate topology.

TABLE III
CYTOSKELETAL PROTEINS INTERACTING WITH RNA VIRUSES

Virus	Family (genome polarity)	Host factor	References
VSV	Rhabdoviridae (−)	Tubulin	Moyer et al. (1986)
MV	Paramyxoviridae (−)	Tubulin	Moyer et al. (1990)
SeV	Paramyxoviridae (−)	Tubulin	Moyer et al. (1986)
		PGK,* enolase	Ogino et al. (2001)
HPIV-3	Paramyxoviridae (−)	Actin	De and Banerjee (1999); Gupta et al. (1998)
HRSV	Paramyxoviridae (−)	Actin Profilin	Burke et al. (1998); Burke et al. (2000); Ulloa et al. (1998)

* Interaction occurs via tubulin.

VSV, vesicular stomatitis virus; MV, measles virus; SeV, Sendai virus; HPIV-3, human parainfluenza virus-3; HRSV, human respiratory syncytial virus; PGK, phosphoglycerate kinase.

Synthesis *in vitro* of VSV and SeV mRNAs requires tubulin (Moyer *et al.*, 1986). In SeV, tubulin and at least two other cellular proteins, phosphoglycerate kinase (PGK) and enolase, both glycolytic enzymes, are required for transcription. Once associated with the transcription complex, tubulin might recruit PGK and enolase into the complex that would become active for elongation of viral mRNA chains (Ogino *et al.*, 2001). In MV-infected cell extracts that support transcription and replication, actin stimulates both these reactions *in vitro* (Moyer *et al.*, 1990).

Actin microfilaments are involved in HPIV-3 replication and transcription *in vivo,* and in transcription *in vitro* (De and Banerjee, 1999; Gupta *et al.*, 1998) as judged by the action of cytochalasin D that depolymerizes actin microfilaments and results in inhibition of viral RNA synthesis. Actin is detected in purified human respiratory syncytial virus (HRSV; *Paramyxoviridae*) (Ulloa *et al.*, 1998); transcription *in vitro* is stimulated by monomeric actin, as well as by actin microfilaments (Burke *et al.*, 1998). In addition to actin, profilin (an actin-modulatory protein) is required for optimal transcription of HRSV (Burke *et al.*, 2000).

VI. HOST FACTORS IDENTIFIED USING THE YEAST MODEL

Among (+) strand RNA viruses, interesting clues have been gathered using BMV, taking advantage of the fact that yeast cells support BMV amplification. Using this approach, several yeast factors have been characterized that stimulate virus amplification (Noueiry and Ahlquist, 2003). Mutations in the small protein Lsm1p involved in mRNA decapping preceding mRNA degradation modulate BMV RNA replication and sg mRNA synthesis in yeast, leading to a selective reduction of RNA3 replication, and hence of sgRNA synthesis (Diez et al., 2000; Noueiry et al., 2003). Mutations in Ydj1p, a chaperone thought to participate in proper polymerase folding to produce an active replication complex, inhibit the formation of such a complex required for (−) RNA strand synthesis (Tomita et al., 2003). Finally, mutations in the Ded1p required for translation initiation (Noueiry et al., 2000) appear to also indirectly decrease BMV RNA replication.

Ded1p was also shown to directly influence replication of the yeast dsRNA L-A virus (Totiviridae) because its addition to extracts of virus-infected yeast cells led to stimulation of (−) but not of (+) strand RNA synthesis. However, the mechanism of action of Ded1p remains only partially elucidated, as only interaction of Ded1p with the capsid protein Gag could be clearly demonstrated (Chong et al., 2004).

A systematic search for host factors involved in BMV replication has been performed using 4500 single-gene yeast deletion mutants transformed with constructs expressing replicase proteins as well as luciferase from a replication template. This led to the identification of 58 and 38 genes whose deletions respectively inhibited and enhanced luciferase expression at least threefold (Kushner et al., 2003). Some of these genes, such as the *LSM* and *SKI* genes, had been previously identified as being implicated in viral replication and/or RNA function and turnover. For a selected group of genes, a more detailed analysis was carried out that further supports the direct effect of the gene products on RNA synthesis. Further studies are needed to elucidate more precisely how these newly implicated host factors influence viral replication, and to uncover possible functions of essential yeast genes in BMV RNA replication.

VII. Involvement of Membrane Components

The replication complexes of all (+) strand eukaryotic RNA viruses are associated with intracellular membranes (Lee et al., 2001), and virus infection frequently leads to membrane proliferation or vesiculation. In addition, proliferation of membrane proteins is linked to lipid biosynthesis—both phenomena appear to be important for the amplification of most viral genomes. Thus, it is to be expected that membrane association is important for viral RNA synthesis.

The polymerase of members of the *Togaviridae* family is active in modified endosomes or lysosomes designated cytoplasmic vacuoles or double-stranded vesicles (Kääriäinen and Ahola, 2002). In equine arteritis virus (*Arteriviridae*) (Pedersen *et al.*, 1999; Snijder *et al.*, 2001), the RNA polymerase is found in presumably endoplasmic reticulum– (ER-) derived double-membrane structures. RNA synthesis of members of the *Flaviviridae* family, such as that of WNV, probably occurs in trans-Golgi membranes (Mackenzie *et al.*, 1999). RNA synthesis of HCV occurs on membranes of the ER or on a membranous matrix designated membranous web (Moradpour *et al.*, 2003) or lipid raft (detergent-resistant membrane) (Gao *et al.*, 2004). Poliovirus infection leads to the appearance of greatly rearranged vesiculated membranes (Suhy *et al.*, 2000), and poliovirus RNA replication occurs on membranes that contain markers of lysosomes, the Golgi apparatus, and the ER (Egger *et al.*, 2000; Schlegel *et al.*, 1996).

Flock house virus (FHV; *Nodaviridae*) is an insect virus that multiplies in diverse host cells such as plant, animal, insect, and yeast cells. Using anti-RdRp antibodies, it was demonstrated that the RdRp is tightly associated with the outer mitochondrial membranes of *Drosophila* cells, implying that FHV replication occurs on these membranes (Miller *et al.*, 2001). This interaction results from the presence of an N-proximal sequence in the RdRp that contains a transmembrane domain. When this domain was replaced by a yeast or HCV ER-targeting sequence, the RdRp chimeras were redirected to the ER membranes in yeast cells, and viral genomic and subgenomic RNA synthesis was enhanced considerably (Miller *et al.*, 2003).

RNA synthesis of plant RNA viruses occurs on cell membranes whose nature depends on the virus such as ER-derived membranes (Carette *et al.*, 2000; den Boon *et al.*, 2001; Dunoyer *et al.*, 2002; Más and Beachy, 1999; Ritzenthaler *et al.*, 2002; Schaad *et al.*, 1997), cytoplasmic inclusions (Dohi *et al.*, 2001), vacuolar membranes (Hagiwara *et al.*, 2003; van der Heijden *et al.*, 2001), invaginations of chloroplast membranes (Prod'homme *et al.*, 2001), or multivesicular bodies derived from peroxisomes, vacuoles, or mitochondria (Dunoyer *et al.*, 2002). Cowpea mosaic comovirus (CPMV) replication is linked to small membranous vesicles that proliferate upon virus infection. CPMV but neither alfalfa mosaic bromovirus nor TMV replication requires ongoing lipid biosynthesis, and replication of CPMV is inhibited by the antibiotic cerulenin, an inhibitor of lipid biosynthesis (Carette *et al.*, 2000; van der Heijden and Bol, 2002).

A yeast mutant defective in BMV RNA replication was characterized; it is defective in the *OLE1* gene (Lee *et al.*, 2001) that encodes a

TABLE IV

Host Cell Membrane Components Involved in Virus Amplification

Virus	Family/Genus (genome polarity)	Host membrane component	References
BMV	Bromoviridae (+)	UFA	Lee et al. (2001)
FHV	Nodaviridae (+)	PGL	Wu et al. (1992)
SFV	Togaviridae (+)	Phosphatidylserine	Ahola et al. (1999)
TMV	Tobamovirus (+)	TOM1	Yamanaka et al. (2000)
		TOM2A	Tsujimoto et al. (2003)
HCV	Flaviviridae (+)	hVAP-33	Gao et al. (2004)

BMV, brome mosaic virus; FHV, flock house virus; SFV, Semliki Forest virus; TMV, tobacco mosaic virus; HCV, hepatitis C virus; UFA, unsaturated fatty acid; PGL, phosphoglycerolipid; TOM1 and 2A, tobamovirus multiplication 1 and 2A proteins; hVAP-33, human homologue of the 33-kDa vesicle-associated membrane protein-associated protein.

Δ9 fatty acid desaturase, essential for the conversion of saturated to unsaturated fatty acids (UFAs) (Table IV). This desaturase is an integral protein of the ER, the site of BMV RNA replication. As such, it is not required for BMV RNA replication, but high levels of UFAs are required at one of the steps leading to RNA replication. UFAs could participate in the assembly of a functional complex for replication.

Isolated membrane complexes can, in certain cases, support the production of replicative intermediates in vitro and the elongation of RNA chains to produce genomic-length viral RNAs. Early work demonstrated that nuclease digestion of the membrane-bound FHV RNA yielded a template-dependent RNA polymerase. However, upon addition of the template RNA, only dsRNA intermediates were produced due to synthesis of only (−) RNA strands. Yet, upon addition of certain phosphoglycerolipids (PGLs), both (−) and (+) strands were produced, and complete replication of the genomic RNA was restored (Buck, 1996; Wu and Kaesberg, 1991; Wu et al., 1992). It was postulated that either interaction of PGLs with the polymerase was required to initiate (+) strand synthesis, or that the PGLs modified the configuration of the membrane to mimic the modifications occurring in virus-infected cells.

In cells infected with viruses such as Semliki Forest virus (SFV; Togaviridae) (Perez et al., 1991) or poliovirus (Guinea and Carrasco, 1991), the level of lipid synthesis increases, and inoculation with cerulenin prevents replication of these viruses. Cerulenin also inhibits poliovirus replication in vitro. Continuous lipid synthesis is required for synthesis of SFV RNA (Perez et al., 1991).

The SFV replicase is composed of the four nonstructural proteins 1 to 4 (NSP1-NSP4). NSP4 is the catalytic subunit of the replicase, and NSP1 contains the capping enzyme activities. NSP1 directly associates with several cell membranes such as the plasma membrane, endosomes, and lysosomes, and its attachment to membranes is required for its activity. Solubilization of NSP1 with various detergents strongly inhibits these activities, but they can be restored by the addition of anionic phospholipids, in particular phosphatidylserine (Ahola *et al.*, 1999). As of yet, the elements in the membrane that are affected by the detergents remain undefined, as does how the lipids affect the membrane-associated replication system.

TOM1 and TOM2A are *Arabidopsis thaliana* transmembrane proteins normally located with vacuolar membranes and other uncharacterized cytoplasmic membranes. TOM1 interacts with TOM2A and with the helicase domain of the TMV polymerase (Tsujimoto *et al.*, 2003; Yamanaka *et al.*, 2000). Inactivation of either of the corresponding genes significantly reduces TMV replication, implying that *in vivo* both proteins are involved in the formation of the replication complex (Ohshima *et al.*, 1998; Yamanaka *et al.*, 2000). Mutations in either the TOM1 or the TOM2A gene decrease TMV multiplication but do not affect replication of three unrelated plant RNA viruses.

It was recently shown (Gao *et al.*, 2004) that binding of the HCV polymerase NS5B as well as of NS5A to the cellular vesicle membrane transport protein hVAP-33 is necessary for the formation of the HCV replication complex on lipid raft and for viral RNA replication. Experiments were carried out using a subgenomic HCV replicon in a hepatocyte cell line transfected with siRNA against hVAP-33 or overexpressing truncated versions of hVAP-33. In both cases, the association of the viral nonstructural proteins with the lipid raft was inhibited, and the overall level of viral proteins was lower than in control cells. In the hVAP-33 mutant-expressing cells, the level of HCV (+) RNA synthesis was correlated with disruption of this association, supporting the significance of interaction of hVAP-33 with NS5B and NS5A for HCV replication.

VIII. PERSPECTIVES

Virus multiplication negatively affects its host as reflected by the appearance of symptoms and diminished performance of the host that are the basis of disease. The organism retorts with a series of programmed responses to counteract the development of virus infection. This situation triggers reiterated counteroffensives between the host and the virus. Whereas the host is indispensable for the virus, the reverse is rarely true. It is only in very few cases that the presence of a virus is beneficial for the host as in the case of yeast strains infected with dsRNA viruses that secrete peptide toxins lethal to virus-free yeast cells (Sesti *et al.*, 2001). Thus, it can appear surprising that so many host factors normally engaged in vital cell functions can be subverted by the virus for its own benefit. The participation of an ever-increasing number of host factors in virus replication/transcription is indeed somewhat unexpected, as is the variety of these factors. More-over, the borrowed factors do not necessarily serve the same function in virus development such as in RNA replication/transcription as they serve in the host, making their identification more laborious. Indeed, some of them are usually associated with functions distantly related to nucleic acid multiplication. In many cases, it still cannot be dis-missed that these factors act by stimulating other factors that are more directly involved in viral RNA genome amplification.

Basically, the viral RdRp constitutes the core of the polymerizing enzyme complex, essentially providing the machinery that will produce complementary RNA copies of the template. It has been shown that relatively pure preparations of several RdRps can replicate certain nat-ural or synthetic RNAs. Specificity is provided in large part by ancillary viral factors and by a cohort of satellite host factors (Klovins and van Duin, 1999; Lai, 1998). Intracellular virus–host interactions are both intimate and varied. Certain viruses are tissue specific, restricted to certain cell types, and this must be ascribed to the presence or absence of adequate host factors anywhere along the path of virus development. Other viruses replicate in diverse hosts, indicating that critical cell proteins required by these viruses must be highly conserved between host species. An interesting example of this situation is provided by experiments showing that the insect virus FHV not only multiplies but also systemically spreads through a transgenic plant expressing the viral movement protein of a plant virus (Dasgupta *et al.*, 2001).

As more is learned about viral genomes, it is becoming clear that interactions of proteins with secondary and tertiary RNA structures allow the RNA to adopt various folding patterns and constitute major elements modulating gene expression. For example, the 5′ and 3′ UTRs as well as RNA structures located within the coding regions are vital for (−) RNA strand synthesis of rhinoviruses, enteroviruses, and cardioviruses (Murray and Barton, 2003).

There is increasing evidence that circularization of mRNAs is an important step in protein synthesis, presumably because it allows ribosomes to be redirected from the 3′ to the 5′ end of the mRNA for a new round of translation and probably because it ensures that only intact mRNA is translated. Circularization involves the participation of cap- and poly(A)-binding proteins. It is also important for translation and replication of RNA viruses. For viral genomes containing a cap structure and a poly(A) tail, circularization could be brought about by a mechanism similar to the one used by cellular mRNAs. For viral genomes devoid of cap and/or poly(A) tail, other mechanisms must be invoked. Host factors participate in bringing close to one another distantly located regions of the RNA in poliovirus (Gamarnik and Andino, 1998; Sadowy *et al.*, 2001; Wimmer, 1982) and Qβ phage (Klovins and van Duin, 1999; Schuppli *et al.*, 2000). There are numerous other cases of viral RNA circularization involving essentially RNA–RNA interactions, and the functional importance of these interactions for viral gene expression has been demonstrated (Khromykh *et al.*, 2001; Shen and Miller, 2004; Yang *et al.*, 2004). It is, however, not unlikely that as for Qβ, viral and/or host factors could be implicated in regulating circularization and triggering a switch between translation and replication. Some of the same host factors could be involved in both mechanisms.

None of the strategies used so far to study the involvement of host factors in viral RNA amplification has been without pitfalls, and frequently the data reported attempt to answer certain but not other questions. What then might be the most appropriate strategy to unambiguously distinguish between factors involved in replication and/or transcription but not in translation, between (−) and (+) strand RNA synthesis, and between those directly involved in a given mechanism and those that affect this mechanism indirectly? It would, for instance, be instructive to investigate by appropriate methods, not only synthesis of the viral genome RNA and of its complementary strand, but at the same time also translation. Such questions must ultimately be addressed both *in vitro* and *in vivo*. Having identified potential factors by a screening process such as the one involving the yeast system (Kushner *et al.*, 2003), a reductionist approach might best be adopted to pinpoint the exact step at which a given factor is required, using various constructs, each allowing only one specific step to be examined. Following these experiments, more intricate models could be envisaged. This strategy demands huge investments in time and effort. However, it can be hoped that the availability of eukaryotic genome sequences and of an ever increasing number of infectious viral clones and replicons, the development of more reliable replication systems, and finally the advent of nanotechnology will facilitate such endeavors. Moreover, the rapidly developing RNA interference methodology and its application to genome-wide screening of gene functions and protein interactions, including in mammals, now enables large-scale identification of host genes whose down regulation (knock-down) will have a severe impact on viral replication (Fraser, 2004; Paddison *et al.*, 2004). An HCV replicon-bearing human hepatocarcinoma cell line that has become one of the favorite models to study host factor influence on viral genome amplification would be a good target for such an experiment.

In conclusion, over the past few years, much has been done to establish which host factors participate in viral genome amplification. Although the exquisite function of many of these protein–protein and RNA–protein interactions must still be established, their identification is a first step towards further understanding of the function of these interactions in the virus life cycle.

ACKNOWLEDGMENTS

We are very grateful to Karla Kirkegaard, Leevi Kääriäinen, Masayuki Ishikawa, Andrzej Palucha, Peter Sarnow, and Eric Snijder for careful reading of the manuscript, fruitful discussions, and constructive comments. This work was partly supported by the Centre National de la Recherche Scientifique (CNRS, France), the Center of Excellence of Molecular Biotechnology (Poland), the French-Polish Center of Plant Biotechnology (CNRS and Polish Academy of Science), and the University of Antioquia and Colciencias (Colombia).

REFERENCES

Ahlquist, P. (2002). RNA-dependent RNA polymerases, viruses and RNA silencing. *Science* **296**:1270–1273.

Ahlquist, P., Noueiry, A. O., Lee, W.-M., Kushner, D. B., and Dye, B. T. (2003). Host factors in positive-strand RNA virus genome replication. *J. Virol.* **77**:8181–8186.

Ahola, T., Lampio, A., Auvinen, P., and Kääriäinen, L. (1999). Semliki Forest virus mRNA capping enzyme requires association with anionic membrane phospholipids for activity. *EMBO J.* **18**:3164–3172.

Anwar, A., Ali, N., Tanveer, R., and Siddiqui, A. (2000). Demonstration of functional requirement of polypyrimidine tract-binding protein by SELEX RNA during hepatitis C virus internal ribosome entry site-mediated translation initiation. *J. Biol. Chem.* **275**:34231–34235.

Atreya, C. D., Lee, N. S., Forng, R. Y., Hofmann, J., Washington, G., Marti, G., and Nakhasi, H. L. (1998). The rubella virus putative replicase interacts with the retinoblastoma tumor suppressor protein. *Virus Genes* **16**:177–183.

Back, S. H., Kim, Y. K., Kim, W. J., Cho, S., Oh, H. R., Kim, J.-E., and Jang, S. K. (2002). Translation of polioviral mRNA is inhibited by cleavage of polypyrimidine tract-binding proteins executed by polioviral 3C^{pro}. *J. Virol.* **76**:2529–2542.

Belov, G. A., Evstafieva, A. G., Rubtsov, Y. P., Mikitas, O. V., Vartapetian, A. B., and Agol, V. I. (2000). Early alteration of nucleocytoplasmic traffic induced by some RNA viruses. *Virology* **275**:244–248.

Blumenthal, T., and Carmichael, G. G. (1979). RNA replication: Function and structure of Q$_\beta$-replicase. *Annu. Rev. Biochem.* **48:**525–548.

Blyn, L. B., Towner, J. S., Semler, B. L., and Ehrenfeld, E. (1997). Requirement of poly (rC) binding protein 2 for translation of poliovirus RNA. *J. Virol.* **71:**6243–6246.

Borden, K. L. B., Campbelldwyer, E. J., Carlile, G. W., Djavani, M., and Salvato, M. S. (1998). Two RING finger proteins, the oncoprotein PML and the Arenavirus Z protein, colocalize with the nuclear fraction of the ribosomal P proteins. *J. Virol.* **72:**3819–3826.

Bouloy, M., Plotch, S. J., and Krug, R. M. (1978). Globin mRNA are primers for the transcription of influenza viral RNA *in vitro. Proc. Natl. Acad. Sci. USA* **75:**4886–4890.

Buck, K. W. (1996). Comparison of the replication of positive-stranded RNA viruses of plants and animals. *Adv. Virus Res.* **47:**159–251.

Burke, E., Dupuy, L., Wall, C., and Barik, S. (1998). Role of cellular actin in the gene expression and morphogenesis of human respiratory syncytial virus. *Virology* **252:**137–148.

Burke, E., Mahoney, N. M., Almo, S. C., and Barik, S. (2000). Profilin is required for optimal actin-dependent transcription of respiratory syncytial virus genome RNA. *J. Virol.* **74:**669–675.

Carette, J. E., Stuiver, M., van Lent, J., Wellink, J., and van Kammen, A. (2000). Cowpea mosaic virus infection induces a massive proliferation of endoplasmic reticulum but not Golgi membranes and is dependent on *de novo* membrane synthesis. *J. Virol.* **74:**6556–6563.

Choi, K. S., Huang, P., and Lai, M. M. C. (2002). Polypyrimidine-tract-binding protein affects transcription but not translation of mouse hepatitis virus RNA. *Virology* **303:**58–68.

Chong, J.-L., Chuang, R.-Y., Tung, L., and Chang, T.-H. (2004). Ded1p, a conserved DExD/H-box translation factor, can promote yeast L-A virus negative-strand RNA synthesis *in vitro. Nucl. Acids Res.* **32:**2031–2038.

Choudhary, S., De, B. P., and Banerjee, A. K. (2000). Specific phosphorylated forms of glyceraldehyde 3-phosphate dehydrogenase associate with human parainfluenza virus type 3 and inhibit viral transcription *in vitro. J. Virol.* **74:**3634–3641.

Das, T., Mathur, M., Gupta, A. K., Janssen, G. M. C., and Banerjee, A. K. (1998). RNA polymerase of vesicular stomatitis virus specifically associates with translation elongation factor-1 $\alpha\beta\gamma$ for its activity. *Proc. Natl. Acad. Sci. USA* **95:**1449–1454.

Dasgupta, R., Garcia, B. H., and Goodman, R. M. (2001). Systemic spread of an RNA insect virus in plants expressing plant viral movement protein genes. *Proc. Natl. Acad. Sci. USA* **98:**4910–4915.

De, B. P., Gupta, S., Zhao, H., Drazba, J. A., and Banerjee, A. K. (1996). Specific interaction *in vitro* and *in vivo* of glyceraldehyde-3-phosphate dehydrogenase and LA protein with cis-acting RNAs of human parainfluenza virus type 3. *J. Biol. Chem.* **271:**24728–24735.

De, B. P., and Banerjee, A. K. (1997). Role of host proteins in gene expression of nonsegmented negative strand RNA viruses. *Adv. Virus Res.* **48:**169–204.

De, B. P., and Banerjee, A. K. (1999). Involvement of actin microfilaments in the transcription/replication of human parainfluenza virus type 3: Possible role of actin in other viruses. *Microsc. Res. Tech.* **47:**114–123.

Dember, L. M., Kim, N. D., Liu, K.-Q., and Anderson, P. (1996). Individual RNA recognition motifs of TIA-1 and TIAR have different RNA binding specificities. *J. Biol. Chem.* **271:**2783–2788.

den Boon, J. A., Chen, J., and Ahlquist, P. (2001). Identification of sequences in brome mosaic virus replicase protein 1a that mediate association with endoplasmic reticulum membranes. *J. Virol.* **75:**12370–12381.

De Nova-Ocampo, M., Villegas-Sepúlveda, N., and del Angel, R. M. (2002). Translation elongation factor-1α, La and PTB interact with the 3′ untranslated region of dengue 4 virus RNA. *Virology* **295:**337–347.

Diez, J., Ishikawa, M., Kaido, M., and Ahlquist, P. (2000). Identification and characterization of a host protein required for efficient template selection in viral RNA replication. *Proc. Natl. Acad. Sci. USA* **97:**3913–3918.

Dohi, K., Mori, M., Furusawa, I., Mise, K., and Okuno, T. (2001). *Brome mosaic virus* replicase proteins localize with the movement protein at infection-specific cytoplasmic inclusions in infected barley leaf cells. *Arch. Virol.* **146:**1607–1615.

Duncan, R. C., and Nakhasi, H. L. (1997). La autoantigen binding to a 5′ *cis*-element of rubella virus RNA correlates with element function *in vivo*. *Gene* **201:**137–149.

Dunoyer, P., Ritzenthaler, C., Hemmer, O., Michler, P., and Fritsch, C. (2002). Intracellular localization of the *Peanut clump virus* replication complex in tobacco BY-2 protoplasts containing green fluorescent protein-labeled endoplasmic reticulum or Golgi apparatus. *J. Virol.* **76:**865–874.

Egger, D., Teterina, N., Ehrenfeld, E., and Bienz, K. (2000). Formation of the poliovirus replication complex requires coupled viral translation, vesicle production, and viral RNA synthesis. *J. Virol.* **74:**6570–6580.

Ehrenfeld, E. (1996). Initiation of translation by picornavirus RNAs. *In* "Translational Control" (J. W. B. Hershey, M. B. Mathews, and N. Sonenberg, eds.), pp. 549–573. Cold Spring Harbor Laboratory Press, Cold Spring Harbor, NY.

Follett, E. A. C., Pringle, C. R., and Pennington, T. H. (1975). Virus development in enucleate cells: Echovirus, poliovirus, pseudorabies virus, reovirus, respiratory syncytial virus and Semliki Forest virus. *J. Gen. Virol.* **26:**183–196.

Forng, R.-Y., and Atreya, C. D. (1999). Mutations in the retinoblastoma protein-binding LXCXE motif of rubella virus putative replicase affect virus replication. *J. Gen. Virol.* **80:**327–332.

Fraser, A. (2004). Human genes hit the big screen. *Nature* **428:**375–378.

Gamarnik, A. V., and Andino, R. (1997). Two functional complexes formed by KH domain containing proteins with the 5′ noncoding region of poliovirus RNA. *RNA* **3:**882–892.

Gamarnik, A. V., and Andino, R. (1998). Switch from translation to RNA replication in a positive-stranded RNA virus. *Genes Dev.* **12:**2293–2304.

Gamarnik, A. V., and Andino, R. (2000). Interactions of viral protein 3CD and poly(rC) binding protein with the 5′ untranslated region of the poliovirus genome. *J. Virol.* **74:**2219–2226.

Gao, L., Tu, H., Shi, S. T., Lee, K.-J., Asanaka, M., Hwang, S. B., and Lai, M. M. C. (2003). Interaction with a ubiquitin-like protein enhances the ubiquitination and degradation of hepatitis C virus RNA-dependent RNA polymerase. *J. Virol.* **77:**4149–4159.

Gao, L., Aizaki, H., He, J.-W., and Lai, M. M. C. (2004). Interactions between viral nonstructural proteins and host protein hVAP-33 mediate the formation of hepatitis C virus RNA replication complex on lipid raft. *J. Virol.* **78:**3480–3488.

Goh, P. Y., Tan, Y.-J., Lim, S. P., Tan, Y. H., Lim, S. G., Fuller-Pace, F., and Hong, W. (2004). Cellular RNA helicase p68 relocalization and interaction with the hepatitis C virus (HCV) NS5B protein and the potential role of p68 in HCV RNA replication. *J. Virol.* **78:**5288–5298.

Gowda, S., Satyanarayana, T., Ayllón, M. A., Albiach-Martí, M. R., Mawassi, M., Rabindran, S., Garnsey, S. M., and Dawson, W. O. (2001). Characterization of the cis-acting elements controlling subgenomic mRNAs of *citrus tristeza virus:* Production of positive- and negative-stranded 3'-terminal and positive-stranded 5'-terminal RNAs. *Virology* **286:**134–151.

Guinea, R., and Carrasco, L. (1991). Effects of fatty acids on lipid synthesis and viral RNA replication in poliovirus-infected cells. *Virology* **185:**473–476.

Gupta, S., De, B. P., Drazba, J. A., and Banerjee, A. K. (1998). Involvement of actin microfilaments in the replication of human parainfluenza virus type 3. *J. Virol.* **72:**2655–2662.

Gustin, K. E., and Sarnow, P. (2001). Effects of poliovirus infection on nucleo-cytoplasmic trafficking and nuclear pore complex composition. *EMBO J.* **20:**240–249.

Gutiérrez-Escolano, A. L., Uribe Brito, Z., del Angel, R. M., and Jiang, X. (2000). Interaction of cellular proteins with the 5' end of Norwalk virus genomic RNA. *J. Virol.* **74:**8558–8562.

Gutiérrez-Escolano, A. L., Vázquez-Ochoa, M., Escobar-Herrera, J., and Hernández-Acosta, J. (2003). La, PTB, and PAB proteins bind to the 3' untranslated region of Norwalk virus genomic RNA. *Biochem. Biophys. Res. Commun.* **311:**759–766.

Hagiwara, Y., Komoda, K., Yamanaka, T., Tamai, A., Meshi, T., Funada, R., Tsuchiya, T., Naito, S., and Ishikawa, M. (2003). Subcellular localization of host and viral proteins associated with tobamovirus RNA replication. *EMBO J.* **22:**344–353.

Herold, J., and Andino, R. (2001). Poliovirus RNA replication requires genome circularization through a protein-protein bridge. *Mol. Cell* **7:**581–591.

Hiscox, J. A. (2002). The nucleolus—a gateway to viral infection? *Arch. Virol.* **147:**1077–1089.

Hiscox, J. A. (2003). The interaction of animal cytoplasmic RNA viruses with the nucleus to facilitate replication. *Virus Res.* **95:**13–22.

Huang, P., and Lai, M. M. C. (1999). Polypyrimidine tract-binding protein binds to the complementary strand of the mouse hepatitis virus 3' untranslated region, thereby altering RNA conformation. *J. Virol.* **73:**9110–9116.

Huang, P., and Lai, M. M. C. (2001). Heterogeneous nuclear ribonucleoprotein A1 binds to the 3'-untranslated region and mediates potential 5'-3' end cross talks of mouse hepatitis virus RNA. *J. Virol.* **75:**5009–5017.

Hunt, S. L., and Jackson, R. J. (1999). Polypyrimidine-tract binding protein (PTB) is necessary, but not sufficient, for efficient internal initiation of translation of human rhinovirus-2 RNA. *RNA* **5:**344–359.

Iseni, F., Garcin, D., Nishio, M., Kedersha, N., Anderson, P., and Kolakofsky, D. (2002). Sendai virus trailer RNA binds TIAR, a cellular protein involved in virus-induced apoptosis. *EMBO J.* **21:**5141–5150.

Isken, O., Grassmann, C. W., Sarisky, R. T., Kann, M., Zhang, S., Grosse, F., Kao, P. N., and Behrens, S.-E. (2003). Members of the NF90/NFAR protein group are involved in the life cycle of a positive-strand RNA virus. *EMBO J.* **22:**5655–5665.

Kääriäinen, L., and Ahola, T. (2002). Functions of Alphavirus nonstructural proteins in RNA replication. *Prog. Nucl. Acid Res. Mol. Biol.* **71:**187–222.

Kaminski, A., and Jackson, R. J. (1998). The polypyrimidine tract binding protein (PTB) requirement for internal initiation of translation of cardiovirus RNAs is conditional rather than absolute. *RNA* **4:**626–638.

Kedersha, N. L., Gupta, M., Li, W, Miller, I., and Anderson, P. (1999). RNA-binding proteins TIA-1 and TIAR link the phosphorylation of eIF-2α to the assembly of mammalian stress granules. *J. Cell Biol.* **147:**1431–1441.

Khromykh, A. A., Meka, H., Guyatt, K. J., and Westaway, E. G. (2001). Essential role of cyclization sequences in Flavivirus RNA replication. *J. Virol.* **75:**6719–6728.

Klovins, J., and van Duin, J. (1999). A long-range pseudoknot in Q_β RNA is essential for replication. *J. Mol. Biol.* **294**:875–884.

Kolakofsky, D., Le Mercier, P., Iseni, F., and Garcin, D. (2004). Viral RNA polymerase scanning and the gymnastics of Sendai virus RNA synthesis. *Virology* **318**:463–473.

Kushner, D. B., Lindenbach, B. D., Grdzelishvili, V. Z., Noueiry, A. O., Paul, S. M., and Ahlquist, P. (2003). Systematic, genome-wide identification of host genes affecting replication of a positive-strand RNA virus. *Proc. Natl. Acad. Sci. USA* **100**:15764–15769.

Lai, M. M. C. (1998). Cellular factors in the transcription and replication of viral RNA genomes: A parallel to DNA-dependent RNA transcription. *Virology* **244**:1–12.

Lamb, R. A., and Horvath, C. M. (1991). Diversity of coding strategies in influenza viruses. *Trends Genetics* **7**:261–266.

Lan, S., Wang, H., Jiang, H., Mao, H., Liu, X., Zhang, X., Hu, Y., Xiang, L., and Yuan, Z. (2003). Direct interaction between α-actinin and hepatitis C virus NS5B. *FEBS Lett.* **554**:289–294.

Lee, W. M., Ishikawa, M., and Ahlquist, P. (2001). Mutation of host $\Delta9$ fatty acid desaturase inhibits brome mosaic virus RNA replication between template recognition and RNA synthesis. *J. Virol.* **75**:2097–2106.

Li, H., Li, W. X., and Ding, S. W. (2002a). Induction and suppression of RNA silencing by an animal virus. *Science* **296**:1319–1321.

Li, H.-P., Zang, X., Duncan, R., Comai, L., and Lai, M. M. C. (1997). Heterogeneous nuclear ribonucleoprotein A1 binds to the transcription-regulatory region of mouse hepatitis virus RNA. *Proc. Natl. Acad. Sci. USA* **94**:9544–9549.

Li, W., Li, Y., Kedersha, N., Anderson, P., Emera, M., Swiderek, K. M., Moreno, G. T., and Brinton, M. A. (2002b). Cell proteins TIA-1 and TIAR interact with the 3′ stem-loop of the West Nile virus complementary minus-strand RNA and facilitate virus replication. *J. Virol.* **76**:11989–12000.

Mackenzie, J. M., Jones, M. K., and Westaway, E. G. (1999). Markers for *trans*-Golgi membranes and the intermediate compartment localize to induced membranes with distinct replication functions in flavivirus-infected cells. *J. Virol.* **73**:9555–9567.

Más, P., and Beachy, R. N. (1999). Replication of tobacco mosaic virus on endoplasmic reticulum and role of the cytoskeleton and virus movement protein in intracellular distribution of viral RNA. *J. Cell Biol.* **147**:945–958.

Miller, D. J., Schwartz, M. D., and Ahlquist, P. (2001). Flock house virus RNA replicates on outer mitochondrial membranes in *Drosophila* cells. *J. Virol.* **75**:11664–11676.

Miller, D. J., Schwartz, M. D., Dye, B. T., and Ahlquist, P. (2003). Engineered retargeting of viral RNA replication complexes to an alternative intracellular membrane. *J. Virol.* **77**:12193–12202.

Miller, W. A., and Koev, G. (2000). Synthesis of subgenomic RNAs by positive-strand RNA viruses. *Virology* **273**:1–8.

Momose, F., Basler, C. F., O'Neill, R. E., Iwamatsu, A., Palese, P., and Nagata, K. (2001). Cellular splicing factor RAF-2p48/NPI-5/BAT1/UAP56 interacts with the influenza virus nucleoprotein and enhances viral RNA synthesis. *J. Virol.* **75**:1899–1908.

Moradpour, D., Gosert, R., Egger, D., Penin, F., Blum, H. E., and Bienz, K. (2003). Membrane association of hepatitis C virus nonstructural poteins and identification of the membrane alteration that harbors the vial replication complex. *Antiviral Res.* **60**:103–109.

Moyer, S. A., Baker, S. C., and Lessard, J. L. (1986). Tubulin: A factor necessary for the synthesis of both Sendai virus and vesicular stomatitis virus RNAs. *Proc. Natl. Acad. Sci. USA* **83**:5405–5409.

Moyer, S. A., Baker, S. C., and Horikami, S. M. (1990). Host cell proteins required for measles virus reproduction. *J. Gen. Virol.* **71**:775–783.

Murray, K. E., and Barton, D. J. (2003). Poliovirus CRE-dependent VPg uridylylation is required for positive-strand RNA synthesis but not for negative-strand RNA synthesis. *J. Virol.* **77**:4739–4750.

Nanda, S. K., and Leibowitz, J. L. (2001). Mitochondrial aconitase binds to the 3′ untranslated region of the mouse hepatitis virus genome. *J. Virol.* **75**:3352–3362.

Nguyen, M., and Haenni, A.-L. (2003). Expression strategies of ambisense viruses. *Virus Res.* **93**:141–150.

Noueiry, A. O., Chen, J., and Ahlquist, P. (2000). A mutant allele of essential, general translation initiation factor *DED1* selectively inhibits translation of a viral mRNA. *Proc. Natl. Acad. Sci. USA* **97**:12985–12990.

Noueiry, A. O., and Ahlquist, P. (2003). Brome mosaic virus RNA replication: Revealing the role of the host in RNA virus replication. *Annu. Rev. Phytopathol.* **41**:77–98.

Noueiry, A. O., Diez, J., Falk, S. P., Chen, J., and Ahlquist, P. (2003). Yeast Lsm1p-7p/Pat1p deadenylation-dependent mRNA-decapping factors are required for brome mosaic virus genomic RNA translation. *Mol. Cell. Biol.* **23**:4094–4106.

Ogino, T., Yamadera, T., Nonaka, T., Imajoh-Ohmi, S., and Mizumoto, K. (2001). Enolase, a cellular glycolytic enzyme, is required for efficient transcription of Sendai virus genome. *Biochem. Biophys. Res. Commun.* **285**:447–455.

Oglesbee, M. J., Liu, Z., Kenney, H., and Brooks, C. L. (1996). The highly inducible member of the 70 kDa family of heat shock proteins increases canine distemper virus polymerase activity. *J. Gen. Virol.* **77**:2125–2135.

Ohshima, K., Taniyama, T., Yamanaka, T., Ishikawa, M., and Naito, S. (1998). Isolation of a mutant *Arabidopsis thaliana* carrying two simultaneous mutations affecting tobacco mosaic virus multiplication within a single cell. *Virology* **243**:472–481.

Osman, T. A. M., and Buck, K. W. (1997). The tobacco mosaic virus RNA polymerase complex contains a plant protein related to the RNA-binding subunit of yeast eIF-3. *J. Virol.* **71**:6075–6082.

Paddison, P. J., Silva, J. M., Conklin, D. S., Schlabach, M., Li, M., Aruleba, S., Balija, V., O'Shaughnessy, A., Gnoj, L., Scobie, K., Chang, K., Westbrook, T., Cleary, M., Sachidanandam, R., McCombie, W. R., Elledge, S. J., and Hannon, G. J. (2004). A resource for large-scale RNA-interference-based screens in mammals. *Nature* **428**:427–431.

Parsley, T. B., Towner, J. S., Blyn, L. B., Ehrenfeld, E., and Semler, B. L. (1997). Poly(rC) binding protein 2 forms a ternary complex with the 5′-terminal sequences of poliovirus RNA and the viral 3CD proteinase. *RNA* **3**:1124–1134.

Pasternak, A. O., van den Born, E., Spaan, W. J. M., and Snijder, E. J. (2001). Sequence requirements for RNA strand transfer during nidovirus discontinuous subgenomic RNA synthesis. *EMBO J.* **20**:7220–7228.

Pedersen, K. W., van der Meer, Y., Roos, N., and Snijder, E. J. (1999). Open reading frame 1a-encoded subunits of the arterivirus replicase induce endoplasmic reticulum-derived double-membrane vesicles which carry the viral replication complex. *J. Virol.* **73**:2016–2026.

Perez, L., Guinea, R., and Carrasco, L. (1991). Synthesis of Semliki Forest virus RNA requires continuous lipid synthesis. *Virology* **183**:74–82.

Prod'homme, D., Le Panse, S., Drugeon, G., and Jupin, I. (2001). Detection and subcellular localization of the turnip yellow mosaic virus 66K replication protein in infected cells. *Virology* **281**:88–101.

Pyper, J. M., Clements, J. E., and Zink, M. C. (1998). The nucleolus is the site of Borna disease virus RNA transcription and replication. *J. Virol.* **72**:7697–7702.

Quadt, R., Kao, C. C., Bowning, K. S., Hershberger, R. P., and Ahlquist, P. (1993). Characterization of a host protein associated with brome mosaic virus RNA-dependent RNA polymerase. *Proc. Natl. Acad. Sci. USA* **90:**1498–1502.

Ritzenthaler, C., Laporte, C., Gaire, F., Dunoyer, P., Schmitt, C., Duval, S., Piéquet, A., Loudes, A. M., Rohfritsch, O., Stussi-Garaud, C., and Pfeiffer, P. (2002). Grapevine fanleaf virus replication occurs on endoplasmic reticulum-derived membranes. *J. Virol.* **76:**8808–8819.

Sadowy, E., Milner, M., and Haenni, A.-L. (2001). Proteins attached to viral genomes are multifunctional. *Adv. Virus Res.* **57:**185–262.

Sawicki, D., Wang, T., and Sawicki, S. (2001). The RNA structures engaged in replication and transcription of the A59 strain of mouse hepatitis virus. *J. Gen. Virol.* **82:**385–396.

Schaad, M. C., Jensen, P. E., and Carrington, J. C. (1997). Formation of plant RNA virus replication complexes on membranes: role of an endoplasmic reticulum-targeted viral protein. *EMBO J.* **16:**4049–4059.

Schaad, M. C., Anderberg, R. J., and Carrington, J. C. (2000). Strain-specific interaction of the tobacco etch virus NIa protein with the translation initiation factor eIF4E in the yeast two-hybrid system. *Virology* **273:**300–306.

Schlegel, A., Giddings, T. H., Ladinsky, M. S., and Kirkegaard, K. (1996). Cellular origin and ultrastructure of membranes induced during poliovirus infection. *J. Virol.* **70:**6576–6588.

Schuppli, D., Georgijevic, J., and Weber, H. (2000). Synergism of mutations in bacterio-phage Qβ RNA affecting host factor dependence of Qβ replicase. *J. Mol. Biol.* **295:**149–154.

Sesti, F., Shih, T. M., Nikolaeva, N., and Goldstein, S. A. N. (2001). Immunity to K1 killer toxin: Internal TOK1 blockade. *Cell* **105:**637–644.

Shen, R., and Miller, W. A. (2004). The 3′ untranslated region of tobacco necrosis virus RNA contains a barley yellow dwarf virus-like cap-independent translation element. *J. Virol.* **78:**4655–4664.

Shen, X., and Masters, P. S. (2001). Evaluation of the role of heterogeneous nuclear ribonucleoprotein A1 as a host factor in murine coronavirus discontinous transcription and genome replication. *Proc. Natl. Acad. Sci. USA* **98:**2717–2722.

Shi, S. T., Huang, P., Li, H.-P., and Lai, M. M. C. (2000). Heterogeneous nuclear ribo-nucleoprotein A1 regulates RNA synthesis of a cytoplasmic virus. *EMBO J.* **19:**4701–4711.

Shi, S. T., Yu, G.-Y., and Lai, M. M. C. (2003). Multiple type A/B heterogeneous nuclear ribonucleoproteins (hnRNPs) can replace hnRNP A1 in mouse hepatitis virus RNA synthesis. *J. Virol.* **77:**10584–10593.

Sit, T. L., Vaewhongs, A. A., and Lommel, S. A. (1998). RNA-mediated trans-activation of transcription from a viral RNA. *Science* **281:**829–832.

Snijder, E. J., van Tol, H., Roos, N., and Pedersen, K. W. (2001). Nonstructural proteins 2 and 3 interact to modify host cell membranes during the formation of the arterivirus replication complex. *J. Gen. Virol.* **82:**985–994.

Song, J., Nagano-Fujii, M., Wang, F., Florese, R., Fujita, T., Ishido, S., and Hotta, H. (2000). Nuclear localization and intramolecular cleavage of N-terminally deleted NS5A protein of hepatitis C virus. *Virus Res.* **69:**109–117.

Suhy, D. A., Giddings, T. H., and Kirkegaard, K. (2000). Remodeling the endoplasmic reticulum by poliovirus infection and by individual viral proteins: an autophagy-like origin for virus-induced vesicles. *J. Virol.* **74:**8953–8965.

Takagi, T., Iwama, M., Seta, K., Kanda, T., Tsukamoto, T., Tominaga, S., and Mizumoto, K. (1996). Positive and negative host factors for Sendai virus transcription and their organ distribution in rats. *Arch. Virol.* **141:**1623–1635.

Tange, T. Ø., Damgaard, C. K., Guth, S., Valcárcel, J., and Kjems, J. (2001). The hnRNP A1 protein regulates HIV-1 tat splicing via a novel intron silencer element. *EMBO J.* **20:**5748–5758.

Taya, Y. (1997). RB kinases and RB-binding proteins: new points of view. *Trends Biochem. Sci.* **22:**14–17.

Tomita, Y., Mizuno, T., Díez, J., Naito, S., Ahlquist, P., and Ishikawa, M. (2003). Mutation of host *dnaJ* homolog inhibits brome mosaic virus negative-strand RNA synthesis. *J. Virol.* **77:**2990–2997.

Tomonaga, K., Kobayashi, T., and Ikuta, K. (2002). Molecular and cellular biology of Borna disease virus infection. *Microbes Infect.* **4:**491–500.

Tsujimoto, Y., Numaga, T., Ohshima, K., Yano, M.-A., Ohsawa, R., Goto, D. B., Naito, S., and Ishikawa, M. (2003). *Arabidopsis TOBAMOVIRUS MULTIPLICATION (TOM) 2* locus encodes a transmembrane protein that interacts with TOM1. *EMBO J.* **22:**335–343.

Ulloa, L., Serra, R., Asenjo, A., and Villanueva, N. (1998). Interactions between cellular actin and human respiratory syncytial virus (HRSV). *Virus Res.* **53:**13–25.

Urcuqui-Inchima, S., Haenni, A.-L., and Bernardi, F. (2001). Potyvirus proteins: a wealth of functions. *Virus Res.* **74:**157–175.

van der Heijden, M. W., Carette, J. E., Reinhoud, P. J., Haegi, A., and Bol, J. F. (2001). Alfalfa mosaic virus replicase proteins P1 and P2 interact and colocalize at the vacuolar membrane. *J. Virol.* **75:**1879–1887.

van der Heijden, M. W., and Bol, J. F. (2002). Composition of alphavirus-like replication complexes: Involvement of virus and host encoded proteins. *Arch. Virol.* **147:**875–898.

Waggoner, S., and Sarnow, P. (1998). Viral ribonucleoprotein complex formation and nucleolar-cytoplasmic relocalization of nucleolin in poliovirus-infected cells. *J. Virol.* **72:**6699–6709.

Walter, B. L., Parsley, T. B., Ehrenfeld, E., and Semler, B. L. (2002). Distinct poly(rC) binding protein KH domain determinants for poliovirus translation initiation and viral RNA replication. *J. Virol.* **76:**12008–12022.

Wang, P., Palese, P., and O'Neill, R. E. (1997). The NPI-1/NPI-3 (karyopherin α) binding site on the influenza A virus nucleoprotein NP is a nonconventional nuclear localization signal. *J. Virol.* **71:**1850–1856.

Wimmer, E. (1982). Genome-linked proteins of viruses. *Cell* **28:**199–201.

Wolff, T., O'Neill, R. E., and Palese, P. (1998). NS1-binding protein (NS1-BP): a novel human protein that interacts with the influenza A virus nonstructural NS1 protein is relocalized in the nuclei of infected cells. *J. Virol.* **72:**7170–7180.

Wolin, S. L., and Cedervall, T. (2002). The La protein. *Annu. Rev. Biochem.* **71:**375–403.

Wu, S.-X., and Kaesberg, P. (1991). Synthesis of template-sense, single-strand flockhouse virus RNA in a cell-free replication system. *Virology* **183:**392–396.

Wu, S.-X., Ahlquist, P., and Kaesberg, P. (1992). Active complete *in vitro* replication of nodavirus RNA requires glycerophospholipid. *Proc. Natl. Acad. Sci. USA* **89:**11136–11140.

Yamanaka, T., Ohta, T., Takahashi, M., Meshi, T., Schmidt, R., Dean, C., Naito, S., and Ishikawa, M. (2000). *TOM1*, an *Arabidopsis* gene required for efficient multiplication of a tobamovirus, encodes a putative transmembrane protein. *Proc. Natl. Acad. Sci. USA* **97:**10107–10112.

Yang, A. D., Barro, M., Gorziglia, and Patton, J. T. (2004). Translation enhancer in the 3′-untranslated region of rotavirus gene 6 mRNA promotes expression of the major capsid protein VP6. *Arch. Virol.* **149:**303–321.

ADVANCES IN VIRUS RESEARCH, VOL 65

TOMATO SPOTTED WILT VIRUS PARTICLE ASSEMBLY AND THE PROSPECTS OF FLUORESCENCE MICROSCOPY TO STUDY PROTEIN–PROTEIN INTERACTIONS INVOLVED

Marjolein Snippe,* Rob Goldbach,[†] and Richard Kormelink[†]

*Department of Asthma, Allergy, and Respiratory Diseases
King's College, London, WC2R 2LS United Kingdom
[†]Laboratory of Virology, Department of Plant Sciences, Wageningen University
Wageningen, 6700 HB The Netherlands

I. Introduction

The site of virus assembly for the animal-infecting members of the *Bunyaviridae* was long ago identified as the Golgi complex. However, for the plant-infecting bunyavirus *Tomato spotted wilt virus* (TSWV) this has long been an enigma, even though most cytopathological structures related to the assembly of TSWV were already observed in infected plants decades past. Later, the site of particle assembly was identified as the Golgi complex. Whereas, for the animal-infecting counterparts, particle assembly occurs by budding of ribonucleoproteins into the vacuolized lumen of the Golgi complex, TSWV particles arise as the result of enwrapment of ribonucleoproteins by Golgi cisternae, leading to the formation of so-called doubly enveloped virus (DEV) particles. Fusion of the latter to each other and likely to ER-derived membranes lead to the formation of large vesicles in which mature TSWV particles accumulate and retain. Unlike what occurs in plant cells, TSWV is secreted from its insect vector (thrips) cells, in which the

63

0065-3527/05 $35.00
DOI: 10.1016/S0065-3527(05)65003-8

virus also replicates. The formation of DEVs has never been observed in thrips cells; the excretion of TSWV from these cells more resembles the infection and release route of animal-infecting bunyaviruses from mammalian cells. This duality in assembly routes of TSWV, retention (plant cells) versus excretion (thrips cells), likely reflects the adaptation of TSWV to infect plant cells while retaining the capacity to infect its insect vector. It is generally assumed that the glycoproteins forming the spikes on the lipid envelope of the virus particle are the major determinants in the process of virus assembly. Hence, various efforts have been made to further characterize these proteins and to study their maturation and interactions compared to other nonstructural proteins that lead to particle assembly.

During the last decade, virological research has slowly shifted from molecular biology to cell biology focused studies. The latter is making increasing use of bioimaging, which is possible due to a greater number of various fluorescence microscopy techniques being developed that not only allow visualization of proteins in a living cell, but also allow investigation of protein-protein interactions, protein-folding studies, etc.

In the first part of this chapter, the current status of TSWV particle assembly and the involvement of structural proteins in this process will be described and compared to what is known of the animal-infecting counterparts. The second part will describe more recently developed fluorescence microscopy techniques that enable the analysis of *in vivo* protein-protein interactions and how some of these techniques have been employed, as exemplified by recently obtained results, to unravel protein-protein interactions during TSWV particle assembly.

A. *Bunyaviridae and the Genus Tospovirus*

Tomats spotted wilt virus (TSWV) is the type species of the genus *Tospovirus* within the family *Bunyaviridae*. Tospoviruses are restricted to plant hosts, whereas all other members of this family infect animals. The latter ones are classified in 4 genera: *Orthobunyavirus*, *Nairovirus*, *Hantavirus*, and *Phlebovirus*. Most members of the *Bunyaviridae* are arthropod-borne viruses, Hantaviruses being the exception (Table I) (Elliott, 1996). At this point, it should be noted that in this review, the term "bunyavirus" will be used to refer to any member of the *Bunyaviridae*, and not only to members of the genus *Orthobunyavirus*.

At this time, 14 different tospovirus species, and one tentative (un-published results) have been recognized, of which TSWV has the

TABLE I

The *Bunyaviridae* Family

Family	Genus	Type species	Arthropod vector
Bunyaviridae	Orthobunyavirus	Bunyamwera	Mosquitoes
	Phlebovirus	Rift Valley Fever	Mosquitoes, Ticks, Phlebotomines
	Nairovirus	Crimean Congo Hemorrhagic fever	Mosquitoes
	Hantavirus	Hantaan	None
	Tospovirus	Tomato spotted wilt	Thrips

broadest host range and has become most widely distributed (Table II). The cumulative host range of tospoviruses encompasses over 1000 different plant species, mostly *Compositae* and *Solanaceae* (Peters, 1998). High yield losses have been reported in many different economically important agricultural and ornamental crops (Goldbach and Peters, 1994). Among these are dicots as well as monocots. Tospoviruses are exclusively transmitted by a limited number of phytophagous thrips (order *Thripidae*) in a propagative manner (Wijkamp *et al.*, 1993). So far, eight different species have been demonstrated to act as a vector, with *Frankliniella occidentalis* (also known as California thrips, Western flower thrips, Alfalfa thrips) as the most important one (Table II).

B. Particle Morphology and Expression Strategy of the Viral Genome

Tospoviruses are relatively easy to identify in infected tissues due to their unique particle morphology. Like all other members of the *Bunyaviridae*, virions are spherical lipid-bound particles, 80–120 nm in diameter, covered with spike projections consisting of two glycoproteins G1 and G2 (Fig. 1, left). The core consists of pseudo-circular ribonucleocapsids (RNPs) (Fig. 1, right), each consisting of a viral RNA segment tightly enclosed by the nucleoprotein (N) and minor amounts of a large protein (L), the putative viral RNA-dependent RNA polymerase (Mohamed, 1981; Mohamed *et al.*, 1973; Tas *et al.*, 1977; Van den Hurk *et al.*, 1977; Van Poelwijk *et al.*, 1993; Verkleij *et al.*, 1982). All members of the *Bunyaviridae*, including the tospoviruses, replicate in the cytoplasm of the cell.

TABLE II
Established *Tospoviruses* and their Geographical Distribution, Host Range, and Vectors

Tospovirus	Geographical distribution	Host range	Reported vectors	References
Tomato spotted wilt virus (TSWV)	Worldwide	Very broad (almost 1000 species)	*Frankliniella fusca*	Francki *et al.*, 1991; Goldbach and Peters, 1994; Murphy *et al.*, 1995; Sakimura, 1963
			F. occidentalis	Gardner *et al.*, 1935
			F. intonsa	Wijkamp *et al.*, 1995
			F. schultzei	Samuel *et al.*, 1930; Wijkamp *et al.*, 1995
			Thrips palmi	Fujisawa *et al.*, 1988
			T. setosus	Fujisawa *et al.*, 1988; Tsuda *et al.*, 1996
			T. tabaci	Pittman, 1927
			F. bispinosa	Webb *et al.*, 1998
Impatiens necrotic spot virus (INSV)	U.S.A., Europe	Mainly ornamentals	*F. occidentalis*	De Ávila *et al.*, 1992; DeAngelis *et al.*, 1994; Law and Moyer, 1990; Louro, 1996; Marchoux *et al.*, 1991; Vaira *et al.*, 1993; Wijkamp and Peters, 1993
Tomato chlorotic spot virus (TCSV)	Brazil	Tomato	*F. occidentalis*	De Ávila *et al.*, 1993; De Ávila *et al.*, 1990; Francki *et al.*, 1991; Granval de Milan *et al.*, 1998; Wijkamp *et al.*, 1995
			F. intonsa	Wijkamp *et al.*, 1995
			F. schultzei	Wijkamp *et al.*, 1995

Virus	Location	Host	Vector	References
Groundnut ringspot virus (GRSV)	Argentine, Brazil, South Africa	Groundnut, tomato	*F. occidentalis*	De Ávila *et al.*, 1993; Dewey *et al.*, 1993; De Ávila *et al.*, 1990; Wijkamp *et al.*, 1995
Watermelon silver mottle virus (WSMV)	Japan, Taiwan	Watermelon, other cucurbits	*F. schultzei* / *Thrips palmi*	Wijkamp *et al.*, 1995; Heinze *et al.*, 1995; Kameya-Iwaki *et al.*, 1984; Yeh and Chang, 1995; Yeh *et al.*, 1992
Groundnut bud necrosis virus (GBNV)*	India, South-East Asia	Groundnut	*F. schultzei* / *Thrips palmi*	Amin *et al.*, 1981; Palmer *et al.*, 1990; Reddy *et al.*, 1992; Satyanarayana *et al.*, 1996a; Vijayalakshmi, 1994; Lakshmi *et al.*, 1995; Palmer *et al.*, 1990; Vijayalakshmi, 1994
Watermelon bud necrosis virus (WBNV)	India	Watermelon	*Thrips palmi*	Jain *et al.*, 1998; Singh and Krishnareddy, 1996; Yeh *et al.*, 1992
Iris yellow spot virus (IYSV)	The Netherlands, Brazil, Israel, USA	Iris, onion, leek	*T. tabaci*	Bezerra *et al.*, 1999; Cortes *et al.*, 1998; Gera *et al.*, 1998a,b; Hall *et al.*, 1993; Resende *et al.*, 1996
Groundnut yellow spot virus (GYSV)*	India, Thailand	Groundnut		Reddy *et al.*, 1991; Satyanarayana *et al.*, 1998

(*Continued*)

TABLE II (Continued)

Tospovirus	Geographical distribution	Host range	Reported vectors	References
Chrysanthemum stem necrosis virus (CSNV)	Brazil	Chrysanthemum, tomato	*F. occidentalis*	Nagata and De Ávila, 2000; Pozzer *et al.*, 1999; Resende *et al.*, 1996
Zucchini lethal chlorosis virus (ZLCV)	Brazil	Zucchini	*F. zucchini* (not confirmed)	Pozzer *et al.*, 1999; Resende *et al.*, 1996
Groundnut chlorotic fan-spot virus (GCFV)*	Taiwan	Groundnut		Chen and Chiu, 1996; Chu *et al.*, 2001
Melon yellow spot virus (MYSV)[†]	Thailand, Taiwan, Japan	Physalis, cucurbits	*T. palmi*	Cortez *et al.*, 2001; Gera *et al.*, 1998b; Kato and Hanada, 2000; Takeuchi *et al.*, 2001
Capsicum chlorosis virus	Australia, Southeast Asia	Capsicum, gloxinia		McMichael *et al.*, 2002
Tentative species:				
Tomato yellow ring virus	Iran (Middle-East)	Tomato, Gazania, Chrysanthemum		

* In literature GBNV is also named peanut bud necrosis (PBNV); GYSV is also peanut yellow spot virus (PYSV); GCFV is also peanut chloritic fan-spot virus (PCFV).

[†] An isolate of MYSV has been characterized and reported in literature as Physalis severe mottle virus (PSMV).

F. schultzei was later correctly identified as *T. palmi* (Palmer *et al.*, 1990). There is no transmission of GBNV by *F. schultzei* (Vijayalakshmi, 1994). (Modified from Wijkamp [1995] and Maris [2002].)

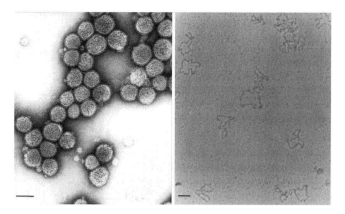

FIG 1. Electron micrograph of purified TSWV particles (left), and TSWV RNPs (right). Size bar indicates 100 nm.

The viral genome consists of three single-stranded, linear RNA segments, which according to their sizes have been denoted small (S) RNA, medium (M) RNA, and large (L) RNA, in analogy to the animal-infecting bunyaviral genomes (De Haan *et al.*, 1989; van den Hurk *et al.*, 1977; Verkleij *et al.*, 1982).

Complete genome sequences have become available for only a few tospovirus species such as TSWV, INSV, and PBNV (De Haan *et al.*, 1990, 1991, 1992; Gowda *et al.*, 1998; Kormelink *et al.*, 1992a; Law *et al.*, 1992; van Poelwijk *et al.*, 1997; Satyanarayana *et al.*, 1996a,b). For other species, only the S and/or M RNA or only the N gene has been sequenced. These sequence data have resulted in a genomic organization and expression strategy depicted for TSWV in Fig. 2, and thus far representative for all tospoviruses.

The first stretch of 8 nucleotides at the 3′ terminal end of all tospoviral RNA segments are conserved and complementary to the 5′ end, a feature that is common for all segmented, negative-strand RNA viruses of the *Bunyaviridae*, *Orthomyxoviridae,* and *Arenaviridae* (De Haan *et al.*, 1989). Moreover, the nucleotide sequence of the L, M, and S RNA segments shows total nucleotide sequence complementarity of the 5′ and 3′ terminal ends for the first 15 nucleotides (De Haan *et al.*, 1990, 1991; Kormelink *et al.*, 1992a). Within each RNA segment, this complementarity extends up to about 65 nucleotides, allowing the formation of a so-called panhandle structure and thereby accounting for the pseudocircular appearance of the RNPs (Fig. 1, right). The

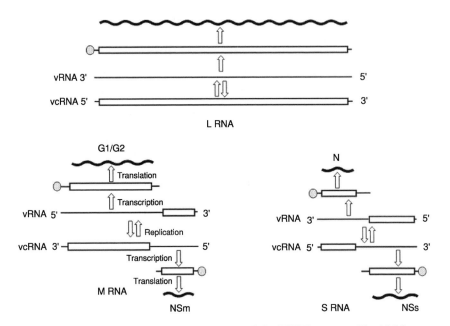

vRNA 3' ——————————————— 5'

vcRNA 5' ——————————————— 3'

L RNA

G1/G2

Translation

Transcription

vRNA 5' ————————————— 3'

Replication

vcRNA 3' ————————————— 5'

Transcription

Translation

M RNA

NSm

N

vRNA 3' ————————————— 5'

vcRNA 5' ————————————— 3'

S RNA   NSs

FIG 2. Organisation and expression strategy of the TSWV genome. The highly conserved terminal sequences are indicated by a black box. The small dotted circles at the 5' end of ([sub]genomic-length) viral transcripts represent non-viral leader sequences and result from transcription initiation by cap-snatching.

terminal sequences are conserved among members of the same genus, but differ between genera of the *Bunyaviridae* (Elliott, 1996).

Purified virus preparations contain all three genomic segments, but these are usually not present in equimolar amounts and the S segment mostly predominates. Moreover, viral (v) RNA is often observed in excess over viral complementary (vc) RNA strands, which may possibly reflect the amounts synthesized during replication (Bouloy *et al.*, 1973/1974; Elliott, 1990; Eshita *et al.*, 1985; Gentsch *et al.*, 1977; Ihara *et al.*, 1985a; Kormelink *et al.*, 1992b; Objieski and Murphy, 1977; Objieski *et al.*, 1976; Pettersson and Kääriäinen, 1973). Encapsidation of viral mRNAs into virus particles did not appear to take place.

The L RNA segment of TSWV is 8897 nucleotides (nt) and of complete negative polarity. It contains one large open reading frame (ORF) in the viral complementary (vc) strand, coding for a protein of 331.5 kDa (De Haan *et al.*, 1990), the putative RNA dependent RNA polymerase. The TSWV M RNA is 4821 nt and of ambisense polarity, containing two

TOMATO SPOTTED WILT VIRUS PARTICLE ASSEMBLY 71

nonoverlapping ORFs, one in the viral (v) strand coding for a nonstructural protein (NSm) of 33.6 kDa, the putative cell-to-cell movement protein, and one in the vc-strand, encoding the precursor to the glycoproteins of 127.4 kDa (Kormelink et al., 1992a, 1994; Storms et al., 1995). The TSWV S RNA is 2918 nt and also of ambisense polarity, encoding a nonstructural protein (NSs) of 52.1 kDa in the v-strand that acts as a suppressor of silencing, and the nucleoprotein (N) of 28.9 kDa in the vc-strand (Bucher et al., 2003; De Haan et al., 1990; Takeda et al., 2002).

Comparison of the genomes of representatives of all genera within the *Bunyaviridae* shows a few differences (Fig. 3). Firstly, only members of the genera *Tospovirus* and *Phlebovirus* have an ambisense S RNA segment that codes for N and NSs proteins. The S RNA of orthobunyaviruses codes for similar proteins, but in overlapping reading frames, whereas the nairo- and hanta-virus S RNA segments only encode the N protein. Secondly, tospoviruses contain one extra gene positioned in ambisense arrangement on the M RNA. The absence from the animal-infecting bunyavirus genomes is easily explained, as the corresponding NSm gene product facilitates cell-to-cell movement of tospoviral RNPs through plasmodesmata, a process that relates to plants as being the hosts of tospoviruses. Thirdly, the L RNA-encoded RdRp shows a relatively large size difference between different genera. Whereas the RdRp of orthobunya-, phlebo-, and hantaviruses are approximately 250 kDa, that of tospoviruses 330 kDa the nairovirus RdRp is almost twice the size (459 kDa). Whether these differences reflect additional biological functions required for certain viruses (genera) remains unclear.

Expression of the viral genes occurs by the synthesis of sub-genomic-sized mRNAs, which differ from full-length antigenomic RNA strands by the additional presence of a nonviral leader sequence at the 5′-ends, generally 10–20 nucleotides in length (Fig. 2) (Duijsings et al., 1999, 2001; Kormelink et al., 1992c; van Knippenberg et al., 2002; Van Poelwijk et al., 1996). This sequence is the result of cap-snatching, a process by which the viral polymerase, encompassing an endonuclease activity, cleaves off capped-RNA leader sequences from cellular mRNA molecules to use these as primers for transcription initiation on the viral genome (Braam et al., 1983; Plotch et al., 1981; Ulmanen et al., 1981). This feature is common for all members of the *Bunyaviridae*, *Arenaviridae*, *Orthomyxoviridae*, and the floating genus *Tenuivirus* (Bishop et al., 1983; Bouloy et al., 1990; Braam et al., 1983; Caton and Robertson, 1980; Collett, 1986; Dhar et al., 1980; Garcin and Kolakofsky, 1990; Huiet et al., 1993; Patterson and Kolakofsky, 1984; Plotch et al., 1981; Raju et al., 1990; Ramirez et al., 1995; Shimizu

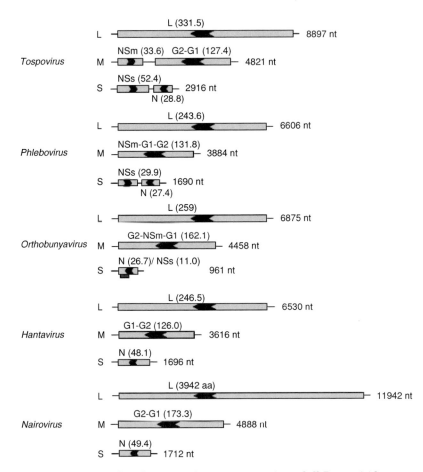

Fɪɢ 3. Comparison of viral genomes from representatives of all *Bunyaviridae* genera. Sizes of the RNA segments (in nucleotides) and encoded gene products (in kilodaltons, except for Nairovirus L) are indicated. Arrows reflect the orientation of the ORF relative to the viral genomic strand. For the tospoviruses and phleboviruses, the NSs gene is located in an ambisense arrangement with the N protein; whereas, for the orthobunya-viruses, the NSs gene is oriented in the same polarity, overlapping the N ORF. (See Color Insert.)

et al., 1996; Simons and Pettersson, 1991; Ulmanen *et al.*, 1981). In contrast to transcription initiation, little is known on transcription termination. Bunyaviral transcripts are not polyadenylated, nor do they share a conserved sequence motif that may act as a transcription termination signal. For both the ambisense TSWV M and S RNA segments, the intergenic region, highly rich in A- and U- stretches and thereby able to form a hairpin structure, has been thought to play a role in this process (De Haan *et al.*, 1991; Kormelink *et al.*, 1992a; Maiss *et al.*, 1991). The 3' ends of TSWV S RNA-specific transcripts have been mapped and indicate that transcription terminated near the 3' end of the intergenic hairpin structure; in other words, viral transcripts contain a predicted stem-loop structure at their 3' end (Van Knippenberg, 2005). It is proposed that this 3'-end hairpin may be involved in stimulation of viral mRNA translation through 5'-3' terminal interaction, as has been observed for the 3'-end structural feature of several plant-infecting RNA viruses that lack a poly(A) tail (Fabian and White, 2004; Gallie, 1998; Gallie and Kobayashi, 1994; Leonard *et al.*, 2004; Matsuda and Dreher, 2004; Meulewaeter *et al.*, 2004; Neeleman *et al.*, 2001).

Few other animal-infecting bunyaviruses with an ambisense S RNA segment (i.e., the Uukuniemi [UUK], Punta Toro [PT], and Toscana [TOS] phleboviruses) having the 3' ends of the N and NSs mRNA transcripts have been mapped (Emery and Bishop, 1987; Grò *et al.*, 1992; Simons and Pettersson, 1991). The termination site has been located at the opposite sides of the hairpin in case of PT (Emery and Bishop, 1987), or downstream from the hairpin structure in the case of UUK (Simons and Pettersson, 1991). In the latter's case, both mRNAs contain the hairpin structure and overlap for about 100 nucleotides. This could indicate a termination mechanism based on the formation of the hairpin structure in the nascent transcript, as also suggested for *Tacaribe virus* (Iapalucci *et al.*, 1991) and *Lymphocytic choriomeningitis virus* (LCV) (Meyer and Southern, 1993), members of the *Arenaviridae* (another virus family of ambisense coding ssRNA viruses). However, in case of Toscana phlebovirus, in which the S RNA contains an intergenic region rich in G residues and lacks a clear secondary structure, the 3' termini of the N and NSs mRNAs have been mapped to a hexanucleotide sequence motif conserved in the S RNA segment of two other phleboviruses (i.e., Rift Valley fever and Sandfly Fever Sicilian). This suggests a possible involvement of this sequence motif in termination of transcription (Grò *et al.*, 1992). Intriguingly, this sequence has also been found at the top of the intergenic hairpin of PTV as well as in the template for transcription of the NSs mRNA of UUK. For the entirely negative strand of RNA-encoded genes, even less is known on possible transcription termination signals, but specific sequences have been suggested to play a role.

The promoters for transcription and replication of the viral genome are embedded within the conserved terminal end sequences, as has been demonstrated with reverse genetics systems by the analyses of replicational and transcriptional activity of minigenomes harboring mutagenized termini for the Bunyamwera orthobunyavirus, Rift Valley fever and Uukuniemi phleboviruses (Dunn *et al.*, 1995; Flick and Pettersson, 2001; Flick *et al.*, 2002, 2004; Lopez *et al.*, 1995; Prehaud *et al.*, 1997).

1. Protein Functions

a. L Protein (RNA-Dependent RNA Polymerase) Based on sequence similarities with putative and proven RNA polymerases from other negative-stranded, segmented RNA viruses (e.g., Influenza viral PB1 protein and the Bunyamwera L RNA encoded RNA polymerase), the TSWV 331.5 kDa protein represents the putative viral RNA-dependent RNA polymerase. The sequence similarity is confined to five types of consensus sequences, which are characteristic of RNA polymerases displaying RNA template specificity, the so-called polymerase motifs (Poch *et al.*, 1989; Tordo *et al.*, 1992). One of these motifs, the "SDD" motif, has been shown to be functionally important in several cases (e.g., Bunyamwera L protein and Influenza A polymerase) (Biswas and Nayak, 1994; Jin and Elliott, 1992). For members of the *Nairovirus* genus (Honig *et al.*, 2004; Kinsella et al., 2004; Marriott and Nuttall, 1996), the complete L RNA sequence shows the additional presence of protease and helicase domains, which led to the suggestion that the nairovirus L protein may act as a polyprotein that becomes cleaved and further processed. The TSWV L protein is not processed and can be detected in purified virus and RNP preparations (Van Poelwijk *et al.*, 1993).

In vitro replicase and transcriptase activity of the RdRp have been demonstrated for purified, detergent-disrupted virus preparations of TSWV, LaCrosse, Lumbo, Germiston, Uukuniemi, and Hantaan and, in some cases, depend on the additional presence of the translation machinery to support ongoing transcription (Bellocq *et al.*, 1986; Bellocq and Kolakofsky, 1987; Bouloy and Hannoun, 1976; Chapman *et al.*, 2003; Gerbaud *et al.*, 1987; Patterson *et al.*, 1984; Raju and Kolakofsky, 1986a, b; Ranki and Pettersson, 1975; Schmaljohn and Dalrymple, 1983; Van Knippenberg *et al.*, 2002, 2004; Vialat and Bouloy, 1992).

The L protein encompasses several enzymatic activities such as polymerase and endonuclease. Depending on the circumstances (e.g., presence or absence of viral/host factors), the polymerase may exhibit transcriptase or replicase activity (Van Knippenberg *et al.*, 2002). A free pool of soluble N protein is assumed to play a role in this, based on observations made for the Influenza virus (Beaton and Krug, 1984, 1986). As the L protein does not possess methyl transferase activity required for capping of viral transcripts, but bunyaviruses instead use cap-snatching to provide their transcripts with a m^7G-ppp cap structure, the L protein likely contains a cap-binding domain to retrieve capped cellular mRNA molecules as cap donors. It is likely that the L protein contains additional domains that play a role in other processes than transcription/replication (e.g., host range determinants or matrix function similar as to the M protein of many other negative strand RNA viruses, in light of the large size differences observed between L proteins of different genera).

b. Glycoproteins   The amino-acid sequence of the TSWV glycoprotein precursor contains eight potential N-linked glycosylation sites and several hydrophobic domains (Fig. 4), some of which act as a signal sequence and others as transmembrane domains. G1 (78 kDa) is located carboxy-terminal and G2 (58 kDa) amino terminal within the precursor. To avoid any confusion in the comparison with functional homologs of the bunyaviral glycoproteins, they are more generally referred to as Gn (amino terminal) and Gc (carboxy-terminal). A low but significant sequence homology has been observed with the glycoprotein precursor of *Orthobunyavirus* members Bunyamwera, Snowshoe hare, La Crosse, and Germiston (Fig. 4) (Kormelink *et al.*, 1992a). This homology is mainly restricted to a stretch of amino acids within Gc. Based on this observation and the additional homology between the respective L proteins, tospoviruses are most closely related to orthobunyaviruses. Both Gn and Gc are type I membrane-spanning proteins and have been shown to be glycosylated when expressed in insect and mammalian cells (Kikkert *et al.*, 2001; Kormelink, 1994).

Although the presence of several N-glycosylation sites was predicted for both Gn and Gc, no N-linked glycosylation was detected in Gn from virus particles purified from infected plants. With respect to O-linked glycosylation, opposing data have been presented (Naidu *et al.*, 2004; Whitfield *et al.*, 2004). To date, no evidence for O-linked oligosaccharides of the animal-infecting members of the *Bunyaviridae* has been obtained, but various reports have appeared on N-linked glycosylation (Antic *et al.*, 1992; Lappin *et al.*, 1994; Shi and Elliott, 2004).

Both glycoproteins mature by cotranslational cleavage from the precursor, a process in which ER residing proteases are assumed to be involved. For TSWV, cleavage of the signal sequence has been predicted to occur at amino-acid residue 35 from the N terminus

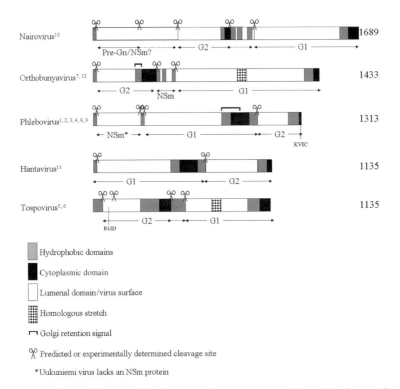

FIG 4. Topology of bunyaviral glycoprotein precursor proteins. Numbers refer to amino acid residues. The sequences of the prototypes of the genera have been used in all cases, except for the *Phlebovirus* genus where the sequence of Punta Toro virus has been used. [1]Andersson *et al*, 1998; [2]Collet *et al.*, 1986; [3]Gerrard and Nichol, 2002; [4]Ihara *et al.*, 1985b; [5]Kikkert *et al.*, 2001; [6]Kormelink *et al*, 1992a; [7]Lees *et al.*, 1986; [8]Matsuoka *et al.*, 1996; [9]Rönnholm and Pettersson, 1987; [10]Sanchez *et al.*, 2002; [11]Schmaljohn *et al.*, 1987; [12]Shi *et al.*, 2004.

(Fig. 4) (von Heijne, 1986). The exact N-terminus of Gc is unknown. For TSWV, as well as its animal-infecting counterparts, cleavage sites have been predicted and some confirmed, but little is known about the signal peptidase involved. For Hantaan virus the glycoprotein precursor is cleaved at a conserved pentapeptide WAASA, which was found absolutely conserved for all hantaviruses (Löber *et al.*, 2001). For Crimean-Congo hemorrhagic fever (CCHF) hantavirus, both Gn and Gc are preceded by tetrapeptides RRLL and RKPL, respectively, which led to the suggestion that endoplasmatic reticulum-residing SKI-1, or related proteases, are involved in their processing (Sanchez *et al.*,

2002). Additional studies revealed that only Gn processing requires subtilase SKI-1, and Gc does not (Vincent *et al.*, 2003). Previously, the surface glycoproteins of Lassa virus, a member of the *Arenaviridae* family, were also shown to be cleaved at an RRLL motif into an N- and C-terminal glycoprotein by the cellular subtilase SKI-1/SP1 (Lenz *et al.*, 2000, 2001).

Analysis of the amino-acid sequence of the TSWV glycoprotein precursor, furthermore, revealed the presence of an Arg-Gly-Asp (RGD)-motif, a putative cell attachment site, within Gn immediately downstream the predicted signal cleavage site (Fig. 4) (Kormelink *et al.*, 1992a; Law *et al.*, 1992). It is recognized that the tospoviral envelope glycoproteins are major determinants for insect transmission and specificity (Wijkamp, 1995), as they may interact with cell-surface receptors on the insect midgut epithelium, the first site of virus infection (Nagata *et al.*, 2002; Tsuda *et al.* 1996; Ullman *et al.*, 1992; Wijkamp *et al.*, 1993), and thereby mediate virus uptake by the thrips (Bandla *et al.*, 1998). Hence, it was speculated that the RGD motif in Gn of TSWV would be involved in the attachment of tospovirus particles to the thrips midgut prior to acquisition (Kormelink *et al.*, 1992a). However, no RGD motif is present in Gn of other tospoviruses, e.g., IYSV, PBNV and WSMV. Furthermore, virus-overlay assays have demonstrated that if the RGD motif in TSWV Gn is responsible for binding to a potential 94 kDa thrips cell receptor, this would probably not occur in the midgut, as the 94 kDa thrips protein was absent from these tissues (Kikkert *et al.*, 1998). On the other hand, studies performed by Bandla *et al.* (1998) showed that both Gn and Gc were involved in virus attachment to the thrips' midgut, and they identified a potential receptor protein of about 50 kDa. More recently, another study again confirmed Gn as the candidate viral ligand for binding to a midgut epithelium receptor (Whitfield *et al.*, 2004).

Also, for the animal-infecting bunyaviruses, only limited information is available on the role of both glycoproteins during the initial infection of a cell. For California encephalitis (CE) virus, neutralization of Gc with monoclonal antibodies, or trypsinization, resulted in loss of infectivity both in mosquitoes and in culture (Hacker *et al.*, 1995). Furthermore, Gc was shown to undergo conformational changes (Pekosz and Gonzalez, 1996) necessary for low pH-mediated entry into cell cultures (Hacker and Hardy, 1997). For LaCrosse virus, a Gc variant was selected that was restricted in its ability to infect mosquitoes when ingested, but not when injected intrathoracically (Sundin *et al.*, 1987), whereas after reversion, the capacity to infect after ingestion was regained, suggesting a Gc requirement for interaction at the midgut level. Other studies by Ludwig *et al.* (1989, 1991) demonstrated the requirement of LaCrosse Gc for infection of cell cultures. Taken together, this indicates an important role for Gc, possibly as the viral attachment protein, in the initial uptake of bunyaviruses in the midgut. Surprisingly, the Gc glycoprotein of all tospoviruses and orthobunya-viruses share a highly conserved peptide sequence in the core of this protein (Fig. 4), which supports the idea of functional homologs and, furthermore, of an involvement of TSWV Gc in the actual attachment during receptor-mediated uptake in vector midgut cells (Pekosz *et al.*, 1995), a function that other studies would attribute to TSWV Gn.

c. NSm Only during the very early stages of a TSWV infection in plant tissues, even prior to the appearance of virus particles, NSm protein is expressed and shown to form tubular structures that ex-tend from plasmodesmata into newly infected cells. These tubules are also formed upon transient expression of NSm in protoplasts in the absence of any other viral protein (Kormelink *et al.*, 1994; Storms *et al.*, 1995). In addition to its capacity to form tubules, NSm is found associated to nonenveloped nucleocapsid aggregates (Kormelink *et al.*, 1994). For these reasons, it has been suggested that the NSm protein is implicated in the cell-to-cell movement of nonenveloped infectious ribonucleocapsid structures of TSWV. This hypothesis is supported by results from microinjection studies of fluorescent dyes into paren-chyma cells of transgenic plants expressing the NSm protein, which reveal a modification of the size exclusion limit of plasmodesmata (Storms *et al.*, 1998). The NSm protein has, furthermore, been shown to interact with N and to reversibly and in a nonspecific manner bind to ssRNA and the panhandle structure (Soellick *et al.*, 2000). The interaction with N was suggested to explain the observed NSm association to RNPs.

For members of the *Orthobunyavirus, Phlebovirus,* and *Nairovirus* genera, the existence of an NSm protein is suggested based on the topology, cleavage products, and maturation of the glycoprotein pre-cursor. However, since tospovirus genomes code for the NSm protein in a separate ORF, while the animal-infecting bunyaviruses encode it as part of the glycoprotein precursor and the tospoviral NSm protein is required for viral movement in plant tissues, it is unlikely that the NSm proteins of the animal-infecting viruses are functionally or struc-turally equivalent to the tospoviral NSm. For Bunyamwera orthobu-nyavirus, the NSm protein has been suggested to play a role in virion formation, based on its Golgi localization (Lappin *et al.*, 1994; Nakitare and Elliott, 1993). For the LaCrosse orthobunyavirus NSm protein, though, such localization profile has not been observed (Bupp *et al.*, 1996).

d. NSs   The distribution of the NSs protein is normally either dispersed throughout the cytoplasm or accumulated in fibrillar inclusion bodies in the cytoplasm of infected cells. A correlation has been observed between the amounts of NSs expression and virulence of tospovirus isolates (Kormelink *et al.*, 1991). However, the function of NSs has long been unknown, due to the absence of any sequence homology to other viral or host proteins. Recently, the NSs protein has been demonstrated to act as a suppressor of RNA silencing (Bucher *et al.*, 2003; Takeda *et al.*, 2002).

Whereas the nairovirus and hantavirus members do not seem to encode an NSs protein, orthobunyaviruses and phleboviruses code for an NSs protein overlapping or in ambisense arrangement with the N ORF, respectively (Fig. 3). For Bunyamwera orthobunyavirus, the NSs protein has been shown to downregulate viral RNA synthesis in a minireplicon system (Weber *et al.*, 2001) and, as was shown for LaCrosse NSs, to be involved in (partial) shutoff of host cell protein synthesis (Bridgen *et al.*, 2001). On the other hand, NSs of RVFV Phlebovirus did not show any effect on viral RNA synthesis (Lopez *et al.*, 1995). Whether this reflects some differences between the orthobunyavirus and phleboviruses, as may seem obvious in light of the observed different genomic locations and sizes of the NSs protein gene, remains a matter of speculation. For the animal-infecting bunyaviruses, NSs has furthermore been shown to be involved in the viral evasion of the host's antiviral interferon response (Bouloy *et al.,* 2001; Bridgen *et al.,* 2001; Haller *et al.,* 2000; Kohl *et al.,* 2003; Weber *et al.,* 2002). Interestingly, LaCrosse nonstructural protein NSs has very recently been shown also to counteract the effect of siRNAs (Soldan *et al.*, 2005), which would support the idea that the mechanism of interferon-induced viral defense, which can be triggered by (large) dsRNA molecules (Iordanov *et al.*, 2001; Zamanian-Daryoush *et al.*, 2000), and the mechanism of RNA silencing, which is triggered by short, 21–24 nt ds RNA molecules (Meister and Tuschl, 2004), may be closely related.

For several bunyaviruses, the NSs protein has furthermore been found associated with polysome fractions or the 40S ribosomal subunit (Di Bonito *et al.*, 1999; Simons *et al.*, 1992; Watkins and Jones, 1993), which would suggest an involvement in translation. This observation has not been made concerning TSWV, but the protein has been hypothesized to play a role in translation, possibly by functioning in a manner homologous to PABP (van Knippenberg, 2005).

e. Nucleocapsid protein (N)   Among tospoviral proteins, the nucleocapsid (N) protein is the least conserved (15–25% sequence identity)
and is used as a discriminative factor for taxonomic classification. For
the animal-infecting bunyaviruses, serological relations are an important criterion for classification. The N protein is likely a multifunctional protein involved in several processes. The protein encapsidates
viral RNA, not viral mRNA, and it is required for transcription and
replication of the viral genome. In the latter processes, the N protein
is hypothesized to be involved in the switch from transcription to
replication and, furthermore, to act as antiterminator of transcription,
similarly as observed for the N protein of Influenza virus (Beaton and
Krug, 1984, 1986) and Vesicular Stomatitis virus (VSV) (Patton
et al., 1984). In relation to these features, RNA-binding properties as
well as homo-oligomerization capacity have been described for the N
protein. Precise details on both features remain unclear. For TSWV,
both N- and C-terminal halves of the protein, as well as the extreme
carboxy-terminal region, exhibited nonspecific RNA affinity in a coop-
erative manner (Fig. 5) (Richmond *et al.*, 1998), but this would not
explain the apparent viral antigenomic RNA-specific encapsidation.
However, no protein-binding studies were performed on RNA constructs including the panhandle structure. For Hantavirus N, an
RNA binding domain that specifically interacted with viral RNA was
mapped to a central conserved region of 42 amino acids within the N
protein (Fig. 5) (Xu *et al.*, 2002). Of this region, the first half contains

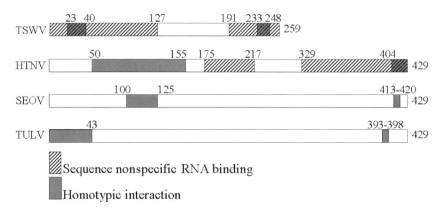

FIG 5. Topology of the N protein of representatives from the different genera of the
Bunyaviridae.

the major determinants for RNA binding, whereas the second half exhibits specificity for viral sense RNA in the presence of the first half. The latter was suggested to explain the specific packaging of viral genomic and not antigenomic RNA into virions. The specific interaction of hantavirus vRNA by the N protein, furthermore, seemed to involve the 5′ noncoding region of viral RNA (Severson *et al.*, 1999, 2001). More recently, trimeric forms of Hantaan virus N were shown to exhibit specific affinity for the panhandle structure, formed by the complementary sequences of the 5′ and 3′ ends of the viral genomic RNA (Mir and Panganiban, 2004), whereas monomeric and dimeric forms of N only showed RNA affinity in a nonspecific manner.

Considering that bunyaviral transcripts are never observed encapsidated by N protein, and transcripts lack the ability to form a panhandle structure, a specific affinity of the (trimeric) N protein for the panhandle seems plausible in order to discriminate between genomic RNA and mRNA, or even sequence nonspecific single stranded RNA. Conversely, Atomic Force microscopy on TSWV RNPs, purified from CsCl gradients revealed that the panhandle structure of RNPs is free from any protein, including N (Kellman *et al.*, 2001). Thus, the panhandle could very well represent the encapsidation signal for (anti)genomic RNA segments, but ultimately remain free of N to some extent. In this way, the terminal sequences of antigenomic RNA molecules would still be accessible for the replicase/transcriptase complex or other (viral) proteins.

Regarding the *in vitro* oligomerization capacity of N, different suggestions have been made for the organization of such interaction. Whereas initially a "head-to-tail" interaction was proposed for TSWV N (Uhrig *et al.*, 1999), more recent data supports the idea of a "tail-to-tail" and "head-to-head" organization (Kainz *et al.* 2004). This has also been suggested for hantavirus N (Alfadhli *et al.*, 2001; Kaukinen *et al.*, 2001, 2003; Yoshimatsu *et al.*, 2003). From these studies, it was clear that neither N-terminal nor C-terminal tags to the N protein seem to interfere in these interactions, suggesting that these sequences are not required in a sterical tight conformation. For TSWV this has recently been confirmed by *in vivo* dimerization of N-fluorophore fusions (Snippe *et al.*, 2005). Most studies on (tospovirus and hantavirus) *in vitro* N multimerization resulted in the identification of an N- and a C-terminal region essential for interaction (Fig. 5). Marked exceptions are Hantaan virus and Seoul virus, for which a C-terminal and a more internal region have been found (Kaukinen *et al.*, 2003).

C. Morphogenesis

1. In the Host

Enveloped viruses usually obtain their lipid membranes by budding through one of the cellular membranes. When budding occurs at the plasma membrane, virion particles are directly released in the extracellular space. Viruses of the *Bunyaviridae* obtain their lipid envelope from intracellular membranes (i.e., the Golgi complex), which requires a final excretion from the cell after transport of the viruses within vesicles to the plasma membrane. For several animal-infecting viruses, the process of virus assembly has been studied more extensively (Booth *et al.*, 1991; Elliott, 1990; Gahmberg *et al.*, 1986; Jäntti *et al.*, 1997; Kuismanen *et al.*, 1982, 1984; Lyons and Heyduk, 1973; Murphy *et al.*, 1973; Rwambo *et al.*, 1996; Salanueva *et al.*, 2003; Smith and Pifat, 1982). Although many of the details of the budding process are still poorly understood, it is generally assumed that the viral glycoproteins accumulate and concentrate at specific foci in smooth-surfaced membranes of the Golgi complex together with the viral ribonucleoproteins (RNPs). Simultaneously, Golgi cisternae vacuolise extensively, allowing RNPs to bud into the lumen of the Golgi and, to a lesser extent, in the ER-Golgi-Intermediate-Cluster (ERGIC) (Fig. 6), and leading to the formation of singly enveloped virus particles in the Golgi complex. The budding process is thought to be triggered by a specific interaction between the cytoplasmic tail of one of the glycoproteins

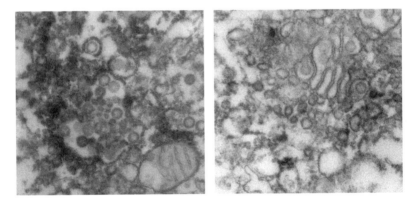

Fɪɢ 6. Presence of Uukuniemi virus particles and budding of RNPs as observed in nucleocapsid-positive 15°C intermediates. Clusters of cytoplasmic vesicles, vacuoles and tubules (ERGIC) are observed, of which the vacuoles contain virus particles and budding virus (left and right panel; with kind permission from Esa Kuismanen). For similar pictures and a more detailed explanation, see Jäntti *et al.* (1997).

(Andersson *et al.*, 1997a) and a structural component of the RNP, for which the nucleoprotein is a candidate protein. The site of virus assembly is dictated by the viral glycoproteins which, during their maturation, accumulate and remain in the Golgi compartment due to the presence of a Golgi retention signal. Final release of virus particles from the cell occurs predominantly by fusion of virus-containing vesicles with the plasma membrane (Murphy *et al.*, 1973; Salanueva *et al.*, 2003; Smith and Pifat, 1982). Animal-infecting bunyaviruses have been observed to bud at the plasma membrane in only a few cases (Anderson and Smith, 1987; Murphy *et al.*, 1973; Ravkov *et al.*, 1997).

For the plant-infecting tospoviruses, *in casu* TSWV, the process of particle morphogenesis has long been an enigma. Early electron microscopic studies (Ie, 1971; Milne, 1970) using infected leaf tissues already revealed most typical structures associated with tospovirus infections that were observed, such as viroplasm (VP), nucleocapsid aggregates (NCA), paired parallel membranes (PPM) doubly-enveloped particles (DEV), and singly enveloped particles (SEV) clustered within ER membranes.

Doubly-enveloped virus particles (DEVs) were found in the cytoplasm, but only during early stages of infection (Francki *et al.*, 1985; Kitajima *et al.*, 1992; Lawson *et al.*, 1996; Milne, 1970). This led to the hypothesis that these structures formed an intermediate stage of particle morphogenesis as a result of budding of nucleocapsids from parallel membranes. Based on their morphology, the paired parallel membranes have been suggested to originate from the Golgi system (Lawson *et al.*, 1996). Subsequent fusion of several DEVs was proposed to finally result in the accumulation of (singly-enveloped) virus particles (SEVs) in clusters within the ER (Kitajima *et al.*, 1992; Milne, 1970). This maturation pathway has been substantiated as the mode of TSWV particle morphogenesis, by a detailed study using a plant-protoplast-infection system (Fig. 7) (Kikkert *et al.*, 1997, 1999). Using ER and Golgi-specific markers, these studies have shown that envelopment of nucleocapsids, leading to the formation of DEVs, occurs by enwrapment of parallel membranes derived from modified Golgi membranes (Kikkert *et al.*, 1999). In a later stage of the maturation, DEVs disappear by fusion to each other and ER-derived membranes, a process possibly facilitated by the presence of an ER retrieval signal that may be exposed after processing of Gn in the Golgi apparatus (Fig. 7) (Kikkert, 1999). Mature virions accumulated and retained within subcellular (ER-derived) compartments serve as the source by which thrips acquire the virus by feeding prior to transmission. Although the maturation of TSWV particles is somewhat different

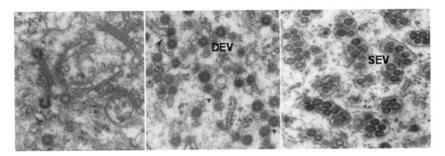

FIG 7. Assembly of TSWV-enveloped particles by enwrapment of RNPs with Golgi cisternae (left panel). Fusion of DEVs to each other (middle panel) and ER-derived membranes leading to accumulation of mature virus particles (SEVs) in large vesicles (right panel).

from that of the animal-infecting bunyaviruses, the involvement of the Golgi system is a common feature of both the plant- and animal-infecting bunyaviruses.

2. In the Arthropod Vector

Tospoviruses are propagatively transmitted by their insect vector thrips (i.e., after acquisition and subsequent dissemination in the thrips' body, the virus not only circulates but multiplies in the vector before being transmitted) (Wijkamp et al., 1993). They only transmit the virus if acquired during the first or second larval instar, but they do not transmit the virus transovarially. They do not seem to suffer from the infection and respond with an immune response upon infection, based on the observed upregulation of several genes coding for antimicrobial peptides, lectins, and receptors that activate the innate immune response (Medeiros et al., 2004). So far, still only limited information is available on the lifecycle of tospoviruses in the thrips vector and particularly, on stages involving virus assembly. Despite this lack of detailed information, an important difference in the morphogenesis pathway of TSWV in thrips cells versus plant cells has emerged. Whereas TSWV particles are clearly retained intracellularly, probably by ER-derived vesicles in plant cells, they are only observed in and secreted from thrips' salivary gland cells, not least to allow transmission of the virus by thrips feeding on plants.

Immunolocalization studies of TSWV in viruliferous thrips, *F. occidentalis,* have shown large amounts of the N and NS_S proteins in the salivary glands, the NSs protein indicating virus replication. Whereas the N protein is found in more confined areas of the gland tissue identified as electron-dense aggregates (viroplasm), the NS_S protein appears to be evenly spread throughout the cytoplasm of the gland cells. Both proteins, moreover, are found in muscle cells associated with the midgut epithelium, the N protein again confined in viroplasm and the NS_S protein dispersed throughout the cytoplasm (Ullman *et al.*, 1992, 1993; Wijkamp *et al.*, 1993). Paracrystalline arrays and amorphous inclusions have also been observed in infected thrips cells. The paracrystalline arrays resemble those previously observed in infected plant cells and have also been shown to contain the NS_S protein (Kitajima *et al.*, 1992; Kormelink *et al.*, 1991), whereas the amorphous inclusions labeled with antiserum are directed to the Gc glycoprotein (Ullman *et al.*, 1995). Mature virus particles are observed only in low amounts in saliva vesicles, but in numerous amounts in the salivary gland ducts, suggesting that the salivary glands are the major site of replication. The presence of virus particles in saliva secretory vesicles supports the idea of virus particles being released during secretion of saliva into the ducts. Although this is an indication that the Golgi system may be involved in the assembly of virus particles in thrips, no actual virus assembly in the Golgi complex has been observed. Instead, budding from the basal plasma membrane of midgut epithelial cells has been suggested to occur (Nagata *et al.*, 2002), but these observations have not been repeated and remain rather vague. In thrips that have acquired TSWV while feeding on infected plants during the adult stage, viral proteins have only been observed in the midgut epithelium, and not beyond these cells, indicating that dissemination of the virus to the hemocoel is blocked, and a midgut escape barrier likely exists in adult thrips (Ullman *et al.*, 1992). Immunofluorescence studies on *Thrips setosus* larvae following acquisition of TSWV have shown that during the initial stages, the virus first accumulates in the anterior part of the midgut, and gradually spreads to the entire midgut before finally spreading to the other organs during later developmental stages of the thrips (Filho *et al.*, 2002; Nagata *et al.*, 2002; Tsuda *et al.*, 1996). A few attempts have been made to study tospovirus infections in primary thrip cells. Neither in primary cells from nontransmitting *T. tabaci*, nor in those from TSWV transmitting *F. occidentalis*, could glycoprotein synthesis or virus accumulation be observed. Although some nonstructural viral protein NSs was detected in these cells, indicating virus replication, passage of culture fluids from these cells to healthy primary cells only resulted in 1% infection 48–72 hours post-introduction (Nagata *et al.*, 1997).

For tospoviruses and all other bunyaviruses, the glycoproteins are the major determinants for infection of the vector and transmission (see section on glycoproteins). No receptors have been identified and characterized to which the glycoproteins bind that lead to receptor-mediated endocytosis and subsequent infection.

3. RNP Formation

In the formation of virus particles, RNPs serve as the first intermediary. This formation is enabled due to the specific affinity of the viral N protein for viral (anti)genomic RNA in concert with the capacity to oligomerize. Viral transcripts, however, are not encapsidated by the N protein (Kormelink *et al.*, 1992b), which can be explained in light of recent observations made for hantavirus N protein, in which a trimeric form of the N protein showed high affinity for the panhandle structure of the viral genomic RNA (Mir and Panganiban, 2004), a structure absent from (bunyavirus) mRNA molecules.

Once viral RNA is encapsidated by N protein, functional RNPs–RNPs containing all components to initiate a virus multiplication cycle are formed. Based on *in vivo* reconstitution analyses of several bunyaviruses (Blakqori *et al.*, 2003; Dun *et al.*, 1995; Flick and Pettersson, 2001; Flick *et al.*, 2003a,b; Jin and Elliott, 1991, 1993; Lopez *et al.*, 1995), a transcriptional/replicational-active, and thereby functional, RNP minimally consists of viral RNA encapsidated by N protein and supplemented with a few copies (catalytic amounts) of the L protein. Whether this RNP represents the entity that becomes membrane-bound or whether a nonstructural viral protein and/or host protein is additionally required to trigger RNP envelopment is not known. It is interesting to mention here that cytoplasmic RNPs of RVFV and TSWV were shown to contain (higher) amounts of NSs protein than RNPs purified from enveloped virus particles (Michele Bouloy, personal communication; Van Knippenberg, 2005). Furthermore, TSWV cytoplasmic and particle RNPs exhibited different replicational and transcriptional activities (Van Knippenberg *et al.*, 2002). This would point toward the existence of RNPs with different compositions. Whether the NSs protein, acting as a suppressor of silencing, is also involved in the selection and subsequent preparation of cytoplasmic RNPs for envelopment, on which moment NSs is released, remains speculative. Also, the RdRp protein has been implicated as a possible matrix protein triggering the envelopment of RNP by Golgi membranes. Very recently, a transcription factor has been identified from the vector thrips *F. occidentalis*, named FoTF, which stimulates

TSWV replication and binds to the TSWV L protein (Medeiros *et al.*, 2005) and also to Gc. This protein was observed to colocalize with L and Gc in ER and Golgi vesicles in thrips cells, suggesting a possible role in particle assembly (e.g., by recruiting the L protein to virus particles) (Medeiros, personal communication).

Next to the cytoplasmic and virus-derived RNPs that are normally distinguished, a subpopulation of cytoplasmic RNPs has been observed during the very early stages of the infection prior to the appearance of virus particles that can be discriminated by the additional association of NSm. As discussed earlier, the NSm protein facilitates the cell-to-cell movement of nonenveloped infectious RNP structures of TSWV through plasmodesmata (Kormelink *et al.*, 1994; Storms *et al.*, 1995, 1998) and allows viral spread. These RNPs do not need to become membrane-bound, as it otherwise would have the contrary effect. After all, enveloped tospovirus particles are 80–120 nm in diameter and thereby too large to pass plasmodesmata, which have an effective size exclusion limit of only 5 nm. In conclusion, it is evident that various compositions of RNP exist, but the exact composition of RNPs to become membrane-bound remains unclear.

Prior to the actual envelopment, RNPs need to locate and concentrate around dilated Golgi membranes where the glycoproteins remain and accumulate. In cells infected with Uukuniemi virus, RNPs are observed in a punctuated pattern throughout the cytoplasm, but also increasingly in the Golgi region (Kuismanen *et al.*, 1984). In Crimean Congo hemorrhagic fever (CCHF) virus, RNPs were observed in perinuclear regions in infected cells (Andersson *et al.*, 2004a). The additional interaction and colocalization of CCHF RNPs with interferon-induced MxA in the perinuclear region (Andersson *et al.*, 2004a) was proposed as the mechanism to make RNPs unavailable for the process of replication and envelopment. Similar observations were earlier made on LaCrosse, Dugbe, and hantaviruses, in which MxA was suggested to inhibit virus replication by sequestering RNPs to the perinuclear region (Bridgen *et al.*, 2004; Flohr *et al.*, 1999; Frese *et al.*, 1996; Kanerva *et al.*, 1996; Kochs *et al.*, 2002). It is unknown whether targeting and final localization of RNPs at the Golgi complex is the result of an early interaction with the glycoproteins. So far, transient expression studies of various bunyavirus N proteins have primarily shown a perinuclear localization. Hantavirus N protein expressed as a membrane-associated protein in the perinuclear region (Ravkov and Compans, 2001) and LaCrosse virus N formed perinuclear complexes, suggested to be caused by MxA (Reichelt *et al.*, 2004). For CCHF and

Black Creek Canal virus (BCCV), transiently expressed N protein colocalized with actin microfilaments in infected cells. The use of inhibitors of actin filament polymerization confirmed the requirement of actin filaments for targeting of CCHF NP to perinuclear regions of mammalian cells (Andersson *et al.*, 2004b). Since a colocalization with the Golgi complex was not observed in any of these expression studies, with the exception of BCCV N protein (Ravkov and Compans, 2001; Ravkov *et al.*, 1998), the presence or interaction with additional viral elements (e.g., RNA, L protein, nonstructural proteins) is required for proper Golgi targeting. The perinuclear accumulation of N/RNPs has in several cases been suggested to represent aggresomes (Andersson *et al.*, 2004b; Kochs *et al.*, 2002), structures that are related to the microtubule organizing center (MTOC) and are possible aggregation sites for misfolded proteins (Johnston *et al.*, 1998). On the other hand, the aggresomes' pathway has also been suggested to be exploited by viruses in order to concentrate structural proteins at virus assembly sites (Heath *et al.*, 2001).

Experiments on transient expression of TSWV N protein in mammalian BHK21 cells partly revealed similar results (Fig. 8A–D). The N protein was not only observed throughout the cytoplasm, but also formed perinuclear aggregates that did not colocalize with markers for the ERGIC or Golgi complex (Fig. 8E,F). Furthermore, upon treatment of the cells with BFA (an inhibitor of ER-export), hardly any difference in the localization pattern of N was observed. In contrast, incubation in the presence of either colchicine or cytochalasin D, inhibitors of microtubule assembly and actin filament polymerization, respectively, rendered an obvious change in the localization pattern (Fig. 9). Whereas small clusters continued to be formed, no perinuclear accumulation was observed, suggesting both microtubules and actin filaments are essential to obtain perinuclear accumulation of N. These results, together with data from the animal-infecting bunyaviruses, could suggest a common antiviral mechanism against bunyaviruses, including tospoviruses, of MxA/GTPase-assisted sequestering and subsequent inactivation of viral N/RNP structures to the perinuclear region. Transgenic tobacco plants constitutively expressing human MxA, however, did not show resistance to TSWV (Frese *et al.*, 2000).

Although several papers have reported the interaction of bunyaviral N proteins with a variety of cellular proteins (Lee *et al.*, 2003; Li *et al.*, 2002), e.g., SUMO-1, Daxx (a Fas-mediated apoptosis enhancer), the biological significance of these still remains to be determined.

4. Maturation of the Glycoproteins

Since the glycoproteins dictate the site of virus assembly, many studies have been performed on expression of glycoproteins to identify and characterize domains within these proteins required for specific steps in the virus assembly. For several bunyaviruses, the glycoproteins were inferred in mammalian cell systems from a precursor or from constructs representing the individual glycoproteins (Fig. 4). For TSWV, it has been shown that the N-terminal protein G2 (Gn) of the glycoprotein precursor could be transported to the Golgi system on its own, though with decreased efficiency, thus suggesting the presence of a Golgi retention signal. The C-terminal protein G1 (Gc) expressed on its own was unable to leave the ER, but in the presence of G2 (Gn) was able to exit the ER and translocate to the Golgi (Kikkert et al., 2001). These data strongly suggested heterodimerization of Gn and Gc for proper targeting and maturation. In the presence of tunicamycin, an inhibitor of glycosylation, the glycoproteins were retained in the ER, suggesting that the absence of N-linked glycans probably resulted in aberrant folding of the proteins leading to a hampered transport from the ER to the Golgi complex. An association with ER-residing chaperone proteins (i.e., calnexin and calreticulin) at this stage is likely, as already observed for Uukuniemi virus glycoproteins during an infection in BHK cells (Veijola and Pettersson, 1999) as well as for most other (viral and nonviral) glycoproteins (Braakman and Van Anken, 2000). More or less similar observations were made during heterologous expression studies of Gn and Gc from Uukuniemi and Punta Toro phleboviruses, and of Bunyamwera orthobunyavirus (Chen and Compans, 1991; Chen et al., 1991; Lappin et al., 1994; Matsuoka et al., 1988; Rönnholm, 1992).

For hantaviruses, the situation seems to be slightly different, as both Gn and Gc retain in the ER upon individual expression, whereas Golgi localization requires coexpression of both glycoproteins (Ruusala et al., 1992; Shi and Elliott, 2002; Spiropoulou et al., 2003).

For Uukuniemi phlebovirus, the Golgi retention signal was mapped to the cytoplasmic tail of Gn, that is, amino-acid residues 10–40 downstream from the transmembrane domain (TMD). The ectodomain and TMD were not required for Golgi retention (Fig. 4) (Andersson et al., 1997b, 1998). For two other phleboviruses, Punta Toro virus and Rift Valley Fever virus, the Golgi retention signal was mapped to the junction region of the TMD and the cytoplasmic tail (Fig. 4) (Gerrard and Nichol, 2002; Matsuoka et al., 1994). On the other hand, for Bunyamwera orthobunyavirus, the Golgi retention signal completely

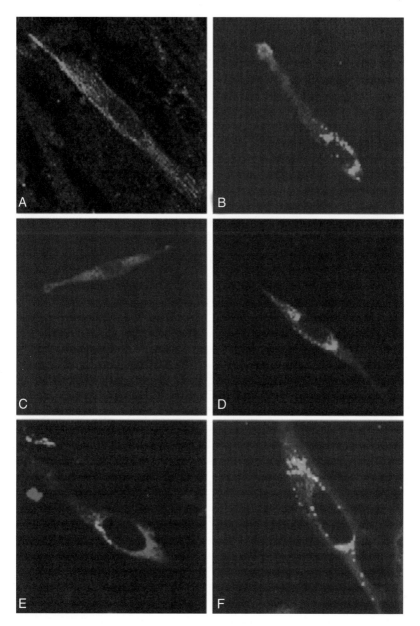

FIG 8. Time course analysis of wild-type TSWV N protein (wt N) and N-YFP expressed in BHK21 cells. Images were taken at 4 (A, C) and 24 (B, D, E, F) hours post transfection. Both N (A, B) and N-YFP (C, D) are observed dispersed thoughout the cytoplasm early in transfection. Upon longer incubation, both proteins are gradually observed to form

resided in the TMD of Gn (Shi *et al.*, 2004), so viruses within the *Bunyaviridae* seem to have developed different strategies for Golgi retention. Alignment of these Golgi retention signals did not reveal any sequence homology, suggesting that the retention signal rather involves a conformational structure. The role of a TMD in Golgi retention in general still remains a matter of debate, as for several Golgi-membrane residing proteins the size of the TMD has been shown to be crucial for Golgi retention, not only in vertebrate cells but also plant cells (Brandizzi *et al.*, 2002; Gerrard and Nichol, 2002; Munro, 1998), whereas in other cases the TMD does not seem to be involved at all (Misumi *et al.*, 2001).

Although there are several discrepancies between the bunyaviruses with respect to Golgi targeting and signals involved, the observation that TSWV glycoproteins contain information necessary and sufficient for their transport to and retention in the Golgi system of mammalian cells not only supports the notion of the putative evolution of an ancestral animal-infecting bunyavirus into the plant-infecting tospoviruses, but also the idea that protein trafficking and targeting machinery is homologous among all eukaryotes. On this point, it is believed that cellular transport and retention signals are not only conserved among closely related organisms, but are similar if not identical for all eukaryotes (Bar *et al.*, 1996; Kermode 1996).

It is still a matter of debate how glycoproteins are incorporated in virus particles after being transported to the Golgi. For UUK, they were first reported to do so as heterodimeric complexes assembled in the ER (Persson and Pettersson, 1991). However, in later studies, Gn and Gc were each observed to associate into homodimeric complexes in mature virus (Rönka *et al.*, 1995). The Gn protein of Sin Nombre virus was also found in monomeric and stable SDS-resistant multimeric forms, with the dimer being the only form present late in infection (Spiropoulou *et al.*, 2003). Conversely, Punta Toro virus Gn was found as a heterodimer with Gc, but not as a homodimer (Matsuoka *et al.*, 1996) Recently, Gn homodimers were also observed in TSWV particles (Whitfield *et al.*, 2004). The biological importance of homodimerization still is unknown.

Although it is not known which of the two (Gn or Gc) cytoplasmic tails interacts with the RNPs prior to envelopment, it has been postulated that for UUK the cytoplasmic tail of Gn is involved. Since of the

------

clusters that are ultimately most abundantly localized near the nucleus. Panels E and F show N (E) and N-YFP (F) in combination with a marker for the trans-Golgi system (E: marker in green, N in red) or for the ERGIC (F: ERGIC in red, N-YFP in green). (See Color Insert.)

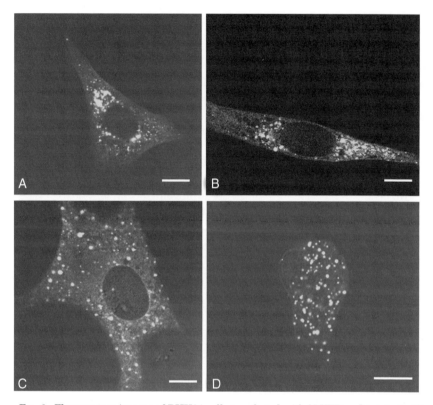

Fɪɢ 9. Fluorescence images of BHK21 cells tranfected with N-YFP and treated with BFA (B), colchicine (C), or cytochalasin D (D). Panel A is an untreated control transfection. The disturbing effect of both colchicine and cytochalasin D on the perinuclear translocation of N-YFP is clearly visible, indicating the involvement of actin filaments as well as micrutubules in the peri-nuclear targeting.

cytoplasmic tail of Gc is rather short (i.e., only 5 amino acids) the tail of Gn was proposed as the target for interaction (Andersson *et al.*, 1997a). In contrast, TSWV Gn and Gc proteins have more or less equal-sized cytoplasmic tails.

II. Fʟᴜᴏʀᴇꜱᴄᴇɴᴄᴇ Mɪᴄʀᴏꜱᴄᴏᴘʏ

During the last decade, bioimaging has slowly gained enormous attention due to the possibilities to study different viral processes in tissue and living cells. Whereas previously, and still, *in vitro* methods such as coimmunoprecipitation, overlay blots, gel shift essays, subcellular fractionation, and sucrose gradient centrifugation have proved valuable to study protein dynamics and localizations (Andersson and Pettersson, 1998; Bandla *et al.*, 1998; Gutiérrez-Escolano *et al.*, 2000; Kikkert *et al.*, 1998; Lappin *et al.*, 1994; Medeiros *et al.*, 2000; Persson and Pettersson, 1991; Rönkä *et al.*, 1995; Ruusala *et al.*, 1992; Veijola and Pettersson, 1999), the spectrum of *in vivo* tools to detect and analyze proteins has long remained more limited, and as a result, there is limited detailed knowledge based on *in vivo* studies. This is slowly changing with the ongoing development of more elegant and sophisticated fluorescence tools, and the significance of these will, no doubt, continue to increase their influence on virus research.

To understand the way viruses infect their host cell, it is very helpful, if not essential, to follow the fate of different viral components during infection. A major advantage of using fluorophore fusion proteins (i.e., proteins fused to a fluorescent group such as green fluorescent protein, GFP) is that the protein of interest can be studied in living cells or even in whole living plants. Another advantage is that it offers the possibility of investigating proteins such as glycoproteins and other membrane proteins that are difficult, if not impossible, to study using, for example, yeast-two-hybrid.

Traditionally, fluorescent tags have been applied to follow the intra- as well as intercellular localization of (viral) proteins to study virus infection (Bosch *et al.*, 2004; Brideau *et al.*, 1999; Silva, 2004; Silva *et al.*, 2002) or expression and localization patterns of individual proteins (Andersson and Pettersson, 1998; Dalton and Rose, 2001; Kochs *et al.*, 2002) as well as, for instance, the effect of a viral infection on the intracellular membranes (Carette *et al.*, 2000). Newer methods have been developed that allow for more detailed investigation of features such as *in vivo* protein-protein interaction (Immink *et al.*, 2002; Larson *et al.*, 2003; Snippe *et al.*, 2005; van Kuppeveld *et al.*, 2002), protein-nucleic acid interaction (Murchie *et al.*, 2003), phosphorylation (Sato *et al.*, 2002), cleavage (Bastiaens and Jovin, 1996; Xu *et al.*, 1998), apoptosis induction (Xu *et al.*, 1998), DNA and RNA folding (Katiliene *et al.*, 2003), protein folding, and conformational changes (Rhoades *et al.*, 2003; Truong and Ikura, 2001). In addition, these properties can be studied *in vivo*, allowing for spatial as well as temporal observation (e.g., where in the cell, and when during a viral infection, does the interaction occur).

Some methods that have recently come into use for virological research will be explained later and supported with work on TSWV aimed at unraveling interactions between structural proteins involved during some stages of particle assembly. A promising method for analyzing processes involving three proteins is three-way or two-step FRET (Galperin *et al.,* 2004; Liu and Lu, 2002; Watrob *et al.,* 2003). Because the method is still being developed, it will not be discussed here. Likewise, other methods such as homotransfer or energy migration FRET (emFRET) (Lidke *et al.,* 2003) and photochromic FRET (pcFRET) (described in Giordano *et al.,* 2002), both efficient techniques to study protein-protein interactions and protein dynamics will not be discussed.

A. *Fluorescence Resonance Energy Transfer*

Fluorescence resonance energy transfer (FRET), or Förster resonance energy transfer, in the case of nonfluorescing molecules, is the phenomenon by which energy is transferred from a donor fluorophore to an acceptor fluorophore (Fig. 10). This energy transfer occurs through dipole-dipole interactions and will only take place if the donor and acceptor fluorophore are in very close proximity (typically 10-100 Å, which is in the same range as protein dimensions) and is therefore a good indicator for direct protein-protein interactions (Pollok and Heim, 1999). Another prerequisite for FRET is that the excitation spectrum of the acceptor has considerable overlap with the emission spectrum of the donor. FRET couples are, for example, CFP–YFP (Cyan and Yellow fluorescent protein, respectively), Cy3-Cy5, or GFP with monomeric DaRed, GFP, or YFP with HcRed and, somewhat less due to the weak fluorescence intrinsic of BFP as well as damaging UV excitation, BFP-GFP (Bastiaens and Jovin, 1996; Erickson *et al.,* 2003; Pollok and Heim, 1999). Different methods have been developed to measure FRET. Since measurement of sensitized emission of acceptor

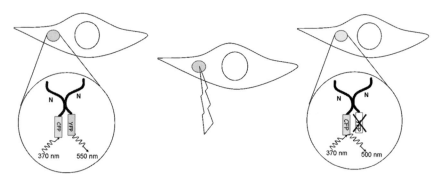

FIG 10. Principles of FRET and bleaching. (See Color Insert.)

fluorescence is very sensitive to cross-talk or bleed-through of the signal, either hardware or software should be employed to correct for this background signal. Hardware can be optimized using narrow filter sets both for excitation and for detection of the fluorophore; software can correct for background signals by using algorithms to compare signals from cells expressing only donor and only acceptor with signals from the sample (Gordon *et al.*, 1998; Nagy *et al.*, 1998; Xia and Liu, 2001). A comparison of different methods can be found in Berney and Danuser, 2003.

A more direct method of direct FRET determination is spectral imaging spectroscopy, or SPIM. In this method, a fluorescence spectrum of the region or whole cell is recorded upon excitation of the donor fluorophore. The spectrum is compared to negative controls. In case FRET occurs, the relative donor emission can be seen to decrease, while acceptor emission increases (Immink *et al.*, 2002; Pouwels, 2004; van Kuppeveld *et al.*, 2002).

Another method to measure FRET is acceptor photobleaching (Fig. 10). This method exploits the phenomenon of the fluorescence of the donor fluorophore being quenched if FRET takes place. Comparing the donor fluorescence before and after irreversibly destructing the fluorescence and energy-absorbing capacity of the acceptor gives a clear indication of FRET. Destruction is carried out by exposing the fluorophore to an intense laser beam.

Still another measuring method is FLIM (fluorescence lifetime imaging) (Fig. 11), in which the fluorescence lifetime of the donor fluorophore is measured on a picosecond scale. If FRET occurs, the fluorescence lifetime of the donor is decreased (Gadella, 1999).

For TSWV, the use of FRET microscopy techniques has recently been used to determine their potential in the identification and analyses of *in vivo* interactions between the structural N, G1, and G2 proteins involved in the lipid envelopment of RNPs. In light of the assembly process and the occurrence of *in vitro* homotypic interactions for N, this protein was selected as the first (cytosolic) candidate protein to analyze *in vivo* protein–protein interactions by means of FRET. To this end, N- and C-terminal fusions of ECFP (donor) and EYFP (acceptor) to the N protein were made and first transiently expressed in BHK21 cells to verify that the fluorophore fusions did not alter the intracellular localization of N. The results demonstrated that 4 hours after transfection, N-YFP (or CFP) appeared throughout the cytoplasm of BHK, but upon longer incubations gradually accumulated as clusters in the perinuclear region, similarly as observed for nonfused N protein

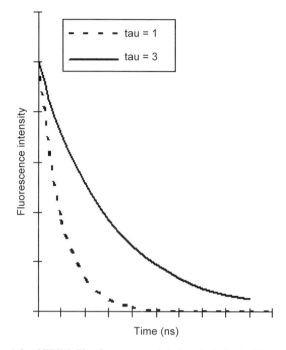

Time (ns)

FIG 11. Principle of FLIM. The fluorescence life time is indicated in nanoseconds (ns). In the additional presence of an acceptor fluorophore, the fluorescence lifetime of the donor decreases (from tau = 3 to tau = 1).

(Fig. 8A–D). These clusters were ultimately most abundant close to the nucleus, but did not colocalize with specific markers for the trans-Golgi system or the ER-Golgi intermediate compartment (ERGIC) (Fig. 8E,F). Since fluorophore fusions did not appear to affect the behavior of the N protein *in vivo*, the fusion proteins were subsequently coexpressed in BHK cells and analyzed for the occurrence of FRET. In all cases, when either N- or C-terminal fusions of CFP (donor) and YFP (acceptor) to the N protein were used, FRET could be observed, although N-CPF fusions rendered lower fluorescence intensities. These results were further substantiated by acceptor photobleaching. A graph in which the CFP-N and YFP-N fluorescence intensities from such experiment were set out clearly showed a drop of YFP signal upon bleaching (indicated by an "X" in Fig. 12, top panel) and a simultaneous increase of CFP signal. After some time, a recovery of the YFP signal could be observed as the result of YFP-N mobility, since the cells in these experiments were not fixed. These results indicated the

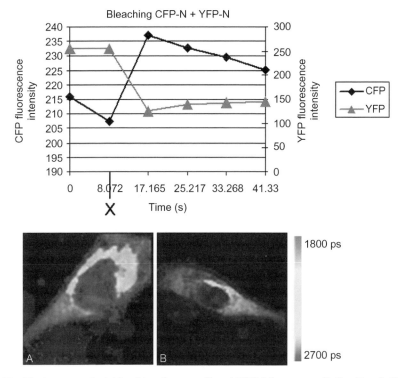

FIG 12. Acceptor photobleaching (top panel) and FLIM (lower panel). Top Panel: CFP and YFP fluorescence intensities (in arbitrary units) of a region in a cell, cotransfected with CFP-N and YFP-N, measured before and after YFP bleaching. The "X" indicates the time point at which the bleaching pulse was active. A decrease in YFP fluorescence caused by bleaching coincides with a significant CFP fluorescence increase, indicating the occurrence of FRET before photobleaching. Lower panel: Fluorescence lifetime of CFP in a cell transfected with (A): CFP-N, and (B): CFP-N and YFP-N. A shorter fluorescence lifetime of CFP(-N) is clearly observed in the presence of YFP-N (B), as indicated by the pseudocolor change into yellow. The legend for the pseudocolors representing CFP fluorescence lifetime is provided in the color scale on the right. The decrease in lifetime is a consequence of fluorescence energy transfer and thus, of interaction between the fusion proteins. (See Color Insert.)

occurrence of (homotypic) N-N interactions, which was further supported by data from FLIM experiments (Fig. 12, lower panel) in which a significant decrease in the N-CFP fluorescence lifetime was observed when YFP-N was present. Furthermore, the results of acceptor photobleaching and FLIM experiments showed that the homotypic interactions not only occurred in the perinuclear region, where large clustering of N was observed, but also throughout the cytoplasm (Snippe *et al.*, 2005). Recently, the use of FRET microscopy has been extended to identify the interactions between N and the viral glyco-proteins and, although preliminary, results suggest that N interacts with Gc and not with Gn (unpublished data).

FRET has been used in the recent past for investigation of (viral) proteins; for example, the homomultimerization of viral proteins (Larson *et al.*, 2003; Snippe *et al.*, 2005; van Kuppeveld *et al.*, 2002), protein-folding dynamics (Rhoades *et al.*, 2003; Truong and Ikura, 2001), and enzyme activity (Bark and Hahn, 2000; Violin *et al.*, 2003; Yoshizaki *et al.*, 2003).

A method related to FRET is bioluminescence resonance energy transfer (BRET) (Xu *et al.*, 1999). In this technique, the donor of a FRET couple is replaced by a luciferase. Energy transfer takes place under the same circumstances as described for FRET, but no potentially damaging laser light is required for excitation. As a consequence, auto-fluorescence and photobleaching no longer occur. In addition, BRET has been reported to be significantly more sensitive than FRET, possi-bly due to much lower background signals from direct excitation of the acceptor fluorophore (Arai *et al.*, 2001).

B. Fluorescence Recovery After Photobleaching and Fluorescence Loss in Photobleaching

Fluorescence recovery after photobleaching (FRAP) is a valuable technique to study the structure of biological membranes and to mea-sure movement (e.g., diffusion) of proteins in a living cell. It is espe-cially useful for gaining insight into the movement of membrane or membrane-bound proteins (Sekar and Periasamy, 2003). In FRAP, the fluorescent molecules in a specific region are bleached (using an intense laser pulse). The nonbleached molecules outside this region might move into the bleached area resulting in a (partial) recovery of the fluorescence. The recovery curve yields information about the diffusion coefficient of the fluorophore and the proportion of mobile fluorophore molecules. Moreover, diffusion can be distinguished be-tween flow-driven processes from the shape of the recovery curve (Sciaky *et al.*, 1997; White and Stelzer, 1999). FRAP may prove helpful in determining the mobility of proteins in membranes such as the cell membrane. The method has been employed to determine the interac-tion between molecules of CPMV viral movement protein and the cell membrane of protoplasts (Pouwels, 2004), to study raft forma-tion in influenza virus infection (Shvartsman *et al.*, 2003) and the mechanisms of protein retention in the ER, and the dynamics of protein-folding complexes in native ER membranes using VSV-G protein (Nehls *et al.*, 2000).

In fluorescence loss in photobleaching (FLIP), a region is repeatedly bleached and the disappearance of fluorescence in the nonbleached areas is monitored. If all proteins can diffuse towards the region subjected to bleaching, this will eventually lead to a loss of fluorescence throughout the cell.

This method can be a especially valuable in gaining insight into ER-Golgi trafficking (White and Stelzer, 1999), demonstrating the continuity of intracellular organelles (Cole *et al.*, 1996; Köhler *et al.*, 1997), and studying nuclear transport and ER retention (Shimi *et al.*, 2004; Imreh *et al.*, 2003).

C. Fluorescence Correlation Spectroscopy

Fluorescence correlation spectroscopy (FCS) is a method that can be used to determine the diffusion rate of fluorescently labeled proteins. This diffusion rate gives information about molecular interactions in a very sensitive way; in principle, molecular interaction of a single fluorescent ligand with a macromolecule can be determined and fully quantified. In addition, transport properties as well as chemical properties such as association-dissociation kinetics can be investigated using FCS (Kinjo and Rigler, 1995; Rauer *et al.*, 1996; Visser and Hink, 1999).

The principle of FCS is that a small open-volume element is created by a focused laser beam. Fluorescent molecules moving through this confocal volume are excited, and fluorescence is measured with a fast photon detector (Fig. 13). The Brownian motion of the molecules results in relative large fluctuations of the fluorescence intensity, especially at low concentrations. The fluctuations can be auto-correlated to give information about the average time needed by the fluorescent molecule to pass through the volume element. This speed of movement

FIG 13. Principle of fluorescence correlation spectroscopy (FCS).

is directly related to the diffusion coefficient of the molecule, which, in turn, is related to its size. Thus, conclusions can be drawn about the state of the protein (i.e., its presence as a free molecule or as part of a large complex). In addition to the diffusion coefficient, analysis of the autocorrelation signal also yields the molecular concentration of fluorescent molecules. Using these parameters together, the state of a binding equilibrium can be calculated. FCS has proven useful in the investigation of protein-protein and protein-lipid interaction as well as DNA dynamics (Berland *et al.*, 1996; Bonnet *et al.*, 1998; Pramanik *et al.*, 2000). The method and applications have been reviewed by Hink *et al.* (2002), Eigen and Rigler (1994), and Hess *et al.* (2002).

D. Bimolecular Fluorescence Complementation

A relatively new method is to tag two ends of one protein (to study folding of that protein) or two proteins (to study their interaction) with two nonfluorescent fragments of a fluorescent protein. When brought together, the fragments complement each other to render a fluorescent protein. Since fluorescence is only generated if the two fragments are allowed to behave as one, the signal is not disturbed by background fluorescence, and therefore, no complex processing of the data is required. The method is sensitive enough to enable studying proteins in a cellular environment. The method was first described in mammalian cells by Hu *et al.* (2002) and has recently been described in plant cells as well, for dimerization of both nuclear proteins and cytoplasmic proteins (Walter *et al.*, 2004). Using several different GFP variants, the technique was extended to multicolor BiFC. With this technique, several protein interactions can be monitored simultaneously in the same cell (Grinberg *et al.*, 2004; Hu and Kerppola, 2003).

III. CONCLUSIONS

Based on the first (preliminary) results on *in vivo* protein-protein interactions between TSWV structural proteins using FRET microscopy, it is clear that this and newly developed fluorescence microscopy techniques may become valuable tools not only to unravel the maturation of TSWV and other bunyavirus particles, but also for virology in general to identify and characterize interactions between viral proteins and/or host factors involved in many different processes like replication, transcription, translation, viral spread, etc. In contrast to more conventional ways to investigate protein-interactions (e.g., coimmunoprecipitation and Yeast-two-hybrid), these newly emerging fluorescence microscopy techniques have the additional advantage that interactions are analyzed in a more natural, cellular environment, and, since monitoring can be done real time in a living cell, it also is possible to determine the initial intracellular site of interaction. With the use of reverse genetics systems that have already been developed for representative members of almost all genera, Bunyamwera (Bridgen and Elliott, 1996; Jin and Elliott, 1991, 1992, 1993) and LaCrosse orthobunyavirus (Blakqori *et al.*, 2003), UUK (Flick and Pettersson, 2001), and RVFV (Lopez *et al.*, 1995) phlebovirus, Hantaan hantavirus (Flick *et al.*, 2003b) and CCHF nairovirus (Flick *et al.*, 2003a), recombinant bunyaviruses can be constructed containing nonstructural proteins in fusion with different fluorophores that may offer the possibility for real time analysis of distinct steps of the virus assembly process *in vivo*.

Acknowledgments

The authors wish to thank Mark Hink for critically reading the manuscript, Ricardo Medeiros for providing (unpublished and in press) data for this review, and Esa Kuismanen for providing an EM-micrograph on UUK virus assembly. The work was supported in part by the Dutch NWO-ALW and EU- RTN grant HPRN-CT-2002–00262.

References

Amin, P. W., Reddy, D. V. R., and Ghanekar, A. M. (1981). Transmission of Tomato spotted wilt virus, causal agent of bud necrosis of peanut, by *Scirtothrips dorsalis* and *Frankliniella schultzei*. *Plant Dis.* **65**:663–665.

Anderson, G. W., Jr., and Smith, J. F. (1987). Immunoelectron microscopy of Rift valley fever viral morphogenesis in primary rat hepatocytes. *Virology* **161**:91–100.

Andersson, A. M., Melin, L., Persson, R., Raschperger, E., Wikstrom, L., and Petterson, R. F. (1997a). Processing and membrane topology of the spike proteins G1 and G2 of Uukuniemi virus. *J. Virol.* **71**:218–225.

Andersson, A. M., Melin, L., Bean, A., and Pettersson, R. F. (1997b). A retention signal necessary and sufficient for Golgi localization maps to the cytoplasmic tail of a *Bunyaviridae* (Uukuniemi virus) membrane glycoprotein. *J. Virol.* **71**:4717–4727.

Andersson, A. M., and Pettersson, R. F. (1998). Targeting of a short peptide derived from the cytoplasmic tail of the G1 membrane glycoprotein of Uukuniemi virus (*Bunyaviridae*) to the Golgi complex. *J. Virol.* **72**:9585–9596.

Andersson, I., Bladh, L., Mousavi-Jazi, M., Magnusson, K. E., Lundkvist, A., Haller, O., and Mirazimi, A. (2004a). Human MxA protein inhibits the replication of Crimean-Congo hemorrhagic fever virus. *J. Virol.* **78**:4323–4329.

Andersson, I., Simon, M., Lundkvist, A., Nilsson, M., Holmstrom, A., Elgh, F., and Mirazimi, A. (2004b). Role of actin filaments in targeting of Crimean Congo hemorrhagic fever virus nucleocapsid protein to perinuclear regions of mammalian cells. *J. Med. Virol.* **72**:83–93.

Antic, D., Wright, K. E., and Kang, C. Y. (1992). Maturation of Hantaan virus glycoproteins G1 and G2. *Virology* **189**:324–328.

Arai, R., Nakagawa, H., Tsumoto, K., Mahoney, W., Kumagai, I., Ueda, H., and Nagamune, T. (2001). Demonstration of a homogeneous noncompetitive immunoassay based on bioluminescence resonance energy transfer. *Anal. Biochem.* **289**:77–81.

Bandla, M. D., Campbell, L. R., Ullman, D. E., and Sherwood, J. L. (1998). Interaction of tomato spotted wilt tospovirus (TSWV) glycoproteins with a thrips midgut protein, a potential cellular receptor for TSWV. *Phytopath.* **88**:98–104.

Bar, P. M., Bassham, D. C., and Raikhel, N. V. (1996). Transport of proteins in eukaryotic cells, more questions ahead. *Plant Mol. Biol.* **32**:223–249.

Bark, S. J., and Hahn, K. M. (2000). Fluorescent indicators of peptide cleavage in the trafficking compartments of living cells, peptides site-specifically labeled with two dyes. *Meth. Orlando* **20**:429–435.

Bastiaens, P. I. H., and Jovin, T. M. (1996). Microspectroscopic imaging tracks the intracellular processing of a signal transduction protein, Fluorescent-labeled protein kinase C beta-I. *Proc. Natl. Acad. Sci USA* **93**:8407–8412.

Beaton, A. R., and Krug, R. M. (1984). Synthesis of the templates for influenza virion RNA replication *in vitro*. *Proc. Natl. Acad. Sci USA* **81**:4682–4686.

Beaton, A. R., and Krug, R. M. (1986). Transcription antitermination during influenza viral template RNA synthesis requires the nucleocapsid protein and the absence of a 5'capped end. *Proc. Natl. Acad. Sci USA* **83**:6282–6286.

Bellocq, C., and Kolakofsky, D. (1987). Translational requirement for LaCrosse virus S messenger RNA synthesis, A possible mechanism. *J. Virol.* **61**:3960–3967.

Bellocq, C., Raju, R., Patterson, J., and Kolakofsky, D. (1986). Translational requirement of LaCrosse virus small messenger RNA synthesis *in vitro* studies. *J. Virol.* **61**:87–95.

Berland, K. M., So, P. T. C., Chen, Y., Mantulin, W. W., and Gratton, E. (1996). Scanning tow-photon fluctuation correlation spectroscopy, Particle counting measurements for detection of molecular aggregation. *Biophys. J.* **71**:410–420.

Berney, C., and Danuser, G. (2003). FRET or no FRET, A quantitative comparison. *Biophys. J.* **84**:3992–4010.

Bezerra, I. C., Resende, R. de O., Pozzer, L., Nagata, T., Kormelink, R., and de Ávila, A. C. (1999). Increase of tospovirus diversity in Brazil with the identification of two new tospovirus species, one from chrysanthemum and one from zucchini. *Phytopath.* **89**:823–830.

Bishop, D. H. L., Gay, M. E., and Matsuoko, Y. (1983). Nonviral heterogeneous sequences are present at the 5'ends of one species of snowshoe hare bunyavirus S complementary RNA. *Nucl. Acids. Res.* **11**:6409–6418.

Biswas, S. K., and Nayak, D. P. (1994). Mutational analysis of the conserved motifs os Influenza A virus polymerase basic protein 1. *J. Virol.* **68**:1819–1826.

Blakqori, G., Kochs, G., Haller, O., and Weber, F. (2003). Functional L polymerase of La Crosse virus allows *in vivo* reconstitution of recombinant nucleocapsids. *J. Gen. Virol.* **84**:1207–1214.

Bonnet, G., Krichevsky, O., and Libchaber, A. (1998). Kinetics of conformational fluctuations in DNA hairpin loops. *Proc. Natl. Acad. Sci USA* **95**:8602–8606.

Booth, T. F., Gould, E. A., and Nuttall, P. A. (1991). Structure and morphogenesis of Dugbe virus *Bunyaviridae* Nairovirus studied by immunogold electron microscopy of ultrathin cryosections. *Virus Res.* **21**:199–212.

Bosch, B. J., de Haan, C. A. M., and Rottier, P. J. M. (2004). Coronavirus spike glycoprotein, extended at the carboxi terminus with green fluorescent protein, is assembly competent. *J. Virol.* **78**:7369–7378.

Bouloy, M., and Hannoun, C. (1976). Studies on Lumbovirus replication, I RNA-dependent RNA polymerase associated with virions. *Virol.* **69**:258–264.

Bouloy, M., Janzen, C., Vialat, P., Khun, H., Pavlovic, J., Huerre, M., and Haller, O. (2001). Genetic evidence for an interferon-antagonistic function of Rift Valley fever virus nonstructural protein NSs. *J. Virol.* **75**:1371–1377.

Bouloy, M., Krans-Ozden, S., Horodniceanu, F., and Hannoun, C. (1973/1974). 3 segment RNA genome of Lumbo virus bunyavirus. *InterVirology* **2**:173–180.

Bouloy, M., Pardigon, N., Vialat, P., Gerbaud, S., and Girard, M. (1990). Characterization of the 5′ and 3′ ends of viral messenger RNA isolated from BHK21 cells infected with Germiston virus bunyavirus. *Virology* **175**:50–58.

Braakman, I., and van Anken, E. (2000). Folding of viral envelope glycoproteins in the endoplasmic reticulum. *Traffic* **1**:533–539.

Braam, J., Ulmanen, I., and Krug, R. M. (1983). Molecular model of a eukaryotic transcription complex functions and movements of Influenza P proteins during capped RNA primed transcription. *Cell* **34**:609–618.

Brandizzi, F., Frangne, N., Marc-Martin, S., Hawes, C., Neuhaus, J. M., and Paris, N. (2002). The destination for single-pass membrane proteins is influenced markedly by the length of the hydrophobic domain. *Plant Cell* **14**:1077–1092.

Brideau, A. D., Del Rio, T., Wolffe, E. J., and Enquist, L. W. (1999). Itracellular trafficking and localization of the pseudorabies virus Us9 type II envelope protein to host and viral membranes. *J. Virol.* **73**:4372–4384.

Bridgen, A., Dalrymple, D. A., Weber, F., and Elliott, R. M. (2004). Inhibition of Dugbe Nairovirus replication by human MxA protein. *Vir. Res.* **99**:47–50.

Bridgen, A., and Elliott, R. M. (1996). Rescue of a negative-stranded RNA virus entirely from cloned complementary DNAs. *Proc. Natl. Acad. Sci USA* **93**:15400–15404.

Bridgen, A., Weber, F., Fazakerley, J. K., and Elliott, R. M. (2001). Bunyamwera bunyavirus nonstructural protein NSs is a nonessential gene product that contributes to viral pathogenesis. *Proc. Natl. Acad. Sci USA* **98**:664–669.

Bucher, E., Sijen, T., de Haan, P., Goldbach, R., and Prins, M. (2003). Negative-strand tospoviruses and tenuiviruses carry a gene for a suppressor of gene silencing at analogous genomic positions. *J. Virol.* **77**:1329–1336.

Bupp, K., Stillmock, K., and Gonzalez, S. F. (1996). Analysis of the intracellular transport properties of recombinant La Crosse virus glycoproteins. *Virology* **220**:485–490.

Carette, J. E., Stuiver, M., van Lent, J., Wellink, J., and van Kammen, A. (2000). Cowpea mosaic virus infection induces a massive proliferation of endoplasmic reticulum but not Golgi membranes and is dependent on *de novo* membrane synthesis. *J. Virol.* **74**:6556–6563.

Caton, A. J., and Robertson, J. S. (1980). Structure of the host-derived sequences present at the 5′ends of Influenza virus mRNA. *Nucl. Acids. Res.* **8**:2591–2603.

Chapman, E. J., Hilson, P., and German, T. L. (2003). Association of L protein and *in vitro* Tomato spotted wilt virus RNA-dependent RNA polymerase activity. *InterVirol.* **46**:177–181.

Chen, C. C., and Chiu, R. J. (1996). A tospovirus infecting peanut in Taiwan. *Acta Hortic.* **431**:57–67.

Chen, S. Y., and Compans, R. W. (1991). Oligomerization, transport, and Golgi retention of Punta Toro virus glycoproteins. *J. Virol.* **65:**5902–5909.

Chen, S. Y., Matsuoka, Y., and Compans, R. W. (1991). Golgi complex localization of the Punta Toro virus G2 protein requires its association with G1 protein. *Virology* **183:**351–365.

Chu, F. H., Chao, C. H., Peng, Y. C., Lin, S. S., Chen, C. C., and Yeh, S. D. (2001). Serological and molecular characterization of Peanut chlorotic fan-spot virus, a new species of the genus Tospovirus. *Phytopath.* **91:**856–863.

Cole, N. B., Smith, C. L., Sciaky, N., Terasaki, M., Edidin, M., and Lippincott-Schwartz, J. (1996). Diffusional mobility of Golgi proteins in membranes of living cells. *Science* **273:**797–801.

Collett, M. S. (1986). Messenger RNA of the M segment RNA of Rift Valley fever virus. *Virology* **151:**151–156.

Cortes, I., Livieratos, I. C., Derks, A., Peters, D., and Kormelink, R. (1998). Molecular and serological characterization of iris yellow spot virus, a new and distinct tospovirus species. *Phytopath.* **88:**1276–1282.

Cortez, I., Saaijer, J., Wongjkaew, K. S., Pereira, A. M., Goldbach, R., Peters, D., and Kormelink, R. (2001). Identification and characterization of a novel tospovirus species using a new RT-PCR approach. *Arch. Virol.* **146:**265–278.

Dalton, K. P., and Rose, J. K. (2001). Vesicular stomatitis virus glycoprotein containing the entire green fluorescent protein on its cytoplasmic domain is incorporated efficiently into virus particles. *Virology* **279:**414–421.

DeAngelis, J. D., Sether, D. M., and Rossignol, P. A. (1994). Transmission of Impatiens necrotic spot virus in peppermint by western flower thrips (Thysanoptera, Thripidae). *J. Econ. Entomol.* **87:**197–201.

de Ávila, A. C., de Haan, P., Kitajima, E. W., Kormelink, R., Resende, R. de O., Goldbach, R., and Peters, D. (1992). Characterization of a distinct isolate of tomato spotted wilt virus (TSWV) from impatiens sp. in The Netherlands. *J. Phytopath.* **134:**133–151.

de Ávila, A. C., de Haan, P., Kormelink, R., Resende, R. de O., Goldbach, R., and Peters, D. (1993). Classification of tospoviruses based on phylogeny of nucleoprotein gene sequences. *J. Gen. Virol.* **74:**153–159.

de Ávila, A. C., Huguenot, C., Resende, R., Kitajima, E. W., Goldbach, R. W., and Peters, D. (1990). Serological differentiation of twenty isolates of tomato spotted wilt virus. *J. Gen. Virol.* **71:**2801–2807.

de Haan, P., De Ávila, A. C., Kormelink, R., Westerbroek, A., Gielen, J. J. L., Peters, D., and Goldbach, R. (1992). The nucleotide sequence of the S RNA of impatiens necrotic spot virus, a novel tospovirus. *FEBS Lett.* **306:**27–32.

de Haan, P., Kormelink, R., Resende, D., van Poelwijk, F., Peters, D., and Goldbach, R. (1991). Tomato spotted wilt virus L RNA encodes a putative RNA polymerase. *J. Gen. Virol.* **72:**2207–2216.

de Haan, P., Wagemakers, L., Peters, D., and Goldbach, R. (1989). Molecular cloning and terminal sequence determination of the S and M RNA species of Tomato spotted wilt virus. *J. Gen. Virol.* **70:**3469–3474.

de Haan, P., Wagemakers, L., Peters, D., and Goldbach, R. (1990). The S RNA segment of Tomato spotted wilt virus has an ambisense character. *J. Gen. Virol.* **71:**1001–1008.

Dewey, R., Semorile, L., Gracia, O., and Grau, O. (1993). TSWV N protein sequence comparison among an Argentine isolate and other tospoviruses. *Abstr. of the IXth Int. Congress Virol.* Glasgow, Scotland, 1993, 357.

Dhar, R., Chanock, R. M., and Lai, C. J. (1980). Nonviral oligonucleotides at the 5′terminus of cytoplasmic influenza viral mRNAs deduced from cloned complete genomic sequences. *Cell* **21:**495–500.

DiBonito, P., Nicoletti, L., Mochi, S., Accardi, L., Marchi, A., and Giorgi, C. (1999). Immunological characterization of Toscana virus proteins. *Arch. Virol.* **144:**1947–1960.

Duijsings, D., Kormelink, R., and Goldbach, R. (1999). Alfalfa mosaic virus RNAs serve as cap donors for tomato spotted wilt virus transcription during coinfection of Nicotiana benthamiana. *J. Virol.* **73:**5172–5175.

Duijsings, D., Kormelink, R., and Goldbach, R. (2001). *In vivo* analysis of the TSWV cap-snatching mechanism, ingle basse complementarity and primer length requirements. *EMBO J.* **20:**2545–2552.

Dunn, E. F., Pritlove, D. C., Jin, H., and Elliott, R. M. (1995). Transcription of a recombinant Bunyavirus RNA template by transiently expressed Bunyavirus proteins. *Virology* **211:**133–143.

Eigen, M., and Rigler, R. (1994). Sorting single molecules, Application to diagnostics and evolutionary biotechnology. *Proc. Natl. Acad. Sci USA* **91:**5740–5747.

Elliott, R. M. (1990). Molecular biology of the *Bunyaviridae*. *J. Gen. Virol.* **71:**501–522.

Elliott, R. M. (1996). *The Bunyaviridae*. Plenum press, New York.

Emery, V. C., and Bishop, D. H. L. (1987). Characterization of Punta toro S messenger RNA species and identification of an inverted complementary sequence in the intergenic region of Punta toro phlebovirus ambisense S RNA that is involved in messenger RNA transcription termination. *Virology* **156:**1–11.

Erickson, M. G., Moon, D. L., and Yue, D. T. (2003). DsRed as a potential FRET partner with CFP and GFP. *Biophys. J.* **85:**599–611.

Eshita, Y., Ericson, B., Romanowski, V., and Bishop, D. H. L. (1985). Analyses of the messenger RNA transcription processes of Snowshoe hare bunyavirus small and medium-sized RNA species. *J. Virol.* **55:**681–689.

Fabian, M. R., and White, K. A. (2004). 5′-3′ RNA–RNA interaction facilitates cap- and poly(A) tail-independent translation of Tomato bushy stunt virus mRNA, a potential common mechanism for *Tombusviridae*. *J. Biol. Chem.* **279:**28862–28872.

Filho, F. M. de Assis., Naidu, R. A., Deom, C. M., and Sherwood, J. L. (2002). Dynamics of tomato spotted wilt virus replication in the alimentary canal of two thrips species. *Phytopath.* **92:**729–733.

Flick, R., and Petterson, R. F. (2001). Reverse genetics system for Uukuniemi virus (*Bunyaviridae*), RNA polymerase I-catalyzed expression of chimeric viral RNAs. *J. Virol.* **75:**1643–1655.

Flick, R., Elgh, F., Hobom, G., and Pettersson, R. F. (2002). Mutational analysis of the Uukuniemi virus (*Bunyaviridae*) promoter reveals two regions of functional importance. *J. Virol.* **76:**10849–10860.

Flick, R., Flick, K., Feldmann, H., and Elgh, F. (2003a). Reverse genetics for Crimean-Congo hemorrhagic fever virus. *J. Virol.* **77:**5997–6006.

Flick, K., Hooper, J. W., Schmaljohn, C. S., Pettersson, R. F., Feldmann, H., and Flick, R. (2003). Rescue of Hantaanvirus minigenomes. *Virology* **306:**219–224.

Flick, K., Katz, A., Överby, A., Feldmann, H., Pettersson, R. F., and Flick, R. (2004). Functional analysis of the noncoding regions of the Uukuniemi virus (*Bunyaviridae*) RNA segments. *J. Virol.* **78:**11726–11738.

Flohr, F., Schneider-Schaulies, S., Haller, O., and Kochs, G. (1999). The central interactive region of human MxA GTPase is involved in GTPase activation and interaction with viral target structures. *FEBS Lett.* **364:**24–28.

Francki, R. I. B., Milne, R. G., and Hatta, T. (1985). Tomato spotted wilt virus group. *In* "Atlas of Plant Viruses" (R. I. B. Franki, R. G. Milne, and T. Hatta, eds.), vol. 1, pp. 101–110. CRC Press, Boca Raton.

Francki, R. I. B., Fauquet, C. M., Knudson, D. L., and Brown, F. (1991). Classification and nomenclature of viruses, Fifth Report of the International Committee on taxonomy of viruses. *Arch. Virol.* Suppl(2):1–450.

Frese, M., Kochs, G., Feldmann, H., Hertkorn, C., and Haller, O. (1996). Inhibition of Bunyaviruses, Phleboviruses and Hantaviruses by human MxA protein. *J. Virol.* **70:**915–923.

Frese, M., Prins, M., Ponten, A., Goldbach, R. W., Haller, O., and Zeltz, P. (2000). Constitutive expression of interferon-induced human MxA protein in transgenic tobacco plants does not confer resistance to a variety of RNA viruses. *Transgen. Res.* **9:**429–438.

Fujisawa, I., Tanaka, K., and Ishii, M (1988). Tomato spotted wilt virus transmisssion by three species of thrips, *Thrips setosus, Thrips tabaci,* and *Thrips palmi* (in Japanese). *Ann. Phytopath. Soc.* Japan **54:**392.

Gadella, T. W. J., Jr. (1999). Fluorescence lifetime imaging microscopy (FLIM), Instrumentation and applications. *In* "Fluorescent and luminescent probes" (W. T. Mason, ed.), 2nd edn. pp. 467–479. Academic Press, London.

Gahmberg, N., Kuismanen, E., Keränen, S., and Pettersson, R. F. (1986). Uukuniemi virus glycoproteins accumulate in and cause morphological changes of the Golgi complex in the absence of virus maturation. *J. Virol.* **57:**899–906.

Gallie, D. R. (1998). A tale of two termini: A functional interaction between termini of an mRNA is a prerequisite for efficient translation initiation. *Gene Amsterdam* **216:**1–11.

Gallie, D. R., and Kobayashi, M. (1994). The role of the 3′ untranslated region of non-polyadenylated plant viral mRNAs in regulating translational efficiency. *Gene Amsterdam* **142:**159–165.

Galperin, E., Verkusha, V. V., and Sorkin, A. (2004). Three-chromophore FRET microscopy to analyze multiprotein interactions in living cells. *Nat. Methods* **1:**209–217.

Garcin, D., and Kolakofsky, D. (1990). A novel mechanism for the initiation of Tacaribe Arenavirus genome replication. *J. Virol.* **64:**6196–6203.

Gardner, M. W., Tompkins, C. M., and Whipple, O. C. (1935). Spotted wilt of truck crops and ornamental plants. *Phytopath.* **25:**17.

Gentsch, J. R., Bishop, D. H. L., and Obijeski, J. F. (1977). The virus particle nucleic acids and proteins of four bunyaviruses. *J. Gen. Virol.* **34:**257–268.

Gera, A., Cohen, J., Salamon, R., and Raccah, B. (1998). Iris yellow spot tospovirus detected in onion (Allium cepa) in Israel. *Plant Dis.* **82:**127.

Gera, A., Kritzman, A., Cohen, J., and Raccah, B. (1998). Tospovirus infecting bulb crops in Israel. *In* "Recent Progress in Tospovirus and Thrips Research" (D. Peters and R. Goldback, eds.), pp. 86–87. Wageningen, The Netherlands.

Gerbaud, S., Pardigon, N., Vialat, P., and Bouloy, M. (1987). The S segment of the Germiston bunyavirus genome, Coding strategy and transcription. *In* "The Biology of Negative Strand Viruses" (B. Mahy and D. Kolakofsky, eds.), pp. 191–198. Elsevier, Amsterdam.

Gerrard, S. R., and Nichol, S. T. (2002). Characterization of the Golgi retention motif of Rift Valley fever virus Gn glycoprotein. *J. Virol.* **76:**12200–12210.

Giordano, L., Jovin, T. M., Masahiro, I., and Jares-Erijman, E. (2002). Diheteroarylethenes as thermally stable photoswichable acceptors in photochromic fluorescence resonance energy transfer (pcFRET). *J. Am. Chem. Soc.* **124:**7481–7489.

Goldbach, R. W., and Peters, D. (1994). Possible causes of the emergence of tospovirus diseases. *Semin. Virol.* **5:**113–120.

Gordon, G. W., Berry, G., Liang, X. H., Levine, B., and Herman, B. (1998). Quantitative fluorescence resonance energy transfer measurements using fluorescence microscopy. *Biophys. J.* **74:**2702–2713.

Gowda, S., Satyanarayana, T., Naidu, R. A., Mushegian, A., Dawson, W. O., and Reddy, D. V. R. (1998). Characterization of the large (L) RNA of peanut bud necrosis tospovirus. *Arch. Virol.* **143:**2381–2390.

Granval de Milan, N., Piccolo, R., and Gracia, O. (1998). Potato (Solanum tuberosum L.) tuber transmission of tospoviruses (TSWV, GRSV and TCSV) in Mendoza province, Argentina. *In* "Recent Progress in Tospovirus and Thrips Research" (D. Peters and R. Goldbach, eds.), pp. 46–47. Wageningen, The Netherlands.

Grinberg, A. V., Hu, C. D., and Kerppola, T. K. (2004). Visualization of Myc/Max/Mad family dimers and the competition for dimerization in living cells. *Mol. Cell. Biol.* **24:**4294–4308.

Grò, M. C., Di Bonito, P., Accardi, L., and Giorgi, C. (1992). Analysis of 3′ and 5′ ends of N and NSs messenger RNAs of Toscana phlebovirus. *Virology* **191:**435–438.

Gutiérrez-Escolano, A. L., Brito, Z. U., del Angel, R. M., and Jiang, X. (2000). Interaction of cellular proteins with the 5′ end of Norwalk virus genomic RNA. *J. Virol.* **74:**8558–8562.

Hacker, J. K., and Hardy, J. L. (1997). Adsorptive endocytosis of California encephalitis virus into mosquito and mammalian cells, a role for G1. *Virol.* **235:**40–47.

Hacker, J. K., Volkman, L. E., and Hardy, J. L. (1995). Requirement for the G1 protein of California encephalitis virus in infection *in vitro* and *in vivo*. *Virol.* **206:**945–953.

Haller, O., Janzen, C., Vialat, P., Huerre, M., Pavlovic, J., and Bouloy, M. (2000). High virulence of attenuated Rift valley fever virus in mice lacking a type I interferon system. *J. Clin. Virol.* **18:**192.

Hall, J. M., Speck, J., Geske, S., Mohan, K., and Moyer, J. W. (1993). A Tospovirus with serologically distinct nucleocapsid isolated from onion (Allium cepa). *Abstr. of the IXth Int. Congress Virol.* Glasgow, Scotland, 1993, 357.

Heath, C. M., Windsor, M., and Wileman, T. (2001). Aggresomes resemble sites specialized for virus assembly. *J. Cell Biol.* **153:**449–455.

Heinze, C., Maiss, E., Adam, G., and Casper, R. (1995). The complete nucleotide sequence of the S RNA of a new Tospovirus species, representing serogroup IV. *Phytopath.* **85:**683–690.

Hess, S.-T., Huang, S. H., Heikal, A. A., and Webb, W. W. (2002). Biological and chemical applications of fluorescence correlation spectroscopy, A review. *Biochem.* **41:**697–705.

Hink, M. A., Bisseling, T., and Visser, A. J. W. G. (2002). Imaging protein-protein interactions in living cells. *Plant Mol. Biol.* **50:**871–883.

Honig, J. E., Osborne, J. C., and Nichol, S. T. (2004). Crimean-Congo hemorrhagic fever virus genome L RNA segment and encoded protein. *Virol.* **321:**29–35.

Hu, C. D., Chinekov, Y., and Kerppola, T. K. (2002). Visualization of interactions among bZIP and Rel family proteins in living cells using bimolecular fluorescence complementation. *Mol. Cell* **9:**789–798.

Hu, C. D., and Kerppola, T. K. (2003). Simultaneous visualization of multiple protein interactions in living cells using multicolor fluorescence complementation analysis. *Nat. Biotech.* **21:**539–545.

Huiet, L., Feldstein, P. A., Tsai, J. H., and Falk, B. W. (1993). The maize stripe virus major noncapsid protein messenger RNA transcripts contain heterogeneous leader sequences at their 5′ termini. *Virol.* **197:**808–812.

Iapalucci, S., Lopez, N., and Franze-Fernandez, M. T. (1991). The 3′ end termini of the Tacaribe arenavirus subgenomic RNAs. *Virology* **182:**269–278.

Ie, T. S. (1971). Electron microscopy of developmental stages of Tomato spotted wilt virus in plant cells. *Virology* **43:**468–479.

Ihara, T., Matsuura, Y., and Bishop, D. H. L. (1985). Analyses of the mRNA transcription processes of Punta Toro phlebovirus (*Bunyaviridae*). *Virol.* **147**:317–325.

Ihara, T., Smith, J., Dalrymple, J. M., and Bishop, D. H. L. (1985). Complete sequences of the glycoproteins and M RNA of Punta Toro phlebovirus compared to those of Rift Valley fever virus. *Virology* **144**:246–259.

Immink, R. G. H., Gadella, T. W. J., Jr., Ferrario, S., Busscher, M., and Angenent, G. C. (2002). Analysis of MADS box protein-protein interactions in living plant cells. *Proc. Natl. Acad. Sci, USA* **99**:2416–2421.

Imreh, G., Maksel, D., de Monvel, J. B., Branden, L., and Hallberg, E. (2003). ER retention may play a role in sorting of the nuclear pore membrane protein POM121. *Exp. Cell Res.* **284**:173–184.

Iordanov, M. S., Wong, J., Bell, J. C., and Magun, B. E. (2001). Activation of NF-kappaB double stranded RNA (dsRNA) in the absence of protein kinase R and Rnase L demonstrates the existence of two separate dsRNA-triggered antiviral programs. *Mol. Cell. Biol.* **21**:61–72.

Jain, R., Pappu, H., Pappu, S., Krishnareddy, M., and Vani, A. (1998). Watermelon bud necrosis tospovirus is a distinct virus species belonging to serogroup IV. Arch. *Virol.* **143**:1637–1644.

Jäntti, J., Hildén, P., Rönkä, H., Mäkiranta, V., Keränen, S., and Kuismanen, E. (1997). Immunocytochemical analysis of Uukuniemi virus budding compartments, role of the intermediate compartment and the Golgi stack in virus maturation. *J. Virol.* **71**:1162–1172.

Jin, H., and Elliott, R. M. (1991). Expression of functional Bunyamwera virus L protein by recombinant vaccinia viruses. *J. Virol.* **65**:4182–4189.

Jin, H., and Elliott, R. M. (1992). Mutagenesis of the L protein encoded by Bunyamwera virus and production of monospecific antibodies. *J. Gen. Virol.* **73**:2235–2244.

Jin, H., and Elliott, R. M. (1993). Characterization of Bunyamwera virus S RNA that is transcribed and replicated by the L protein expressed from recombinant vaccinia virus. *J. Virol.* **67**:1396–1404.

Johnston, J. A., Ward, C. L., and Kopito, R. R. (1998). Aggresomes, A cellular response to misfolded proteins. *J. Cell Biol.* **143**:1883–1898.

Kainz, M., Hilson, P., Sweeney, L., DeRose, E., and German, T. L. (2004). Interaction between Tomato spotted wilt virus N protein monomers involves nonelectrostatic forces governed by multiple distinct regions in the primary structure. *Phytopath.* **94**:759–765.

Kameya-Iwaki, M., Honda, Y., Hanada, K., Tochihara, H., Yonaha, T., Hokama, K., and Yokoyama, T. (1984). Silver mottle disease of watermelon caused by tomato spotted witt virus. *Plant Dis.* **68**:1006–1008.

Kanerva, M., Melen, K., Vaheri, A., and Julkunen, I. (1996). Inhibition of puumula and tula hantaviruses in Vero cells by MxA protein. *Virol.* **224**:55–62.

Katiliene, Z., Katilius, E., and Woodbury, N. W. (2003). Single molecule detection of DNA looping by NgoMIV restriction endonuclease. *Biophys. J.* **84**:4053–4061.

Kato, K., and Hanada, K. (2000). Characterization of the S RNA segment of melon yellow spot virus. Jap. *J. Phytopath.* **66**:252–254.

Kaukinen, P., Koistinen, V., Vapalahti, O., Vaheri, A., and Plyusnin, A. (2001). Interaction between molecules of Hantavirus nucleocapsid protein. *J. Gen. Virol.* **82**:1845–1853.

Kaukinen, P., Vaheri, A., and Plyusnin, A. (2003). Mapping the regions involved in homotypic interactions of Tula hantavirus N protein. *J. Virol.* **77**:10910–10916.

Kellman, J. F., Liebisch, P., Schmitz, K. P., and Piechulla, B. (2001). Visual representation by atomif force microscopy (AFM) of tomato spotted wiltvirus ribonucleoproteins. *Biol. Chem.* **382**:1559–1562.

Kermode, A. R. (1996). Mechanisms of intracellular protein transport and targeting in plant cells. *Crit. Rev. Plant Sci* **15**:285–423.

Kikkert, M. (1999). Role of the envelope glycoproteins in the infection cycle of Tomato spotted wilt virus. PhD thesis, Wageningen Agricultural University, The Netherlands.

Kikkert, M., Meurs, C., van de Wetering, F., Kormelink, R., and Goldbach, R. (1998). Binding of Tomato spotted wilt virus to a 94 kDa thrips protein. *Phytopath.* **88**:63–69.

Kikkert, M., van Lent, J., Storms, M., Bodegom, P., Kormelink, R., and Goldbach, R. (1999). Tomato spotted wilt virus particle morphogenesis in plant cells. *J. Virol.* **73**:2288–2297.

Kikkert, M., van Poelwijk, F., Storms, M., Bloksma, H., Karsies, W., Kormelink, R., and Goldbach, R. (1997). A protoplast system for studying tomato spotted wilt virus infection. *J. Gen. Virol.* **78**:1755–1763.

Kikkert, M., Verschoor, A., Kormelink, R., Rottier, P., and Goldbach, R. (2001). Tomato spotted wilt virus glycoproteins exhibit trafficking and localization signals that are functional in mammalian cells. *J. Virol.* **75**:1004–1012.

Kinjo, M., and Rigler, R. (1995). Ultrasensitive hybridization analysis using fluorescence correlation spectroscopy. *Nucleic Acids Res.* **23**:1795–1799.

Kinsella, E., Martin, S. G., Grolla, A., Czub, M., Feldmann, H., and Flick, R. (2004). Sequence determination of the Crimean-Congo hemorrhagic fever virus L segment. *Virol.* **321**:23–28.

Kitajima, E. W., de Ávila, A. C., Resende, R. de O., Goldbach, R. W., and Peters, D. (1992). Comparative cytological and immunogold labelling studies on different isolates of Tomato spotted wilt virus. *J. Submicrosc. Cytol. Pathol.* **24**:1–14.

Kochs, G., Janzen, C., Hohenberg, H., and Haller, O. (2002). Antivirally active MxA protein sequesters La Crosse virus nucleocapsid protein into perinuclear complexes. *Proc. Natl. Acad. Sci USA* **99**:3153–3158.

Kohl, A., Clayton, R. F., Weber, F., Bridgen, A., Randall, R. E., and Elliott, R. M. (2003). Bunyamwera virus nonstructural protein NSs counteracts interferon regulatory factor 3-mediated induction of early cell death. *J. Virol.* **77**:7999–8008.

Köhler, R. H., Cao, J., Zipfel, W. R., Webb, W. W., and Hanson, M. R. (1997). Exchange of protein molecules through connections between higher plant plastids. *Science* **276**:2039–2042.

Kormelink, R. (1994). Structure and expression of the Tomato spotted wilt virus genome, a plant-infecting Bunyavirus. PhD thesis, Wageningen Agricultural, The Netherlands.

Kormelink, R., De Haan, P., Meurs, C., Peters, D., and Goldbach, R. (1992). The nucleotide sequence of the M RNA segment of Tomato spotted wilt virus, A bunyavirus with two ambisense RNA segments. *J. Gen. Virol.* **73**:2795–2804.

Kormelink, R., de Haan, P., Peters, D., and Goldbach, G. (1992). Viral RNA synthesis in Tomato spotted wilt virus-infected *Nicotiana rustica* plants. *J. Gen. Virol.* **73**:687–693.

Kormelink, R., Kitajima, E. W., de Haan, P., Zuidema, D., Peters, D., and Goldbach, R. (1991). The nonstructural protein (NSs encoded by the ambisense S RNA of Tomato spotted wilt virus is associated with fibrous structures in infected plant cells. *Virology* **181**:459–468.

Kormelink, R., Storms, M., van Lent, J., Peters, D., and Goldbach, R. (1994). Expression and subcellular location of the NSm protein of Tomato spotted wilt virus (TSWV), a putative viral movement protein. *Virology* **200**:56–65.

Kormelink, R., Van Poelwijk, F., Peters, D., and Goldbach, R. (1992). Nonviral heterogeneous sequences at the 5′ ends of tomato spotted wilt virus (TSWV) mRNAs. *J. Gen. Virol.* **73:**2125–2128.

Kuismanen, E., Bang, B., Hurme, M., and Pettersson, R. F. (1984). Uukuniemi virus maturation, immunofluorescence microscopy with monoclonal glycoprotein-specific antibodies. *J. Virol.* **51:**137–146.

Kuismanen, E., Hedman, K., Saraste, J., and Pettersson, R. F. (1982). Uukuniemi virus maturation, accumulation of virus particles and viral antigens in the Golgi complex. *Mol. Cell Biol.* **2:**1444–1458.

Lakshmi, K. V., Wightman, J. A., Reddy, D. V. R., Ranga Rao, G. V., Buiel, A. A. M., and Reddy, D. D. R. (1995). Transmission of peanut bud necrosis virus by Thrips palmi in India. *In* "Thrips Biology and Management" (B. L. Parker and T. Lewis, eds.), pp. 170–184. Plenum Press, New York.

Lappin, D. F., Nakitare, G. W., Palfreyman, J. W., and Elliott, R. M. (1994). Localization of Bunyamwera bunyavirus G1 glycoprotein to the Golgi requires association with G2 but not with NSm. *J. Gen. Virol.* **75:**3441–3451.

Larson, D. R., Ma, Y. M., Vogt, V. M., and Webb, W. W. (2003). Direct measurement of Gag-Gag interaction during retrovirus assembly with FRET and fluorescence correlation spectroscopy. *J. Cell Biol.* **162:**1233–1244.

Law, M. D., and Moyer, J. W. (1990). A tomato spotted wilt-like virus with a serologically distinct N protein. *J. Gen. Virol.* **71:**933–938.

Law, M. D., Speck, J., and Moyer, J. W. (1992). The M RNA of Impatiens necrotic spot Tospovirus (*Bunyaviridae*) has an ambisense genomic organization. *Virology* **188:**732–741.

Lawson, R. H., Dienelt, M. M., and Hsu, H. T. (1996). Ultrastructural comparisons of defective, partially defective, and nondefective isolates of Impatiens necrotic spot virus. *Phytopath.* **86:**650–661.

Lee, B.-H., Yoshimatsu, K., Maeda, A., Ochiai, K., Morimatsu, M., Araki, K., Ogino, M., Morikawa, S., and Arikawa, J. (2003). Association of the nucleocapsid protein of the Seoul and Hantaan hantaviruses with small ubiquitin-like modifier-1-related molecules. *Vir. Res.* **98:**83–91.

Lees, J. F., Pringle, C. R., and Elliott, R. M. (1986). Nucleotide sequence of the Bunyamwera virus M RNA segment, conservation of structural features in the Bunyavirus glycoprotein gene product. *Virology* **148:**1–14.

Leonard, S., Viel, C., Beauchemin, C., Daigneault, N., Fortin, M. G., and Laliberte, J. F. (2004). Interaction of VPg-pro of Turnip mosaic virus with the translation initiation factor 4E and the poly(A)-binding protein *in planta*. *J. Gen. Virol.* **85:**1055–1063.

Lenz, O., ter Meulen, J., Feldmann, H., Klenk, H. D., and Garten, W. (2000). Identification of a novel consensus sequence at the cleavage site of the lassa virus glycoprotein. *J. Virol.* **74:**11418–11421.

Lenz, O., ter Meulen, J., Klenk, H. D., Seidah, N. G., and Garten, W. (2001). The Lassa virus glycoprotein precursor GP-C is proteolytically processed by subtilase SKI-1/S1P. *Proc. Natl. Acad. Sci USA* **98:**12701–12705.

Li, X. D., Makela, T. P., Guo, D., Soliymani, R., Koistinen, V., Vapalahti, O., Vaheri, A., and Lankinen, H. (2002). Hantavirus nucleocapsid protein interacts with the Fas-mediated apoptosis enhancer Daxx. *J. Gen. Virol.* **83:**759–766.

Lidke, D. S., Nagy, P., Barisas, B. G., Heintzmann, R., Post, J. N., Lidke, K. A., Clayton, A. H. A., Arndt-Jovin, D. J., and Jovin, T. M. (2003). Imaging molecular interactions in cells by dynamic and static fluorescence anisotropy (rFLIM and rFRET). *Biochem. Soc. Trans.* **31:**1020–1027.

Liu, J., and Lu, Y. (2002). FRET study of a trifluorophore-labeled DNAzyme. *J. Am. Chem. Soc.* **124:**15208–15216.

Löber, C., Anheier, B., Lindow, S., Klenk, H. D., and Feldmann, H. (2001). The Hantaan virus glycoprotein precursor is cleaved at the conserved pentapeptide WAASA. *Virology* **289:**224–229.

Lopez, N., Muller, R., Prehaud, C., and Bouloy, M. (1995). The L protein of Rift valley fever virus can rescue viral ribonucleoproteins and transcribe synthetic genome-like RNA molecules. *J. Virol.* **69:**3972–3979.

Louro, D. (1996). Detection and identification of tomato spotted wilt virus and impatiens necrotic spot virus in Portugal. *Act. Horticult.* **431:**99–105.

Ludwig, G. V., Christensen, B. M., Yuill, T. M., and Schultz, K. T. (1989). Enzyme processing of LaCrosse virus glycoprotein G1, a bunyavirus-vector infection model. *Virology* **171:**108–113.

Ludwig, G. V., Israel, B. A., Christensen, B. M., Yuill, T. M., and Schultz, K. T. (1991). Role of La Crosse virus glycoproteins G1 in attachment of virus to host cells. *Virol.* **181:**564–571.

Lyons, M., and Heyduk, J. (1973). Aspects of the developmental morphology of California encephalitis virus in cultured vertebrate and arthropod cells and in mouse brain. *Virology* **54:**37–52.

Maiss, E., Ivanova, L., Breyel, E., and Adam, G. (1991). Cloning and sequencing of the S RNA from a Bulgarian isolate of tomato spotted wilt virus. *J. Gen. Virol.* **72:**461–464.

Marchoux, G., Gebreselassie, K., and Villevieille, M. (1991). Detection of tomato spotted wilt virus and transmission by Frankliniella occidentalis in France. *Plant Pathol.* **40:**347–380.

Marriott, A. C., and Nuttall, P. A. (1996). Large RNA segment of DUG Nairovirus encodes the putative RNA polymerase. *J. Gen. Virol.* **77:**1775–1789.

Matsuda, D., and Dreher, T. W. (2004). The tRNA-like structure of Turnip yellow mosaic virus RNA is a 3'-translational enhancer. *Virology* **321:**36–46.

Matsuoka, Y., Ihara, T., Bishop, D. H. L., and Compans, R. W. (1988). Intracellular accumulation of punta toro glycoproteins expressed from cloned cDNAs. *Virol.* **167:**251–260.

Matsuoka, Y., Chen, S. Y., and Compans, R. W. (1994). A signal for Golgi retention in the Bunyavirus G1 glycoprotein. *J. Biol. Chem.* **269:**22565–22573.

Matsuoka, Y., Chen, S. Y., Holland, C. E., and Compans, R. W. (1996). Molecular determinants of Golgi retention in the Punta Toro virus G1 protein. *Arch. Biochem. Biophys.* **336:**184–189.

McMichael, L. A., Persley, D. M., and Thomas, J. E. (2002). A new tospovirus serogroup IV species infecting capsicum and tomato in Queensland, Australia. *Australasian. Plant Pathol.* **31:**231–239.

Medeiros, R. B., Figueiredo, J., Resende, R. de O., and Ávila, A. C. (2005). Expression of a viral polymerase –bound host factor turns human cell lines permissive to a plant- and insect infecting virus. *Proc. Natl. Acad. Sci USA,* **102:**1175–1180.

Medeiros, R. B., Resende, R. de O., and de Avila, A. C. (2004). The plant virus *Tomato Spotted Wilt Tospovirus* activates the immune system of its main insect vector, *Frankliniella occidentalis*. *J. Virol.* **78:**4976–4982.

Medeiros, R. B., Ullman, D. E., Sherwood, J. L., and German, T. L. (2000). Immunoprecipitation of a 50-kDa protein, a candidate receptor component for tomato spotted wilt tospovirus (*Bunyaviridae*) in its main vector, *Frankliniella occidentalis*. *Vir. Res.* **67:**109–118.

Meister, G., and Tuschl, T. (2004). Mechanisms of gene silencing by double-stranded RNA. *Nature* **431:**343–349.

Meulewaeter, F., van Lipzig, R., Gultayev, A. P., Pleij, C. W., van Damme, D., Cornelissen, M., and van Eldik, G. (2004). Conservation of RNA structures enables TNV and BYDV 5′ and 3′ elements to cooperate synergistically in cap-independent translation. *Nucl. Acids. Res.* **32:**1721–1730.

Meyer, B. J., and Southern, P. J. (1993). Concurrent sequence analysis of 5′ and 3′ RNA termini by intramolecular circularization reveals 5′ nontemplated bases and 3′ terminal heterogeneity for Lymphocytic choriomeningitis virus mRNAs. *J. Virol.* **67:**2621–2627.

Milne, R. G. (1970). An electron microscopic study of Tomato spotted wilt virus in sections of infected cells and in negative stain preparations. *J. Gen. Virol.* **6:**267–276.

Mir, M., and Panganiban, A. (2004). Trimeric hantavirus nucleocapsid protein binds specifically to the viral RNA panhandle. *J. Virol.* **78:**8281–8288.

Misumi, Y., Sohda, M., Tashiro, A., Sato, H., and Ikehara, Y. (2001). An essential cytoplasmic domain for the Golgi localization of coiled-coil proteins with a COOH-terminal membrane anchor. *J. Biol. Chem.* **276:**6867–6873.

Mohamed, N. A. (1981). Isolation and characterisation of subviral structures from Tomato spotted wilt virus. *J. Gen. Virol.* **53:**197–206.

Mohamed, N. A., Randles, J. W., and Francki, R. I. B. (1973). Protein composition of Tomato spotted wilt virus. *Virology* **56:**12–21.

Munro, S. (1998). Localization of proteins to the Golgi apparatus. *Trends Cell Biol.* **8:**11–15.

Murchie, A. I. H., Davis, B., Isel, C., Afshar, M., Drysdale, M. J., Bower, J., Potter, A. J., Starkey, I. D., Swarbrick, T. M., Mirza, S., Prescott, C. D., Vaglio, P., Aboul-ela, F., and Karn, J. (2003). Structure-based drug design targeting an inactive RNA conformation, Exploiting the flexibility of HIV-1 TAR RNA. *J. Mol. Biol.* **336:**625–638.

Murphy, F. A., Fauquet, C. M., Bishop, D. H. L., Ghabrial, S. A., Jarvis, A. W., Martelli, G. P., Mayo, M. A., and Summers, M. D. (1995). Virus Taxonomy, Sixth Report of the International Committee on taxonomy of viruses. *Arch. Virol.* **10:**1–586.

Murphy, F. A., Harrison, A. K., and Whitfield, S. G. (1973). Bunyaviridae, Morphologic and morphogenetic similarities of Bunyamwera serologic supergroup viruses and several other arthropod-borne viruses. *InterVirol.* **1:**297–316.

Nagata, T., and de Ávila, A. C. (2000). Transmission of chrysanthemum stem necrosis virus, a receently discovered Tospovirus, by two thrips species. *J. Phytopath. Berlin* **148:**123–125.

Nagata, T., Inoue-Nagata, A. K., van Lent, J., Goldbach, R., and Peters, D. (2002). Factors determining vector competence and specificity for transmission of Tomato spotted wilt virus. *J. Gen. Virol.* **83:**663–671.

Nagata, T., Storms, M. M. H., Goldbach, R., and Peters, D. (1997). Multiplication of Tomato spotted wilt virus in primary cell cultures derived from two thrips species. *Virus Res.* **49:**59–66.

Nagy, P., Vamosi, G., Bodnar, A., Lockett, S. J., and Szollosi, J. (1998). Intensity-based energy transfer measurements in digital imaging microscopy. *Europ. Biophys. J.* **27:**377–389.

Naidu, R. A., Ingle, C. J., Deom, C. M., and Sherwood, J. L. (2004). The two membrane glycoproteins of Tomato spotted wilt virus show differences in lectin-binding properties and sensitivities to glycosidases. *Virology* **319:**107–117.

Nakitare, G. W., and Elliot, R. M. (1993). Expression of the Bunyamwera virus M genome segment and intracellular localization of NSm. *Virol.* **195:**511–520.

Neeleman, L., Olsthoorn, R. C. L., Linthorst, H. J. M., and Bol, J. F. (2001). Translation of a nonpolyadenylated viral RNA is enhanced by binding of viral coat protein or polyadenylation of the RNA. *Proc. Natl. Acad. Sci USA* **98**:14286–14291.

Nehls, S., Snapp, E. L., Cole, N. B., Zaal, K. J. M., Kenworthy, A. K., Roberts, T. H., Ellenberg, J., Presley, J. F., Siggia, E., and Lippincott-schwartz, J. (2000). Dynamics and retention of misfolded proteins in native ER membranes. *Nat. Cell Biol.* **2**:288–295.

Objieski, J. F., Bishop, D. H. L., Palmer, E. L., and Murphy, F. A. (1976). Segmented genome and nucleocapsid of La Crosse virus. *J. Gen. Virol.* **20**:664–675.

Palmer, J. M., Reddy, D. V. R., Wightman, J. A., and Ranga Rao, G. V. (1990). New information on the thrips vectors of tomato spotted wilt virus in groundnut crops in India. *Int. Arachis Newsl.* **7**:24–25.

Patterson, J. L., and Kolakofsky, D. (1984). Characterization of La Crosse virus small-genome transcripts. *J. Virol.* **49**:680–685.

Patterson, J. L., Holloway, B., and Kolakofsky, D. (1984). La Crosse virions contain a primer-stimulated RNA polymerase and a methylated cap-dependent endonuclease. *J. Virol.* **52**:215–222.

Patton, J. T., Davis, N. L., and Wertz, G. W. (1984). Nucleocapsid protein alone satisfies the requirement for protein synthesis during RNA replication of Vesicular stomatitis virus. *J. Virol.* **49**:303–309.

Pekosz, A., Griot, C., Nathanson, N., and Gonzalez, S. F. (1995). Tropism of bunyaviruses, Evidence for a G1 glycoprotein-mediated entry pathway common to the California serogroup. *Virology* **214**:339–348.

Pekosz, A., and Gonzalez, S. F. (1996). The extracellular domain of La Crosse virus G1 forms oligomers and undergoes pH-dependent conformational changes. *Virology* **225**:243–247.

Persson, R., and Pettersson, R. F. (1991). Formation and intracellular transport of a heterodimeric viral spike protein complex. *J. Cell Biol.* **112**:257–266.

Peters, D. (1998). An updates list of plant species susceptible to tospoviruses. *In* "Recent Progress in Tospovirus and Thrips Research" (D. Peters and R. Goldbach, eds.), pp. 107–110. Wageningen, The Netherlands.

Pittman, H. A. (1927). Spotted wilt of tomatoes. Preliminary note concerning the transmission of the "spotted wilt" of tomatoes by an insect vector (Thrips tabaci Lind.) *Counc. Scien. Ind. Res. Bull.* **1**:74–77.

Plotch, S. J., Bouloy, M., Ulmanen, I., and Krug, R. M. (1981). A unique cap 7 methyl guanosine 5′ tri phosphoryl 5′-2-O methyl nucleoside dependent Influenza virion endo nuclease cleaves capped RNA to generate the primers that initiate viral RNA transcription. *Cell* **23**:847–858.

Poch, O., Sauvaget, I., Delarue, M., and Tordo, N. (1989). Identification of four conserved motifs among the RNA-dependent polymerase encoding elements. *EMBO J.* **8**:3867–3874.

Pollok, B. A., and Heim, R. (1999). Using GFP in FRET-based applications. *Trends Cell Biol.* **9**:57–60.

Pouwels, J. (2004). Functional analysis of the Cowpea mosaic virus movement protein. PhD thesis, Wageningen University, The Netherlands.

Pozzer, L., Bezerra, I., Kormelink, R., Prins, M., Peters, D., Resende, R. de O., and de Ávila, A. C. (1999). Characterization of a distinct tospovirus isolate of iris yellow spot virus associated with a disease in onion fields in Brazil. *Plant Dis.* **83**:345–350.

Pramanik, A., Thyberg, P., and Rigler, R. (2000). Molecular interactions of peptides with phospholipid vesicle membranes as studied by fluorescence correlation spectroscopy. *Chem. Phys. Lipids* **104**:35–47.

Prehaud, C., Lopez, N., Blok, M. J., Obry, V., and Bouloy, M. (1997). Analysis of the 3′ terminal sequence recognized by the Rift Valley fever virus transcription complex in its ambisense S segment. *Virology* **227**:189–197.

Raju, R., and Kolakofsky, D. (1986a). Inhibitors of protein synthesis inhibit both La-Crosse virus S messenger and S genome synthesis *in vivo*. *Virus Res.* **5**:1–10.

Raju, R., and Kolakofsky, D. (1986b). Translational requirement of LaCrosse virus small messenger RNA synthesis *in vivo* studies. *J. Virol.* **61**:96–103.

Raju, R., Raju, L., Hacker, D., Garcin, D., Compans, R., and Kolakofsky, D. (1990). Nontemplated bases at the 5′ ends of Tacaribe virus messenger RNA. *Virology* **174**:53–59.

Ramirez, B.-C., Garcin, D., Calvert, L. A., Kolakofsky, D., and Haenni, A.-L. (1995). Capped nonviral sequences at the 5′ end of the mRNAs of rice hoja blanca virus RNA4. *J. Virol.* **69**:1951–1954.

Ranki, M., and Pettersson, R. F. (1975). Uukuniemi virus contains an RNA polymerase. *J. Virol.* **16**:1420–1425.

Rauer, B., Neumann, E., Widegren, J., and Rigler, R. (1996). Fluorescence correlation spectrometry of the interaction kinetics of tetramethylrhodamine alpha-bungarotoxin with *Torpedo californica* acetylcholine receptor. *Biophys. Chem.* **58**:3–12.

Ravkov, E. V., and Compans, R. W. (2001). Hantavirus nucleocapsid protein is expressed as a membrane-associated protein in the perinuclear region. *J. Virol.* **75**:1808–1815.

Ravkov, E. V., Nichol, S. T., and Compans, R. W. (1997). Polarized entry and release in epithelial cells of Black Creek Canal virus, a new world hantavirus. *J. Virol.* **71**:1147–1154.

Ravkov, E. V., Nichol, S. T., Peters, C. J., and Compans, R. W. (1998). Role of actin microfilaments in Black Creek Canal virus morphogenesis. *J. Virol.* **72**:2865–2870.

Reddy, D. V. R., Ratna, A. S., Sudarshana, M. R., Poul, F., and Kumar, I. K. (1992). Serological relationships and purification of bud necrosis virus, a tospovirus occurring in peanut (Arachis hypogaea L.) in India. *Ann. Appl. Biol.* **120**:279–286.

Reddy, D. V. R., Sudarshana, M. R., Ratna, A. S., Reddy, A. S., Amin, P. W., Kumar, I. K., and Murphy, A. K. (1991). The occurrence of yellow spot virus, a member of tomato spotted wilt group, on peanut (Arachis hypogaea L.) in India. *In* "Virus-Thrips-Plant Interactions of TSWV" (H.-T. Hsu and R. H. Lawson, eds.), pp. 77–78. *Proc. USDA Workshop*, Beltsville, MD.

Reichelt, M., Stertz, S., Krijnse-Locker, J., Haller, O., and Kochs, G. (2004). Missorting of LaCrosse virus nucleocapsid protein by the interferon-induced MxA GTPase involves smooth ER membranes. *Traffic* **5**:772–784.

Resende, R. de O., Posser, L., Nagata, I., Bezerra, I. C., Lima, M. I., Kitajima, E. W., and de Ávila, A. C. (1996). New tospoviruses found in Brazil. *Act. Horticult.* **431**:78–79.

Rhoades, E., Gussakovsky, E., and Haran, G. (2003). Watching proteins fold one molecule at a time. *Proc. Natl. Acad. Sci USA* **100**:3197–3202.

Richmond, K. E., Chenault, K., Sherwood, J. L., and German, T. L. (1998). Characterization of the nucleic acid binding properties of tomato spotted wilt virus nucleocapsid protein. *Virology* **248**:6–11.

Rönkä, H., Hildén, P., von Bonsdorff, C. H., and Kuismanen, E. (1995). Homodimeric association of the spike glycoproteins G1 and G2 of Uukuniemi virus. *Virology* **211**:241–250.

Rönnholm, R. (1992). Localization to the Golgi complex of Uukuniemi virus glycoproteins G1 and G2 expressed from cloned cDNAs. *J. Virol.* **66**:4525–4531.

Rönnholm, R., and Pettersson, R. F. (1987). Complete nucleotide sequence of the M RNA segment of Uukuniemi virus encoding the membrane glycoproteins G1 and G2. *Virology* **160**:191–202.

Ruusala, A., Persson, R., Schmaljohn, C. S., and Pettersson, R. F. (1992). Coexpression of the membrane glycoproteins G1 and G2 of Hantaan virus is required for targeting to the Golgi complex. *Virology* **186**:53–64.

Rwambo, P. M., Shaw, M. K., Rurangirwa, F. R., and deMartini, J. C. (1996). Ultrastructural studies on the replication and morphogenesis of Nairobi sheep disease virus, a Nairovirus. *Arch. Virol.* **141**:1479–1492.

Sakimura, K. (1963). Frankliniella fusca, an additional vector for the tomato spotted wilt virus, with notes on Thrips tabacii, another vector. *Phytopath.* **53**:412–415.

Salanueva, I. J., Novoa, R. R., Cabezas, P., Lopez-Iglesias, C., Carrascosa, J. L., Elliott, R. M., and Risco, C. (2003). Polymorphism and structural maturation of Bunyamwera virus in golgi and post-golgi compartments. *J. Virol.* **77**:1368–1381.

Samuel, G., Bald, J. G., and Pittman, H. A. (1930). Investigations of "spotted wilt" of tomatoes *Scien. Ind. Res. Bull.* **44**:1–64.

Sanchez, A. J., Vincent, M. J., and Nichol, S. T. (2002). Characterization of the glycoproteins of Crimean-Congo hemorrhagic fever virus. *J. Virol.* **76**:7263–7275.

Sato, M., Ozawa, T., Inukai, K., Asano, T., and Umezawa, Y. (2002). Fluorescent indicators for imaging protein phosphorylation in single living cells. *Nat. Biotechnol.* **20**:287–294.

Satyanarayana, T., Gowda, S., Lakshminarayana Reddy, K., Mitchell, S. E., Dawson, W. O., and Reddy, D. V. R. (1998). Peanut yellow spot virus is a member of a new serogroup of Tospovirus genus based on small (S) RNA sequence and organisation. *Arch. Virol.* **143**:353–364.

Satyanarayana, T., Mitchell, S. E., Reddy, D. V. R., Brown, S., Kresovich, S., Jarret, R., Naidu, R. A., and Demski, J. W. (1996a). Peanut bud necrosis tospovirus S RNA, Complete nucleotide sequence, genome organization and homology to other tospoviruses. *Arch. Virol.* **141**:85–98.

Satyanarayana, T., Mitchell, S. E., Reddy, D. V. R., Kresovich, S., Jarret, R., Naidu, R. A., Gowda, S., and Demski, J. W. (1996b). The complete nucleotide sequence and genome organization of the M RNA segment of peanut bud necrosis tospovirus and comparison with other tospoviruses. *J. Gen. Virol.* **77**:2347–2352.

Schmaljohn, C., and Dalrymple, J. M. (1983). Analysis of Hantaan virus RNA, Evidence for a new genus of Bunyaviridae. *Virol.* **131**:482–491.

Schmaljohn, C. S., Schmaljohn, A. L., and Dalrymple, J. M. (1987). Hantaan virus M RNA coding strategy, nucleotide sequence and gene order. *Virology* **1987**:31–39.

Sciaky, N., Presley, J., Smith, C. L., Zaal, K. J. M., Cole, N., Moreira, J. E., Terasaki, M., Siggia, E., and Lippincott Schwartz, J. (1997). Golgi tubule traffic and the effects of Brefeldin A visualized in living cells. *J. Cell Biol.* **139**:1137–1155.

Sekar, R. B., and Periasamy, A. (2003). Fluorescence resonance energy transfer (FRET) microscopy imaging of live cell protein localizations. *J. Cell Biol.* **160**:629–633.

Severson, W., Partin, L., Schmaljohn, C., and Jonsson, C. B. (1999). Characterization of the Hantaan virus nucleocapsid protein-ribonucleic acid interaction. *J. Biol. Chem.* **274**:33732–33739.

Severson, W. E., Xu, X., and Jonsson, C. B. (2001). Cis-acting signals in encapsidation of Hantaan virus S-segment viral genomic RNA by its N protein. *J. Virol.* **75**:2646–2652.

Shi, X. H., and Elliott, R. M. (2002). Golgi localization of Hantaan virus glycoproteins requires coexpression of G1 and G2. *Virology* **300**:31–38.

Shi, X. H., and Elliott, R. M. (2004). Analysis of N-linked glycosylation of Hantaan virus glycoproteins and the role of oligosaccharide side chains in protein folding and intracellular trafficking. *J. Virol.* **78:**5414–5422.

Shi, X. H., Lappin, D. F., and Elliott, R. M. (2004). Mapping the golgi targeting and retention signal of Bunyamwera virus glycoproteins. *J. Virol.* **78:**10793–10802.

Shimi, T., Koujin, T., Segura-Totten, M., Wilson, K. L., Haraguchi, T., and Hiraoka, Y. (2004). Dynamic interaction between BAF and emerin revealed by FRAP, FLIP, and FRET analyses in living HeLa cells. *J. Stuct. Biol.* **147:**31–41.

Shimizu, T., Toriyama, S., Takahashi, M., Akutsu, K., and Yoneyama, K. (1996). Nonviral sequences at the 5′ termini of mRNAs derived from virus-sense and virus-complementary sequences of the ambisense RNA segments of rice stripe Tenuivirus. *J. Gen. Virol.* **77:**541–546.

Shvartsman, D. E., Kotler, M., Tall, R. D., Roth, M. G., and Henis, Y. (2003). Differently anchored influenza hemagglutinin mutants display distinct interaction dynamics with mutual rafts. *J. Cell Biol.* **163:**879–888.

Silva, M. S. (2004). Mechanisms underlying Cowpea mosaic virus systemic infection. PhD thesis,Wageningen University, The Netherlands.

Silva, M. S., Wellink, J., Goldbach, R. W., and van Lent, J. W. M. (2002). Phloem loading and unloading of Cowpea mosaic virus in *Vigna unguiculata*. *J. Gen. Virol.* **83:**1493–1504.

Simons, J. F., Persson, R., and Pettersson, R. F. (1992). Association of the nonstructural protein NSs of Uukuniemi virus with the 40S ribosomal subunit. *J. Virol.* **66:**4233–4241.

Simons, J. F., and Pettersson, R. F. (1991). Host-derived 5′ ends and overlapping complementary 3′ ends of the two messenger RNAs transcribed from the ambisense S segment of Uukuniemi virus. *J. Virol.* **65:**4741–4748.

Singh, S. J., and Krishna Reddy, M. (1996). Watermelon bud necrosis, a new tospovirus disease. *Acta Horticult.* **431:**68–77.

Smith, J. F., and Pifat, D. Y. (1982). Morphogenesis of sandfly fever viruses *Bunyaviridae* family. *Virology* **121:**61–81.

Snippe, M., Borst, J. W., Goldbach, R., and Kormelink, R. (2005). The use of fluorescence microscopy to visualise homolypic interactions of Tomato Spotted Wilt Virus nucleocapsid protein in living cells. *J. Virol.* Methods, in press.

Soellick, T. R., Uhrig, J. F., Bucher, G. L., Kellmann, J. W., and Schreier, P. H. (2000). The movement protein NSm of Tomato spotted wilt tospovirus (TSWV), RNA binding, interaction with the TSWV N protein, and identification of interacting plant proteins. *Proc. Natl. Acad. Sci USA* **97:**2373–2378.

Soldan, S. S., Plassmeyer, M. L., Matukonis, M. K., and Gonzalez-Scarano, F. (2005). LaCrosse virus nonstructural protein NSs counteracts the effects of short interfering RNA. *J. Virol.* **79:**234–244.

Spiropoulou, C. F., Goldsmith, C. S., Shoemaker, T. R., Peters, C. J., and Compans, R. W. (2003). Sin Nombre virus glycoprotein trafficking. *Virol.* **308:**48–63.

Storms, M. M. H., Kormelink, R., Peters, D., van Lent, J. W. M., and Goldbach, R. W. (1995). The nonstructural NSm protein of Tomato spotted wilt virus induces tubular structures in plant and insect cells. *Virology* **214:**485–493.

Storms, M. M. H., van der schoot, C., Prins, M., Kormelink, R., van Lent, J. W. M., and Goldbach, R. W. (1998). A comparison of two methods of microinjection for assessing altered plasmodesmal gating in tissues expressing viral movement proteins. *Plant J.* **13:**131–140.

Sundin, D. R., Beaty, B. J., Nathanson, N., and Gonzalez-Scarano, F. (1987). A G1 glycoprotein epitope of La Crosse virus, a determinant of infection of Aedes triseriatus. *Science* **235**:591–592.

Takeda, A., Sugiyama, K., Nagano, H., Mori, M., Kaido, M., Mise, K., Tsuda, S., and Okuno, T. (2002). Identification of a novel RNA silencing suppressor, NSs protein of Tomato spotted wilt virus. *FEBS lett.* **532**:75–79.

Takeuchi, S., Okuda, M., Hanada, K., Kawada, Y., and Kameya-Iwaki, M. (2001). Spotted wilt disease of cucumber (Cucumis sativus) caused by melon yellow spot virus. *Jap. J. Phytopath.* **67**:46–51.

Tas, P. W. L., Boerjan, M. L., and Peters, D. (1977). The structural proteins of Tomato spotted wilt virus. *J. Gen. Virol.* **36**:267–279.

Tordo, N., de Haan, P., Goldbach, R., and Poch, O. (1992). Evolution of negative – stranded RNA genomes. *Sem. Virol.* **3**:341–357.

Truong, K., and Ikura, M. (2001). The use of FRET imaging microscopy to detect protein-protein interactions and protein conformational changes *in vivo. Curr. Opin. Struct. Biol.* **11**:573–578.

Tsuda, S., Fujisawa, I., Ohnishi, J., Hosokawa, D., and Tomaru, K. (1996). Localization of Tomato spotted wilt tospovirus in larvae and pupae of the insect vector *Thrips setosus. Phytopath.* **86**:1199–1203.

Uhrig, J. F., Soellick, T. R., Minke, C. J., Philipp, C., Kellmann, J. W., and Schreier, P. H. (1999). Homotypic interaction and multimerization of nucleocapsid protein of Tomato spotted wilt tospovirus, identification and characterization of two interacting domains. *Proc. Natl. Acad. Sci USA* **96**:55–60.

Ullman, D. E., Cho, J. J., Mau, R. F. L., Westcoc, D. M., and Custer, D. M. (1992). A midgut barrier to Tomato spotted wilt virus acquisition by adult Western flower thrips. *Phytopath.* **82**:1333–1342.

Ullman, D. E., German, T. L., Sherwood, J. L., Westcot, D. M., and Cantone, F. A. (1993). Tospovirus replication in insect vector cells, Immunocytochemical evidence that the nonstructural protein encoded by the S RNA of Tomato spotted wilt tospovirus is present in thrips vector cells. *Phytopath.* **83**:456–463.

Ullman, D. E., Westcot, D. M., Chenault, K. D., Sherwod, J. L., German, T. L., Bandla, M. D., Cantone, F. A., and Duer, H. L. (1995). Compartmentalization, intracellular transport, and autophagy of tomato spotted wilt tospovirus proteins in infected thrips cells. *Phytopath.* **85**:644–654.

Ulmanen, I., Broni, B. A., and Krug, R. M. (1981). Role of 2 of the Influenza virus core P proteins in recognizing cap 1 structures on RNA and in initiating viral RNA transcription. *Proc. Natl. Acad. Sci USA* **78**:7355–7359.

Vaira, A. M., Roggero, P., Luisoni, E., Milne, R. G., and Lisa, V. (1993). Characterization of two tospoviruses in Italy, tomato spotted wilt virus and Impatiens necrotic spot virus. *Plant Pathol.* **42**:530–542.

van den Hurk, J., Tas, P. W. L., and Peters, D. (1977). The RNA of Tomato spotted wilt virus. *J. Gen. Virol.* **36**:81–91.

van Knippenberg, I. (2005). Analysis of Tomato spotted wilt virus genome transcription. PhD thesis, Wageningen University, The Netherlands.

van Knippenberg, I., Goldbach, R., and Kormelink, R. (2002). Purified Tomato spotted wilt virus particles support both genome replication and transcription *in vitro. Virology* **303**:278–286.

van Knippenberg, I., Goldbach, R., and Kormelink, R. (2004). *In vitro* transcription of TSWV is independent of translation. *J. Gen. Virol.* **85**:1335–1338.

van Kuppeveld, F. J. M., Melchers, W. J. G., Willems, P. H. G. M., and Gadella, T. W. J., Jr. (2002). Homomultimerization of the coxsackievirus 2B protein in living cells visualized by fluorescence resonance energy transfer microscopy. *J. Virol.* **76:**9446–9456.

van Poelwijk, F., Boye, K., Oosterling, R., Peters, D., and Goldbach, R. (1993). Detection of the L protein of Tomato spotted wilt virus. *Virology* **197:**468–470.

van Poelwijk, F., Kolkman, J., and Goldbach, R. (1996). Sequence analysis of the 5′ ends of Tomato spotted wilt virus N mRNAs. *Arch. Virol.* **141:**177–184.

van Poelwijk, F., Prins, M., and Goldbach, R. (1997). Completion of the Impatiens necrotic spot virus genome sequence and genetic comparison of the L proteins within the family *Bunyaviridae*. *J. Gen. Virol.* **78:**543–546.

Veijola, J., and Petterson, R. F. (1999). Transient association of calnexin and calreticulin with newly synthesized G1 and G2 glycoproteins of Uukuniemi virus (family *Bunyaviridae*). *J. Virol.* **73:**6123–6127.

Verkleij, F. N., de Vries, P., and Peters, D. (1982). Evidence that Tomato spotted wilt virus RNA is a positive strand. *J. Gen. Virol.* **58:**329–338.

Verkley, F. N., and Peters, D. (1983). Characterization of a defective form of Tomato spotted wilt virus. *J. Gen. Virol.* **64:**677–682.

Vialat, P., and Bouloy, M. (1992). Germiston virus transcriptase requires active 40S riboosomal subunits and utilizes capped cellular RNAs. *J. Virol.* **66:**685–693.

Vijayalakshmi, K. (1994). Transmission and ecoology of Thrips palmi Karny, the vector of peanut bud necrosis virus. PhD thesis, Andrha Pradesh Agricultural University, Hyderabad.

Vincent, M. J., Sanchez, A. J., Erickson, B. R., Basak, A., Chretien, M., Seidah, N. G., and Nichol, S. T. (2003). Crimean Congo Hemorrhagic fever virus glycoprotein proteolytic processing by subtilase SKI-1. *J. Virol.* **77:**8640–8649.

Violin, J. D., Zhang, J., Tsien, R. Y., and Newton, A. C. (2003). A genetically encoded fluorescent reporter reveals oscillatory phosphorylation by protein kinase C. *J. Cell Biol.* **161:**899–909.

Visser, A. J. W. G., and Hink, M. (1999). New perspectives of Fluorescence Correlation Spectroscopy. *J. Fluorescence* **9:**81–87.

von Heijne, G. (1986). A new method for predicting signal sequence cleavage sites. *Nucl. Acids. Res.* **14:**4683–4690.

Walter, M., Chaban, C., Schutze, K., Batistic, O., Weckermann, K., Nake, C., Blazevic, D., Grefen, C., Schumacher, K., Oecking, C., Harter, K., and Kudla, J. (2004). Visualization of protein interactions in living plant cells using bimolecular fluorescence complementation. *Plant J.* **40:**428–438.

Watkins, C. A., and Jones, I. M. (1993). Association of the 40S ribosomal subunit with the NSs nonstructural protein of Punta Toro virus. *Abstr. Int. Congres Virol.* p. 136-Glasgow, Scotland.

Watrob, H. M., Pan, C. P., and Barkley, M. D. (2003). Two-step FRET as a structural tool. *J. Am. Chem. Soc.* **125:**7336–7343.

Webb, S., Tsai, J., and Mitchell, F. (1998). Bionomics of Frankliniella bispinosa and its transmission of tomato spotted wilt virus. *In* "Recent Progress in Tospovirus and Thrips Research" (D. Peters and R. Goldbach, eds.), p. 67. Wageningen, Netherlands.

Weber, F., Bridgen, A., Fazakerley, J. K., Stretenfeld, H., Kessler, N., Randall, R. E., and Elliott, R. M. (2002). Bunyamwera Bunyavirus nonstructural protein NSs counteracts the induction of alpha/beta interferon. *J. Virol.* **76:**7949–7955.

Weber, F., Dunn, E. F., Bridgen, A., and Elliott, R. M. (2001). The Bunyamwera virus nonstructural protein NSs inhibits viral RNA synthesis in a minireplicon system. *Virology* **281**:67–74.

White, J., and Stelzer, E. (1999). Photobleaching GFP reveals protein dynamics inside live cells. *Trends Cell Biol.* **9**:61–65.

Whitfield, A. E., Ullman, D. E., and German, T. L. (2004). Expression and characterization of a soluble form of Tomato spotted wilt virus glycoprotein GN. *J. Virol.* **78**:13197–13206.

Wijkamp, I. (1995). Virus-vector relationships in the transmission of tospoviruses. PhD thesis, Wageningen University, The Netherlands.

Wijkamp, I., Almarza, N., Goldbach, R., and Peters, D. (1995). Distinct levels of specificity in thrips transmission of tospoviruses. *Phytopath.* **85**:1069–1074.

Wijkamp, I., and Peters, D. (1993). Determination of the median latent period of two tospoviruses in *Frankliniella occidentalis*, using a novel leaf disk assay. *Phytopath.* **83**:986–991.

Wijkamp, I., van Lent, J., Kormelink, R., Goldbach, R., and Peters, D. (1993). Multiplication of Tomato spotted wilt virus in its insect vector, *Frankliniella occidentalis*. *Phytopath.* **83**:986–991.

Xia, Z. P., and Liu, Y. H. (2001). Reliable and global measurement of fluorescence resonance energy transfer using fluorescence microscopes. *Biophys. J.* **81**:2395–2402.

Xu, X., Gerard, A. L. V., Huang, B. C. B., Anderson, D. C., Payan, D. G., and Luo, L. (1998). Detection of programmed cell death using fluorescence energy transfer. *Nucleic Acids Res.* **26**:2034–2035.

Xu, X., Severson, W., Villegas, N., Schmaljohn, C., and Jonsson, C. B. (2002). The RNA binding domain of the Hantaan virus N protein maps to a central, conserved region. *J. Virol.* **76**:3301–3308.

Xu, Y., Piston, D. W., and Johnson, C. H. (1999). A bioluminescence resonance energy transfer (BRET) system, Application to interacting circadian clock proteins. *Proc. Natl. Acad. Sci USA* **96**:151–156.

Yeh, S. D., and Chang, T. F. (1995). Nucleotide sequence of the N gene of watermelon silver mottle virus, a proposed new member of the genus Tospovirus. *Phytopath.* **85**:58–64.

Yeh, S. D., Lin, Y. C., Cheng, Y. H., Jih, C. L., Chen, M. J., and Chen, C. C. (1992). Identification of tomato spotted wilt-like virus on watermelon in Taiwan. *Plant Dis.* **76**:835–840.

Yoshimatsu, K., Lee, B. H., Araki, K., Morimatsu, M., Ogino, M., Ebihara, H., and Arikawa, J. (2003). The multimerization of hantavirus nucleocapsid protein depends on type-specific epitopes. *J. Virol.* **77**:943–952.

Yoshizaki, H., Ohba, Y., Kurokawa, K., Itoh, R. E., Nakamura, T., Mochizuki, N., Nagashima, K., and Matsuda, M. (2003). Activity of Rho-family GTPases during cell division as visualized with FRET-based probes. *J. Cell Biol.* **162**:223–232.

Zamanian-Daryoush, M., Mogensen, T. H., Di Donato, J. A., and Williams, B. R. (2000). NF-kappaB activation by ouble-stranded-RNA-activated protein kinase (PKR) is mediated through NF-kappaB-inducing kinase and IkappaB kinase. *Mol. Cell. Biol.* **20**:1278–1290.

FURTHER READING

Alfadhi, A., Love, Z., Arvidson, B., Seeds, J., Willey, J., and Barklis, E. (2001). Hantavirus nucleocapsid oligomerization. *J. Virol.* **75:**2019–2023.

Collett, M. S., Purchio, A. F., Keegan, K., Frazier, S., Hays, W., Anderson, D. K., Parker, M. D., Schmaljohn, S., Schmidt, J., and Dalrymple, J. M. (1985). Complete nucleotide sequence of the M RNA segment of Rift Valley fever virus. *Virology* **144:**228–245.

Maris, P. (2004). Evaluation of thrips resistance in pepper to control Tomato spotted wilt virus infection. PhD thesis, Wageningen University, The Netherlands.

Obijeski, J. F., and Murphy, F. A. (1977). *Bunyaviridae*, recent biochemical developments. *J. Gen. Virol.* **37:**1–14.

Pekosz, A., Griot, C., Nathanson, N., and Gonzalez Scarano, F. (1995). Tropism of bunyaviruses, Evidence for a G1 glycoprotein-mediated entry pathway common to the California serogroup. *Virol.* **214:**339–348.

Petterson, R., and Kääriäinen, L. (1973). The ribonucleic acids of Uukuniemi virus, a non-cubal tick-borne arbovirus. *Virology* **56:**608–619.

ADVANCES IN VIRUS RESEARCH, VOL 65

INFLUENZA VIRUS VIRULENCE AND ITS MOLECULAR DETERMINANTS

Diana L. Noah*,† and Robert M. Krug‡

†Institute for Cellular and Molecular Biology, University of Texas at Austin
Austin, Texas 78712
‡Institute for Cellular and Molecular Biology, Section of Molecular Genetics and
Microbiology, University of Texas at Austin, Austin, Texas 78712

I. INTRODUCTION: THE HUMAN DISEASE CAUSED BY INFLUENZA VIRUSES

Influenza A and B viruses cause a highly contagious respiratory disease in humans, resulting in approximately 36,000 deaths in the United States annually (Prevention, 2005). The vast majority of these deaths occur in individuals over the age of 65 and/or have underlying disease-lung, heart, and kidney disease, and diseases that result in immunosuppression. These annual epidemics also have a large economic impact and cause more than 100,000 hospitalizations per year in the United States alone. Influenza A viruses, which infect a wide number of avian and mammalian species, are responsible for the periodic widespread epidemics, or pandemics, that have caused high mortality rates. The most devastating pandemic occurred in 1918, which caused an estimated 20 million to 40 million deaths worldwide. Less devastating pandemics occurred in 1957 and 1968. Many investigators feel that a

---

* Present address: Department of Emerging Pathogens, Southern Research Institute, Birmingham, Alabama 35205-5305.

0065-3527/05 $35.00
DOI: 10.1016/S0065-3527(05)65004-X

new pandemic may occur soon, as discussed later. Influenza B virus infections comprise about 20% of the yearly cases, but influenza B virus does not cause pandemics because it does not have an animal reservoir (see following).

Influenza virus is highly infectious because it is effectively transmitted by aerosol. Clinical symptoms associated with an uncomplicated infection include fever up to 41°C, headaches, nonproductive cough, and muscle aches that resolve within several days. Complications of annual influenza virus infections include viral pneumonia, secondary bacterial pneumonia, exacerbated airway inflammation in asthma patients, and an increase risk of myocardial infarction and stroke post influenza infection. Additional complications can occur in children, including inflammation of the middle ear and Reye's syndrome, which is characterized by fatty deposits in the liver and acute encephalopathy (Linnemann et al., 1975). The latter syndrome is usually associated with infections by influenza B virus. The 1918 influenza A virus was much more virulent than the influenza A viruses that cause current annual epidemics (Reid et al., 2001). In 1918, infection often resulted in death, particularly among young adults (15–45 year olds), who did not appear to have any underlying disease. The avian influenza A viruses currently circulating in Asia are also more virulent, resulting in fatal pneumonia and multiple organ failure (Claas et al., 1998; Suarez et al., 1998; Subbarao et al., 1998; To et al., 2001; Tran et al., 2004). Mortality among avian virus-infected individuals has been high, ranging from about 30% in 1997 to 75% in 2004. Less than 100 total individuals have been infected. Although most infections have occurred via avian-to-human transmission, there is some evidence for inefficient human-to-human transmission (Ungchusak et al., 2005). If this avian virus acquires the ability to be transmitted efficiently from humans to humans, it would be expected to cause the next human pandemic, potentially resulting in millions of deaths.

At present, the primary means for controlling influenza virus epidemics is vaccination. The major surface protein of the virus, the hemagglutinin (HA), elicits an immune response that neutralizes virus infectivity (Lamb and Krug, 2001). The antigenic structure of the HA undergoes two types of changes. Antigenic drift, which occurs with both influenza A and B viruses, results from the selection of mutant viruses that evade antibodies directed against the major antigenic type of the HA circulating in the human population. Mutant viruses are readily generated because the viral RNA-dependent RNA polymerase has no proofreading function. Because of antigenic drift, the vaccine has to be reformulated each year. Influenza A, but not influenza B, viruses also undergo a process called antigenic shift that generates pandemic viruses (as discussed later). Currently, two classes of antivirals are used to treat influenza virus infections. These antivirals and their protein targets are discussed later in this review. In light of the fact that it takes approximately six months to produce a new influenza vaccine, antivirals, rather than vaccination, would serve as the first line of defense against a rapidly spreading pandemic.

II. Assays for Virulence of Influenza A Viruses

Despite intensive research, the understanding of the characteristics of influenza A virus that determine its virulence is incomplete. Virulence has been assayed primarily in two animal model systems, mice and ferrets. In addition, the virulence of avian influenza A viruses has also been assayed in chickens.

Most virulence studies have been carried out in mice (Wright and Webster, 2001), Knockout mice with specific gene disruptions are available, and therefore mouse studies have the potential of elucidating the role of specific host genes in influenza A virus-induced disease. However, it can be argued that mice do not provide an appropriate model system. Mice are not naturally infected with influenza A (or B) viruses, and human isolates of influenza A viruses usually result in mild, or essentially inapparent, symptoms in mice. Consequently, most mouse studies employ influenza A virus strains (e.g., A/WSN/33 and A/PR8/34) that have been mouse-adapted via serial passages in mice. It is not clear whether the mutations that accompany mouse adaptation alter the properties of these viruses so that they differ significantly from those of naturally occurring human isolates of influenza A virus. In contrast to other human influenza A isolates, the currently circulating avian influenza A viruses do not require mouse adaptation (Bright et al., 2003; Mase et al., 2005). Another caveat is that the mice employed in these studies lack the interferon (IFN)-inducible Mx1 gene, which provides the major IFN-induced protection against influenza A virus (Staeheli et al., 1984). In the absence of the Mx1 gene, IFN treatment results in only a modest reduction in virus replication, whereas IFN treatment in the presence of the Mx1 gene severely reduces virus replication (Krug et al., 1985; Staeheli et al., 1984). Consequently, the Mx1-minus mice that have been employed have a crippled innate immune response against influenza A virus, suggesting that studies using these mice may overestimate the virulence of influenza A viruses. Finally, the disease caused by mouse-adapted influenza viruses often differs from the human disease caused by most naturally occurring influenza A virus strains (Wright and Webster, 2001).

In contrast to mice, ferrets are naturally infected by influenza A viruses (Wright and Webster, 2001). In fact, because of these natural infections, it is often difficult to obtain ferrets that have not experienced a prior influenza A virus infection. Ferrets are the only small animals that have been found to develop a human-like illness when infected with human isolates of influenza A virus (Zitzow et al., 2002). For these reasons, ferrets may provide the best animal model currently available to assess the virulence of influenza A viruses in humans. (However, see the later discussion concerning the role of the viral polymerase in influenza A virus virulence.) It should be noted that studies in ferrets cannot at present provide information about the roles of specific host genes in influenza A virus-induced disease because ferret genetics has not yet been developed.

Experiments using tissue culture cells can contribute information about virus–cell interactions that may provide important insights into virus virulence. For example, mutations in viral genes that affect steps in virus replication and/or the ability of the virus to counter cellular antiviral response may be expected to influence virus virulence. In addition, viral mutations that alter the ability of the virus to induce apoptosis, or programmed cell death, may also affect virulence. Viruses containing the above mutations could be evaluated in ferrets and/or mice.

The consensus from most virulence studies is that influenza virus virulence is polygenic (Florent et al., 1977; Oxford et al., 1978; Rott et al., 1979). In other words, virulence cannot be ascribed to a single specific viral gene, but rather requires a combination of several viral genes. These results indicate that more than one virus-induced effect on the host may be required for virulence and/or virulence may require that two or more of viral proteins undergo specific interactions with each other.

In this review we will discuss what is currently known about how individual influenza A virus-encoded proteins and their interactions with host factors contribute to human virulence. We will also discuss several functional interactions between influenza A virus proteins, although it has not yet been established whether any or all of these interactions play a role in virulence. Even less is known about the molecular determinants of the virulence of influenza B viruses. One result is probably pertinent: the influenza B virus-encoded nonstructural protein 1 (NS1B protein) interacts with the cellular IFN system in a novel way that has implications for influenza B virus-induced human disease (Yuan and Krug, 2001). However, in this review, we will limit our discussion to the molecular determinants of the virulence of influenza A viruses.

III. The Genes and Proteins of Influenza A Viruses

Influenza A viruses contain negative-stranded, segmented RNA genomes (Lamb and Krug, 2001) (Fig. 1). As is the case for other negative-stranded RNA viruses, the virion contains the polymerase that carries out the initial copying of the single-stranded virion RNA (vRNA) into viral mRNAs. The polymerase, which is comprised of three subunits (PB1, PB2, and PA), is associated with each of the vRNA segments. Because the vRNA segments also contain nucleocapsid (NP) protein molecules bound at regular intervals along their entire length, the

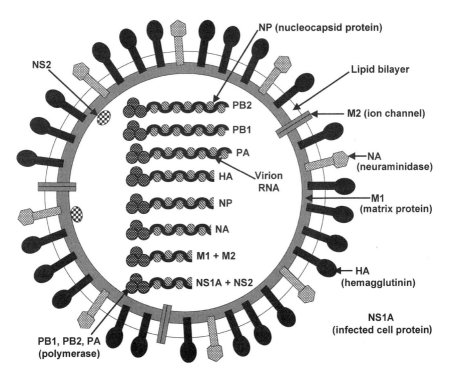

Fig 1. Diagram of the structure of the influenza A virion, indicating the location of the nine virion proteins, the viral envelope and the eight genomic RNA segments. The NS1A protein is found only in infected cells, and not in virions.

vRNAs are in the form of ribonucleoproteins, or RNPs. The viral polymerase initiates viral mRNA synthesis using capped RNA primers that are cleaved from cellular pre-mRNAs in the nucleus (Krug *et al.*, 1989). This process has been called cap snatching. Viral mRNAs are not full-length copies of the vRNAs: transcription terminates at a stretch of uridines, 15–22 bases before the 5′ ends of the vRNA templates, followed by the addition of poly(A) by the viral polymerase. During infection the viral polymerase catalyzes not only viral mRNA synthesis but also vRNA replication, which occurs in two steps: the synthesis of full-length copies of the vRNAs (often termed complementary RNAs, or cRNAs) and the copying of cRNAs into vRNAs. The synthesis of both the cRNAs and the vRNAs is initiated *de novo* without a primer. It has not yet been established how the polymerase switches from capped RNA-primed to unprimed RNA synthesis. All virus-specific RNA synthesis takes place in the nucleus.

Influenza A viruses contain eight vRNA segments (Fig. 1) (Lamb and Krug, 2001). The three largest vRNA segments encode the three subunits of the polymerase. The middle-sized vRNA segments encode the HA, NP, and the neuraminidase (NA). HA, the major surface protein of the virus, binds to sialic acid-containing receptors on host cells and is the protein against which neutralizing antibodies are produced. The NA viral surface protein removes sialic acid from glycoproteins. One of its likely functions is to remove sialic acid during virus budding from the cell surface and from the HA and NA of the newly assembled virions, thereby obviating aggregation of the budding virions on the cell surface. The seventh genomic RNA segment encodes two proteins, M1 (matrix protein) and M2 (influenza A). The M1 protein underlies the viral lipid membrane and is thought to interact with the virion RNPs (vRNPs) and the inner (cytoplasmic) tails of the surface proteins (e.g., HA and NA). The M2 (influenza A) protein is an ion channel protein that is essential for the uncoating of the virus. The smallest segment encodes two proteins, NS1A and NS2. The NS2 protein mediates the export of newly synthesized vRNPs from the nucleus to the cytoplasm. The NS1A protein is not incorporated into virions and is a multifunctional protein that will be discussed in detail later in this review.

IV. The Emergence of New Influenza A Viruses in Humans

Because the genome is segmented, influenza A and B viruses can form reassortant viruses between members of the same type (A or B) but not between types (Lamb and Krug, 2001; Wright and Webster,

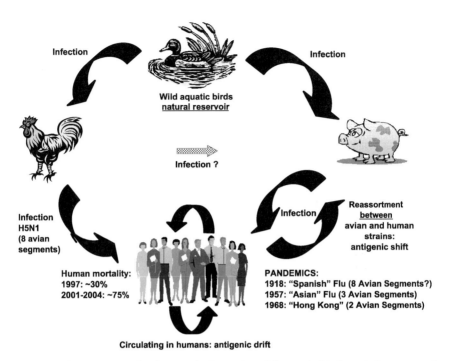

FIG 2. The reservoir of influenza A viruses in wild aquatic birds and the routes of transmission of these viruses to humans. The infection of wild aquatic birds with most strains of influenza A virus is asymptomatic, so these viruses can be maintained and propagated in aquatic birds without significant mortality. The routes of transmission are described in the text. (See Color Insert.)

2001). Influenza B viruses appear to infect only humans. In contrast, influenza A viruses have been isolated from a wide variety of avian and mammalian species, and wild aquatic birds are thought to be the reservoir for all influenza A viruses that infect avian and mammalian species (Fig. 2). In addition, with influenza A virus, reassortants between human and avian strains can result in new pandemic viruses that encode a novel HA that is immunologically distinct from the HA of the previous circulating strain, an event known as antigenic shift. As a consequence, the human population has little or no immunological protection against the new virus. The H1 subtype of HA (found in the 1918 pandemic virus and its descendents) was replaced by the H2 subtype in the 1957 pandemic virus, followed by the H3 subtype in the 1968 pandemic virus. The H1 subtype also reemerged in 1977 and is currently circulating. It is thought that the avian–human reassortants in 1919, 1957, and 1968 were produced in an intermediate host, probably the pig, which possesses both avian and human receptors for influenza A virus (Fig. 2) (Wright and Webster, 2001). Human cell receptors contain sialic acid linked to galactose by $\alpha2$, 6 linkages, whereas avian receptors contain sialic acid linked to galactose by $\alpha2$, 3 linkages. These naturally occurring reassortants also contain other avian genes in addition to HA: the 1957 virus also contains avian NA and PB1 genes; and the 1968 virus also contains the avian PB1 gene.

The virulent avian virus currently circulating in Asia was transmitted directly from chickens to humans without an intermediate host and contains only avian genes (Fig. 2) (Wright and Webster, 2001). This virus has a new H5 HA subtype that is not present in circulating human strains, and is thus a candidate for a new pandemic virus. Because this virus contains a N1 NA subtype, it is classified as a H5N1 virus. However, the ability of this virus to cause a human pandemic is limited at least in part because the H5 HA currently in this virus has retained its specificity for avian ($\alpha2,3$) receptors (Matrosovich et al., 2004; Puthavathana et al., 2005). Whereas human influenza A viruses normally infect nonciliated lung epithelial cells containing $\alpha2,6$ receptors, the avian H5N1 virus infects ciliated epithelial cells which contain avian-like $\alpha2,3$ receptors (Matrosovich et al., 2004). Infection via these ciliated lung cells is probably not optimal for virus replication, which would explain why exposure to high levels of the avian H5N1 virus from infected chickens (and other avians) is required for human infection and why human-to-human transmission is inefficient. Key mutations in the receptor binding site of the H5 HA (e.g., Q226 changed to L and G228 changed to S) have been shown to be sufficient to change its binding specificity from avian $\alpha2,3$ to human $\alpha2,6$ (Harvey et al., 2004). This result suggests that future human infections with the H5N1 viruses have the potential for selecting similar naturally occurring HA mutants, thereby facilitating human-to-human transmission.

V. The Role of the Major Surface Proteins HA and NA in Influenza A Virus Virulence

The influenza A virus HA protein, which binds to sialic acid receptors on the cell surface, mediates the fusion of viral and cellular endosomal membranes, resulting in the release of the viral RNPs into the cytoplasm of infected cells (Lamb and Krug, 2001). Prior to mediating fusion, HA first has to be cleaved into two disulfide-linked subunits,

HA1 and HA2. The resulting N-terminus of HA2 mediates the fusion of viral and endosomal membranes. Consequently, cleavage of HA is a prerequisite for initiating infection, and is a crucial determinant of virulence. H1, H2, and H3 influenza A viruses, including the strains that caused previous pandemics, contain a single arginine at the cleavage site, and cleavage of the HA of these viruses is carried out by extracellular trypsin-like proteases that are secreted only by certain cell types (e.g., cells in the respiratory tract). In contrast, highly pathogenic H5N1 viruses, as well as H7 viruses, contain a stretch of basic residues adjacent to the HA cleavage site (Bender *et al.*, 1999; Govorkova *et al.*, 2005; Puthavathana *et al.*, 2005). The presence of these basic amino acids allows these HAs to be cleaved by ubiquitous intracellular proteases including furin and other subtilisin-like proteases (Stieneke-Grober *et al.*, 1992). As a consequence, cleavage of these HAs is more efficient and can occur in many different cell types. Recombinant H5N1 viruses lacking these basic amino acids are no longer virulent in mice (Hatta *et al.*, 2001), demonstrating that both the presence of these amino acids and the consequent cleavage by intracellular furin-like proteases are required for the virulence of these viruses.

However, the degree of virulence, as assayed in mice and chickens, can be modulated by amino acids in other regions of the HA of H5N1 viruses, and the presence of a polybasic sequence at the HA cleavage site does not always result in a virulent H5N1 virus (Hulse *et al.*, 2004). The HAs of another type of H5 avian viruses (H5N2 viruses) contain a polybasic cleavage site, but a carbohydrate side chain limits the accessibility of the cleavage site to furin-like proteases, resulting in an avirulent virus (Kawaoka and Webster, 1989). Furthermore, properties of a HA other than a polybasic region at the cleavage site can enhance virulence, as exemplified by the HA of the 1918 pandemic virus. This HA, which lacks such a polybasic region, greatly enhances virulence in mice (Kobasa *et al.*, 2004). When the original HA of two human influenza A viruses of low virulence was replaced by the 1918 HA, highly virulent viruses were generated. It is not known what properties of the 1918 HA confer high virulence.

HA cleavage can also be enhanced by NA, the other major influenza virus surface glycoprotein, as occurs with influenza A/WSN/33, which is virulent in mice (Goto and Kawaoka, 1998; Goto, Wells, Takada, and Kawaoka, 2001). The NA of this virus binds and sequesters cellular plasminogen, which is activated to form plasmin, a protease that cleaves the WSN HA. Plasminogen binding to the NA requires that the NA contain a C-terminal Lys. At present, no NA other than the WSN NA has been found to bind plasminogen. However, NA can affect virulence in other ways. For example, the presence of an additional glycosylation site in the NA of a H5N1 virus enhances virulence, although the mechanism of this enhancement has not been determined (Hulse *et al.*, 2004). In general, a great deal of evidence indicates that there needs to be an optimal balance between the activities of the HA and NA (i.e., a balance between the strength of HA needed for binding the virus to cells to initiate infection with the efficiency of NA-mediated release of progeny virus).

NA is the target of a major class of antivirals, which were designed based on the NA X-ray crystal structure (Colman *et al.*, 1983; Varghese *et al.*, 1983; von Itzstein *et al.*, 1993). NA is a tetramer, and each monomer has a sialic acid-binding pocket lined by amino acids that are invariant in all influenza A and B virus strains. A sialic acid derivative in which the 4-hydroxyl group of sialic acid is replaced with a guanidinyl group binds to this pocket tighter than sialic acid itself. The resulting compound, zanamivir (commercially called Relenza), effectively inhibits the NA of influenza A and B viruses, and is orders of magnitude less active against nonviral neuraminidases (Gubareva *et al.*, 1995; Ryan *et al.*, 1994; von Itzstein *et al.*, 1993; Woods *et al.*, 1993). Zanamivir inhibits influenza virus replication *in vitro* and *in vivo*, but must be delivered by the intranasal route. Subsequently, a different compound containing a lipophilic side chain was synthesized (oseltamivir, commercially called Tamiflu), which is well absorbed when administered orally and achieves stable, efficacious levels in the serum while maintaining low toxicity (Kim *et al.*, 1997; Mendel *et al.*, 1998). These neuraminidase inhibitors are most effective when administered prophylactically or within the first 30 hours of symptom onset (Hayden *et al.*, 1996, 1997; McNicholl and McNicholl, 2001). Because influenza viruses resistant to neuraminidase inhibitors were not readily generated in tissue culture experiments, it had been hoped that such resistant viruses would not appear when these inhibitors were used for treatment of humans during an influenza epidemic. This hope is likely unfounded, as a recent study of children treated with oseltamivir showed that about 20% of the children shed drug-resistant viruses (Kiso *et al.*, 2004).

VI. The Role of the Viral Polymerase in Influenza A Virus Virulence

The influenza A virus polymerase, which consists of the PB1, PB2, and PA proteins, is implicated in the virulence of H5N1 avian viruses (Hatta *et al.*, 2001; Shinya *et al.*, 2004). Two of the H5N1 viruses isolated from humans in 1997 possess the polybasic sequence at the HA cleavage site, but differ substantially in virulence in mice. The influenza A/Hong Kong/483/97 (HK483) virus causes lethal systemic infection, whereas the A/Hong Kong/486/97 (HK486) virus causes a nonlethal respiratory infection. Mutational analysis demonstrated that this difference in virulence is due to the identity of the amino acid at position 627 in the PB2 protein (Hatta et al., 2001). The presence of a Lys at this position enables the virus to replicate more efficiently in the mouse, and it was concluded that this efficient replication is the reason that host cell defenses are prevented from confining infection to respiratory organs (Shinya et al., 2004). In addition, tissue culture experiments show that viruses with Lys at this position replicate more efficiently than viruses with a Glu at this position in mouse cells. However, in avian cells, no difference in replication is observed. A H5N1 virus isolated from an acutely infected human in 2004 (A/Vietnam/1203/04) also has Lys at position 627, providing further evidence that this amino acid influences virulence in humans (Govorkova et al., 2005). These results are consistent with earlier results linking the ability of influenza A viruses to replicate in mammalian cells with the presence of a Lys rather than a Glu at position 627 in the PB2 protein (Crescenzo-Chaigne et al., 2002).

Experiments using ferrets revealed that both the 1997 H5N1 viruses (HK483 and HK486) were highly virulent with no significant difference in the level of virulence (Zitzow et al., 2002). Again, using 2004 H5N1 viruses, it was found that Lys 627 of the PB2 protein is not required for high virulence in ferrets (Govorkova et al., 2005). Furthermore, the human isolate (A/Vietnam/1203/04), which contained a Lys at this position, but also a virus isolated from quails, which contained a Glu at this position, caused severe disease in ferrets. These results indicate that ferrets are highly susceptible to virulent avian as well as human influenza A viruses, and as a consequence cannot detect the host range restriction mediated by amino acid 627 in the PB2 protein.

The amino acid at position 627 in the PB2 protein has not been implicated in known functions of the viral RNA polymerase (Lamb and Krug, 2001; Ohtsu et al., 2002). Figure 3 denotes the functions of the PB1, PB2, and PA proteins that have been mapped to specific regions of these proteins. The region of the PB2 protein that interacts with the PB1 protein has been mapped to amino acids 51–259 (Ohtsu et al., 2002). The PB2 protein has a crucial function in capped RNA-primed initiation of viral mRNA synthesis (Li et al., 2001; Rao et al., 2003). As a result of the binding of the 5′ terminal sequence of virion

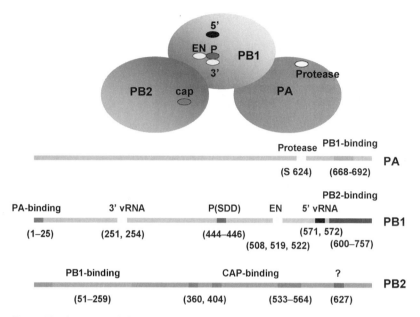

FIG 3. The functions of the three protein subunits (PB1, PB2, and PA) of the influenza A virus polymerase have been mapped to specific regions of these proteins. The amino acid at position 627 in the PB2 protein has been implicated in the virulence of H5N1 avian viruses, but it has not been associated with known functions of the viral RNA polymerase. Two functions of the PB2 protein have been reported: the region (amino acids 51–259) that binds to PB1 and the cap-binding site (either amino acids 360 and 404 or the region from amino acids 533–564). EN, the active site for endonuclease cleavage. P (SDD), the polymerization active site which has the sequence SDD. (See Color Insert.)

RNA to a sequence in the PB1 protein, the cap-dependent endonuclease of the viral polymerase is activated. The endonuclease is comprised of two active sites, one for cap-binding (on the PB2 subunit) and one for endonuclease cleavage (on the PB1 subunit). Cleavage of the capped RNA preferentially occurs at the 3' side of a CA sequence located 10–13 nucleotides from the cap. The 3' end of vRNA then binds to a site in the PB1 subunit, and the CA-terminated capped fragment is used as a primer to initiate transcription. The crucial cap-binding site of the PB2 protein has been mapped to either the 533–564 amino-acid region (Li et al., 2001) or the nonadjacent amino acids 360 and 404 (Fechter et al., 2003). Neither of these regions, nor any other defined functional region, includes amino acid 627.

Two roles for amino acid 627 in the PB2 protein have been postulated. One hypothesis is that this amino acid is crucial for the interaction of the viral polymerase with essential host factors, which differ between avians and mammals (Crescenzo-Chaigne et al., 2002). A second hypothesis is that this amino acid determines the temperature sensitivity of viral RNA polymerase activity (Massin et al., 2001). Virus replication occurs at different temperatures in avians and mammals. Human influenza A viruses replicate in the upper respiratory tract at a temperature of about 30°C, whereas avian influenza A viruses replicate in the intestinal tract at a temperature of about 41°C. Evidence in support of the second hypothesis has been obtained. Polymerase complexes from avian viruses, but not those from mammalian viruses, are less active in mammalian cells at 30°C compared to 41°C, and this cold sensitivity of the avian polymerases is determined primarily by the presence of Glu rather than Lys at position 627 of the PB2 protein (Massin et al., 2001).

VII. The Role of the NS1A Protein in Influenza A Virus Virulence

Evidence has been obtained for two roles of the NS1A protein in the virulence of H5N1 viruses. First, infection of monocyte-derived macrophages in tissue culture with 1997 H5N1 viruses induced a greater amount of cytokine mRNAs (including IFN-β and tumor necrosis factor (TNF)-α) compared to other influenza A viruses isolated from humans during the same period (Cheung et al., 2002). It was postulated that this high cytokine production contributes to the virulence of these H5N1 viruses in humans, particularly because autopsies of two humans indicated that elevated levels of cytokines might have been a major cause of death. By generating recombinant viruses, this virus trait was attributed to the NS1A gene of the 1997 H5N1 virus. Second, the 1997 H5N1 viruses were more resistant to the antiviral effects of IFN and TNF-α, and this resistance required the presence of Glu rather than Asp at position 92 in the NS1A protein (Seo et al., 2002). This property of these H5N1 would enable them to multiply to high titers in the presence of the high cytokine levels that they themselves induce. However, results obtained with a 2001 H5N1 virus are not consistent with resistance to IFN and TNF-α playing a role in virulence in mice and pigs (Lipatov et al., 2005). In addition, 2004 H5N1 viruses with high virulence in humans (e.g., A/Vietnam/1203/04) had Asp, and not Glu, at position 92 in the NS1A protein (Govorkova et al.,

2005). Nonetheless, the H5N1 viruses isolated from acutely infected humans in 2003 retained the ability to induce high cytokine levels, indicating that this property is independent of the presence of Glu rather than Asp at position 92 in the NS1A protein (Guan *et al.*, 2004).

It has not been established how the NS1A protein of H5N1 viruses mediates high cytokine production without attenuating these viruses. The NS1A protein is a multifunctional protein (Falcon *et al.*, 2004; Krug *et al.*, 2003). One of its best characterized functions is the inhibition of the 3' end processing of cellular pre-mRNAs, which occurs in virus-infected cells as well as in transient transfection assays (Chen *et al.*, 1999; Li *et al.*, 2001; Nemeroff *et al.*, 1998; Noah *et al.*, 2003; Shimuzu *et al.*, 1999). This inhibition is mediated by the binding of the NS1A protein to two crucial 3' end processing factors, the 30 kDa subunit of CPSF (cleavage and polyadenylation specificity factor) and PABII (poly (A)-binding protein II) (Chen *et al.*, 1999; Nemeroff *et al.*, 1998). As a consequence, in cells infected by influenza A virus, the processing of the pre-mRNAs encoded by cellular genes, including those encoding cytokines, is inhibited, resulting in a profound reduction of the production of mature cellular mRNAs in the cytoplasm. For example, the transcription of the cellular IFN-β gene occurs in human cells infected with a human influenza A virus (A/Udorn/72) because the virus activates the transcription factors, IRF-3 and NF-κB, that are required for the transcription of the IFN-β gene (Kim *et al.*, 2002; Noah *et al.*, 2003). Both IFN-β mRNA and the IFN-α/β-induced MxA mRNA are synthesized. IFN-α/β-induced mRNAs, including MxA mRNA, are also synthesized in cells infected by the mouse-adapted influenza A/WSN/33 virus (Geiss *et al.*, 2002), demonstrating that IRF-3 and NF-κB are also activated by this virus. Recent experiments have shown that activation of these transcription factors by several RNA viruses is mediated by the RIG-I RNA helicase (Sumpter *et al.*, 2005; Yoneyama *et al.*, 2004). As is the case with these RNA viruses, influenza A virus triggers the activation of IRF-3 and NF-κB through RIG-I-dependent signaling (Loo *et al.*, 2005). Consequently, the NS1A protein does not block the activation of IRF-3 and NF-κB, contrary to some reports (Garcia-Sastre, 2001; Talon *et al.*, 2000). Because 3' end processing of the newly synthesized IFN-β pre-mRNA is inhibited by influenza A virus, only a low amount of mature IFN-β mRNA is produced (Noah *et al.*, 2003). The level of mature IFN-β mRNA is substantially increased when cells are infected with a recombinant influenza A/Udorn/72 virus that encodes an NS1A protein containing a mutated binding site for the 30 kDa subunit of CPSF (centered around amino acid 186 of the 237-amino acid-long protein) (Noah *et al.*, 2003). This mutant virus is substantially attenuated, most likely because of the increase in IFN-β synthesis.

The relationship of these results to the increased cytokine synthesis in cells infected by the H5N1 viruses is unclear for several reasons: (1) the sequences centered around amino acid 186 (the region required for binding 30 kDa CPSF) in the NS1A proteins of H5N1 viruses and wild-type A/Udorn/72 virus are virtually identical; and (2) the increased cytokine synthesis in cells infected by H5N1 viruses does not result in an attenuated phenotype, unlike the situation with the mutant influenza A/Udorn/72 virus described previously. Based on the results obtained with this A/Udorn/72 mutant virus, it would be expected that an influenza A virus, which reduced, rather than increased, the amount of cytokine synthesis, particularly IFN-β synthesis, would render the virus more virulent. In fact, this appears to be the case for the mouse-adapted influenza A/PR8/34 virus, which has been reported to activate IRF-3 inefficiently (Talon *et al.*, 2000) and to induce much lower amounts of IFN-β than other influenza A viruses (Marcus, Rojek, and Sekellick, 2005). Perhaps the generation of this trait by multiple mouse passages is crucial to the high virulence of the A/PR8/34 virus in mice.

VIII. The Molecular Determinants of Influenza A Virus-Induced Cell Death: Possible Role in Virulence

Influenza A viruses have been reported to induce apoptosis, or programmed cell death, in tissue culture cells (Hinshaw *et al.*, 1994; Zhirnov *et al.*, 2002). Apoptosis is characterized by cytoskeletal disintegration and cellular DNA fragmentation, and is mediated by the caspase family of proteases (Green, 2000). Because influenza A virus is an obligate parasite, apoptosis would be expected to be detrimental for the virus unless apoptotic processes occur predominantly after the virus has utilized cellular machinery for its replication. On the other hand, an increase in influenza A virus-induced apoptosis would be expected to contribute to the virulence of the virus.

Three influenza A virus-encoded proteins have been ascribed apoptotic functions. NA has been reported to induce apoptosis both directly and indirectly by activating transforming growth factor β (TGF-β), which then functions as the apoptosis inducer (Schultz-Cherry and Hinshaw, 1996). In contrast, antiapoptotic activity has been attributed to the NS1A protein, based on the observation that a recombinant influenza A/PR8/34 virus lacking the entirety of its NS1A protein coding sequence was much more effective in inducing apoptosis than

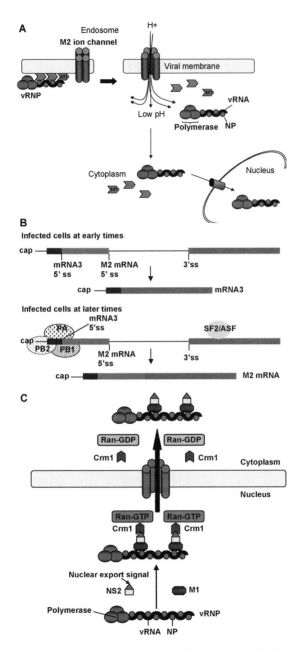

Fɪɢ 4. (A) Role of the M2 ion channel protein in the uncoating of influenza A virus. By promoting the influx of H$^+$ ions into the interior of the virion, the M2 protein causes the

the wild-type virus (Zhirnov *et al.*, 2002). The increase in apoptosis required that the cells possess functional IFN genes, indicating that apoptosis was caused at least in part by the increased synthesis of IFN by infected cells due to the absence of the NS1A protein. These conclusions are reminiscent of those described previously for the H5N1 avian influenza viruses, whose virulence has been attributed at least in part to the substantially increased levels of cytokines, including IFNs, that these viruses induce in infected cells (Cheung *et al.*, 2002; Guan *et al.*, 2004). The third virus-encoded protein involved in apoptosis is a small 87-amino-acid protein that is encoded in the PB1 gene, but in a different (+1) reading frame from that of the large PB1 protein subunit of the viral polymerase (Chen *et al.*, 2001). This small protein (PB1-F2) is not required for virus replication in embryonated eggs and in several epithelial cell lines, but enhances apoptosis in a cell-specific manner, specifically in monocytes but not in epithelial cells. Based on these results, it was postulated that the function of PB1-F2 is to kill host immune cells that respond to influenza A virus infection (Chen *et al.*, 2001). Such a function would be expected to contribute to the virulence of the virus.

IX. Functional Interactions Between Influenza A Virus-Encoded Proteins: Do These Interactions Play a Role in Virulence?

Because the virulence of influenza A viruses is polygenic, specific functional interactions between two or more virus-encoded proteins may be required for, or at least enhance, virulence. One type of functional interaction has been tied to virulence, namely between certain NA and HA proteins, as described previously for influenza A/WSN/33 virus and for H5N1 viruses. It might be argued that any protein–protein interactions that optimize packaging of vRNPs into virus particles and virus budding would facilitate virus spread and hence potentially would enhance virulence. It is not known if other functional interactions between

---

disruption of protein–protein interactions, including that between the M1 protein and vRNPs. The dissociation of M1 from the vRNPs enables the vRNPs to be transported into the nucleus. (B) The viral polymerase and the cellular SF2/ASF splicing enhancer regulate the alternative splicing of M1 mRNA, thereby regulating M2 protein synthesis. See text for the description of the mechanism. Redrawn with permission from Shih and Krug (1996). (C) A M1-NS2 protein complex mediates the nuclear export of influenza vRNPs. The nuclear export signal (NES) interacts with Crm1, the protein that mediates RanGTP-dependent nuclear export. (See Color Insert.)

virus-encoded proteins play significant roles in virulence. We will describe three such functional interactions later (Fig. 4).

The M2 protein, which possesses low pH-activated ion channel activity (Chizhmakov et al., 1996; Lamb and Krug, 2001; Mould et al., 2000;
Pinto et al., 1992), affects crucial interactions between several viral
components during the uncoating of an influenza A virion in endosomes
(Fig. 4A). The M2 protein promotes the influx of H^+ ions into the interior
of the virion (Lamb and Krug, 2001). This influx disrupts protein–
protein interactions, including that between the M1 protein and
vRNPs. After the RNPs are released into the cytoplasm, they are transported to the nucleus, where they catalyze the initial transcription of
the viral genome (Lamb et al., 1994; Martin and Helenius, 1991; Skehel
et al., 1978). The antiviral drugs amantadine and rimantadine inhibit
M2 ion channel activity, and as a result, the M1 protein remains associated with the vRNPs, thereby inhibiting their nuclear import. Treatment with these drugs results in the rapid selection of drug-resistant
virus mutants that replicate as well as the drug-sensitive parent virus
(Cox and Subbarao, 1999; Suzuki et al., 2003). Surprisingly, the 2004
H5N1 avian viruses contain one of the M2 protein mutations that
render these viruses resistant to these two drugs (Puthavathana
et al., 2005). It is unclear how such a mutant H5N1 virus was selected
and continues to circulate in the absence of these drugs. It is conceivable
that, for reasons that we do not currently understand, the drug-resistant form of the M2 ion channel protein is the best fit for H5N1 viruses.

The synthesis of the M2 ion channel protein is regulated by alternative splicing of M1 mRNA, a process that involves both the viral
polymerase and a specific host cell splicing enhancer protein (Shih
and Krug, 1996; Shih et al., 1995) (Fig. 4B). M1 mRNA contains two
5' splice sites, at nucleotide 11 (distal site) and at nucleotide 51 (proximal site), and a single 3' splice site. Usage of the distal 5' splice site
produces a mRNA (mRNA$_3$) that encodes only a 9-amino-acid peptide,
the production of which has not been detected, whereas usage of the
proximal 5' splice site produces the mRNA for the M2 protein. The
distal (mRNA$_3$) 5' splice site, which fits the consensus 5' splice site
more closely than the proximal 5' splice, is the 5' splice site that is
exclusively used when a DNA plasmid expressing the M1 mRNA is
transfected into uninfected cells. In infected cells, the mRNA$_3$ 5' splice
site is blocked by the viral polymerase, which binds to the 5' end of M1
mRNA in a sequence-specific and cap-dependent fashion (Shih et al.,
1995). After the 5' mRNA$_3$ splice site is blocked by the viral polymerase,
the splicing machinery can switch to the M2 5' splice site. Consequently,
the production of M2 mRNA is delayed until sufficient amounts of the

polymerase complex are synthesized, so that the time course of the synthesis and assembly of functional polymerase complexes controls the time at which the M2 5' splice site can be utilized in infected cells. Because this control system is designed to delay the synthesis of the M2 protein, it is quite possible that the presence of increased amounts of the M2 ion channel protein at early times of infection is deleterious for virus replication. If this is the case, a recombinant virus encoding a M1 mRNA with a mutated 5' mRNA$_3$ splice site would be expected to be attenuated. The switch to the M2 5' splice site requires not only the viral polymerase but also a specific SR (serine-rich splicing factor), SF2/ASF, which binds to a specific splicing enhancer sequence in the 3' exon of M1 mRNA (Shih and Krug, 1996). In fact, the amount of the M2 mRNA and hence of the M2 protein is controlled by the amount of SF2/ASF produced by the cell, which was shown to vary among several cell lines. It is not known how different levels of the M2 protein affect virus replication, nor whether the control of M2 levels by a specific cellular protein (SF2/ASF) impacts tissue tropism or the virulence of virus infection.

Nuclear export of newly synthesized vRNPs is a necessary step in the packaging of new virions and is mediated by a M1-NS2 protein complex (Cros and Palese, 2003; O'Neill *et al.*, 1998) (Fig. 4C). The M1 protein binds to the nuclear vRNPs, which may involve both M1-NP and M1-vRNA interactions. The NS2 protein, which binds to M1, provides the nuclear export signal (NES) that enables the vRNP-M1-NS2 complex to be exported from the nucleus via the Crm1-dependent pathway. It is not known whether the nuclear export of vRNPs is coordinately regulated with other steps in virion packaging.

X. Conclusions

The virulence of influenza A viruses is polygenic. Several virus-specific proteins, by their individual actions and/or by their functional interactions with other virus-specific and/or host proteins, likely have important roles in determining how virulent an influenza A virus is in different hosts. As outlined in this review, virulence is a complicated trait, and at present there are more questions than answers about the molecular determinants of the virulence of influenza A viruses.

Acknowledgments

Research in the authors' laboratory was supported by National Institutes of Health grant AI11772 and Welch Foundation grant F-1468. (to R. M. K.)

References

Bender, C., Hall, H., Huang, J., Klimov, A., Cox, N., Hay, A., Gregory, V., Cameron, K., Lim, W., and Subbarao, K. (1999). Characterization of the surface proteins of influenza A (H5N1) viruses isolated from humans in 1997–1998. *Virology* **254:**115–123.

Bright, R. A., Cho, D. S., Rowe, T., and Katz, J. M. (2003). Mechanisms of pathogenicity of influenza A (H5N1) viruses in mice. *Avian Dis.* **47:**1131–1134.

Chen, W., Calvo, P. A., Malide, D., Gibbs, J., Schubert, U., Bacik, I., Basta, S., O'Neill, R., Schickli, J., Palese, P., Henklein, P., Bennink, J. R., and Yewdell, J. W. (2001). A novel influenza A virus mitochondrial protein that induces cell death. *Nat. Med.* **7:**1306–1312.

Chen, Z., Li, Y., and Krug, R. M. (1999). Influenza A virus NS1 protein targets poly (A)-binding protein II of the cellular 3′-end processing machinery. *EMBO J.* **18:**2273–2283.

Cheung, C. Y., Poon, L. L., Lau, A. S., Luk, W., Lau, Y. L., Shortridge, K. F., Gordon, S., Guan, Y., and Peiris, J. S. (2002). Induction of proinflammatory cytokines in human macrophages by influenza A (H5N1) viruses: A mechanism for the unusual severity of human disease? *Lancet* **360:**1831–1837.

Chizhmakov, I. V., Geraghty, F. M., Ogden, D. C., Hayhurst, A., Antoniou, M., and Hay, A. J. (1996). Selective proton permeability and pH regulation of the influenza virus M2 channel expressed in mouse erythroleukaemia cells. *J. Physiol.* **494:**329–336.

Claas, E. C., Osterhaus, A. D., van Beek, R., De Jong, J. C., Rimmelzwaan, G. F., Senne, D. A., Krauss, S., Shortridge, K. F., and Webster, R. G. (1998). Human influenza A H5N1 virus related to a highly pathogenic avian influenza virus. *Lancet* **351:**472–477.

Colman, P. M., Varghese, J. N., and Laver, W. G. (1983). Structure of the catalytic and antigenic sites in influenza virus neuraminidase. *Nature* **303:**41–44.

Cox, N. J., and Subbarao, K. (1999). Influenza. *Lancet* **354:**1277–1282.

Crescenzo-Chaigne, B., van der Werf, S., and Naffakh, N. (2002). Differential effect of nucleotide substitutions in the 3′ arm of the influenza A virus vRNA promoter on transcription/replication by avian and human polymerase complexes is related to the nature of PB2 amino acid 627. *Virology* **303:**240–252.

Cros, J. F., and Palese, P. (2003). Trafficking of viral genomic RNA into and out of the nucleus: Influenza, Thogoto and Borna disease viruses. *Virus Res.* **95:**3–12.

Falcon, A. M., Marion, R. M., Zurcher, T., Gomez, P., Portela, A., Nieto, A., and Ortin, J. (2004). Defective RNA replication and late gene expression in temperature-sensitive influenza viruses expressing deleted forms of the NS1 protein. *J Virol.* **78:**3880–3888.

Fechter, P., Mingay, L., Sharps, J., Chambers, A., Fodor, E., and Brownlee, G. G. (2003). Two aromatic residues in the PB2 subunit of influenza A RNA polymerase are crucial for cap binding. *J. Biol. Chem.* **278:**20381–20388.

Florent, G., Lobmann, M., Beare, A. S., and Zygraich, N. (1977). RNAs of influenza virus recombinants derived from parents of known virulence for man. *Arch. Virol.* **54:**19–28.

Garcia-Sastre, A. (2001). Inhibition of interferon-mediated antiviral responses by influenza A viruses and other negative-strand RNA viruses. *Virology* **279:**375–384.

Geiss, G. K., Salvatore, M., Tumpey, T. M., Carter, V. S., Wang, X., Basler, C. F., Taubenberger, J. K., Bumgarner, R. E., Palese, P., Katze, M. G., and Garcia-Sastre, A. (2002). Cellular transcriptional profiling in influenza A virus-infected lung epithelial cells: The role of the nonstructural NS1 protein in the evasion of the host innate defense and its potential contribution to pandemic influenza. *Proc. Natl. Acad. Sci. USA* **99:**10736–10741.

Goto, H., and Kawaoka, Y. (1998). A novel mechanism for the acquisition of virulence by a human influenza A virus. *Proc. Natl. Acad. Sci. USA* **95**:10224–10228.

Goto, H., Wells, K., Takada, A., and Kawaoka, Y. (2001). Plasminogen-binding activity of neuraminidase determines the pathogenicity of influenza A virus. *J. Virol.* **75**:9297–9301.

Govorkova, E. A., Rehg, J. E., Krauss, S., Yen, H. L., Guan, Y., Peiris, M., Nguyen, T. D., Hanh, T. H., Puthavathana, P., Long, H. T., Buranathai, C., Lim, W., Webster, R. G., and Hoffmann, E. (2005). Lethality to ferrets of H5N1 influenza viruses isolated from humans and poultry in 2004. *J. Virol.* **79**:2191–2198.

Green, D. R. (2000). Apoptotic pathways: Paper wraps stone blunts scissors. *Cell* **102** (1):1–4.

Guan, Y., Poon, L. L., Cheung, C. Y., Ellis, T. M., Lim, W., Lipatov, A. S., Chan, K. H., Sturm-Ramirez, K. M., Cheung, C. L., Leung, Y. H. C., Yuen, K. Y., Webster, R. G., and Peiris, J. S. M. (2004). H5N1 influenza: A protean pandemic threat. *Proc. Natl. Acad. Sci. USA* **101**:8156–8161.

Gubareva, L. V., Penn, C. R., and Webster, R. G. (1995). Inhibition of replication of avian influenza viruses by the neuraminidase inhibitor 4-guanidino-2,4-dideoxy-2,3-dehydro-N-acetylneuraminic acid. *Virology* **212**:323–330.

Harvey, R., Martin, A. C., Zambon, M., and Barclay, W. S. (2004). Restrictions to the adaptation of influenza a virus H5 hemagglutinin to the human host. *J. Virol.* **78**:502–507.

Hatta, M., Gao, P., Halfmann, P., and Kawaoka, Y. (2001). Molecular basis for high virulence of Hong Kong H5N1 influenza A viruses. *Science* **293**:1840–1842.

Hayden, F. G., Osterhaus, A. D., Treanor, J. J., Fleming, D. M., Aoki, F. Y., Nicholson, K. G., Bohnen, A. M., Hirst, H. M., Keene, O., and Wightman, K. (1997). Efficacy and safety of the neuraminidase inhibitor zanamivir in the treatment of influenzavirus infections. GG167 Influenza Study Group. *N. Engl. J. Med.* **337**:874–880.

Hayden, F. G., Treanor, J. J., Betts, R. F., Lobo, M., Esinhart, J. D., and Hussey, E. K. (1996). Safety and efficacy of the neuraminidase inhibitor GG167 in experimental human influenza. *JAMA* **275**:295–299.

Hinshaw, V. S., Olsen, C. W., Dybdahl-Sissoko, N., and Evans, D. (1994). Apoptosis: A mechanism of cell killing by influenza A and B viruses. *J. Virol.* **68**:3667–3673.

Hulse, D. J., Webster, R. G., Russell, R. J., and Perez, D. R. (2004). Molecular determinants within the surface proteins involved in the pathogenicity of H5N1 influenza viruses in chickens. *J. Virol.* **78**:9954–9964.

Kawaoka, Y., and Webster, R. G. (1989). Interplay between carbohydrate in the stalk and the length of the connecting peptide determines the cleavability of influenza virus hemagglutinin. *J. Virol.* **63**:3296–3300.

Kim, C. U., Lew, W., Williams, M. A., Liu, H., Zhang, L., Swaminathan, S., Bischofberger, N., Chen, M. S., Mendel, D. B., Tai, C. Y., Laver, W. G., and Stevens, R. C. (1997). Influenza neuraminidase inhibitors possessing a novel hydrophobic interaction in the enzyme active site: Design, synthesis and structural analysis of carbocyclic sialiac acid analogues with potent anti-influenza activity. *J. Amer. Chem. Soc.* **119**:681–690.

Kim, M. J., Latham, A. G., and Krug, R. M. (2002). Human influenza viruses activate an interferon-independent transcription of cellular antiviral genes: Outcome with influenza A virus is unique. *Proc. Natl. Acad. Sci. USA* **99**:10096–10101.

Kiso, M., Mitamura, K., Sakai-Tagawa, Y., Shiraishi, K., Kawakami, C., Kimura, K., Hayden, F. G., Sugaya, N., and Kawaoka, Y. (2004). Resistant influenza A viruses in children treated with oseltamivir: Descriptive study. *Lancet* **364**:759–765.

Kobasa, D., Takada, A., Shinya, K., Hatta, M., Halfmann, P., Theriault, S., Suzuki, H., Nishimura, H., Mitamura, K., Sugaya, N., Usui, T., Murata, T., Maeda, Y., Watanabe, S., Suresh, M., Suzuki, T., Suzuki, Y., Feldmann, H., and Kawaoka, Y. (2004). Enhanced virulence of influenza A viruses with the haemagglutinin of the 1918 pandemic virus. *Nature* **431**:703–707.

Krug, R. M., Alonso-Caplen, F. V., Julkunen, I., and Katze, M. G. (1989). Expression and replication of the influenza virus genome. *In* "The Influenza Viruses" (R. M. Krug, ed.), pp. 89–152. Plenum Press, New York.

Krug, R. M., Shaw, M., Broni, B., Shapiro, G., and Haller, O. (1985). Inhibition of influenza viral mRNA synthesis in cells expressing the interferon-induced Mx gene product. *J. Virol.* **56**:201–206.

Krug, R. M., Yuan, W., Noah, D. L., and Latham, A. G. (2003). Intracellular warfare between human influenza viruses and human cells: The roles of the viral NS1 protein. *Virology* **309**:181–189.

Lamb, and Krug, R. (2001). Orthomyxoviridae: The viruses and their replication. *In* "Fields Virology" (D. M. Knipe and P. M. Howley, eds.), pp. 1487–1531. Lippincott, Williams, and Wilkins, Philadelphia.

Lamb, R. A., Holsinger, L. J., and Pinto, L. H. (1994). The influenza A virus M2 ion channel protein and its role in the influenza virus life cycle. *In* "Receptor-Mediated Virus Entry into Cells" (E. Wimmer, ed.), pp. 303–312. Cold Spring Harbor Laboratory Press, Cold Spring Harbor, NY.

Li, M. L., Rao, P., and Krug, R. M. (2001). The active sites of the influenza cap-dependent endonuclease are on different polymerase subunits. *EMBO J.* **20**:2078–2086.

Li, Y., Chen, Z. Y., Wang, W., Baker, C. C., and Krug, R. M. (2001). The 3′-end-processing factor CPSF is required for the splicing of single-intron pre-mRNAs *in vivo*. *RNA* **7**:920–931.

Linnemann, C. C., Jr., Shea, L., Partin, J. C., Schubert, W. K., and Schiff, G. M. (1975). Reye's syndrome: Epidemiologic and viral studies, 1963–1974. *Am. J. Epidemiol.* **101**:517–526.

Lipatov, A. S., Andreansky, S., Webby, R. J., Hulse, D. J., Rehg, J. E., Krauss, S., Perez, D. R., Doherty, P. C., Webster, R. G., and Sangster, M. Y. (2005). Pathogenesis of Hong Kong H5N1 influenza virus NS gene reassortants in mice: The role of cytokines and B- and T-cell responses. *J. Gen. Virol.* **86**:1121–1130.

Loo, Y.-M., Fredericksen, B., and Gale, M. J. (2005). Requirement of RIG-I in signaling the host response to virus infection. Submitted for publication.

Marcus, P. I., Rojek, J. M., and Sekellick, M. J. (2005). Interferon induction and/or production and its suppression by influenza A viruses. *J. Virol.* **79**:2880–2890.

Martin, K., and Helenius, A. (1991). Nuclear transport of influenza virus ribonucleoproteins: The viral matrix protein (M1) promotes export and inhibits import. *Cell* **67**:117–130.

Mase, M., Tsukamoto, K., Imada, T., Imai, K., Tanimura, N., Nakamura, K., Yamamoto, Y., Hitomi, T., Kira, T., Nakai, T., Kiso, M., Horimoto, T., Kawaoka, Y., and Yamaguchi, S. (2005). Characterization of H5N1 influenza A viruses isolated during the 2003–2004 influenza outbreaks in Japan. *Virology* **332**:167–176.

Massin, P., van der Werf, S., and Naffakh, N. (2001). Residue 627 of PB2 is a determinant of cold sensitivity in RNA replication of avian influenza viruses. *J. Virol.* **75**:5398–5404.

Matrosovich, M. N., Matrosovich, T. Y., Gray, T., Roberts, N. A., and Klenk, H. D. (2004). Human and avian influenza viruses target different cell types in cultures of human airway epithelium. *Proc. Natl. Acad. Sci. USA* **101**:4620–4624.

McNicholl, I. R., and McNicholl, J. J. (2001). Neuraminidase inhibitors: Zanamivir and oseltamivir. *Ann. Pharmacother.* **35:**57–70.

Mendel, D. B., Tai, C. Y., Escarpe, P. A., Li, W., Sidwell, R. W., Huffman, J. H., Sweet, C., Jakeman, K. J., Merson, J., Lacy, S. A., Lew, W., Williams, M. A., Zhang, L., Chen, M. S., Bischofberger, N., and Kim, C. U. (1998). Oral administration of a prodrug of the influenza virus neuraminidase inhibitor GS 4071 protects mice and ferrets against influenza infection. *Antimicrob. Agents Chemother.* **42:**640–646.

Mould, J. A., Drury, J. E., Frings, S. M., Kaupp, U. B., Pekosz, A., Lamb, R. A., and Pinto, L. H. (2000). Permeation and activation of the M2 ion channel of influenza A virus. *J. Biol. Chem.* **275:**31038–31050.

Nemeroff, M. E., Barabino, S. M., Li, Y., Keller, W., and Krug, R. M. (1998). Influenza virus NS1 protein interacts with the cellular 30 kDa subunit of CPSF and inhibits 3′ end formation of cellular pre-mRNAs. *Mol. Cell* **1:**991–1000.

Noah, D. L., Twu, K. Y., and Krug, R. M. (2003). Cellular antiviral responses against influenza A virus are countered at the posttranscriptional level by the viral NS1A protein via its binding to a cellular protein required for the 3′ end processing of cellular pre-mRNAS. *Virology* **307:**386–395.

O'Neill, R. E., Talon, J., and Palese, P. (1998). The influenza virus NEP (NS2 protein) mediates the nuclear export of viral ribonucleoproteins. *EMBO J.* **17:**288–296.

Ohtsu, Y., Honda, Y., Sakata, Y., Kato, H., and Toyoda, T. (2002). Fine mapping of the subunit binding sites of influenza virus RNA polymerase. *Microbiol. Immunol.* **46:**167–175.

Oxford, J. S., McGeoch, D. J., Schild, G. C., and Beare, A. S. (1978). Analysis of virion RNA segments and polypeptides of influenza A virus recombinants of defined virulence. *Nature* **273:**778–779.

Pinto, L. H., Holsinger, L. J., and Lamb, R. A. (1992). Influenza virus M2 protein has ion channel activity. *Cell* **69:**517–528.

Prevention (2005). Background on Influenza. http://www.cdc.gov/flu/background/professionals/background.

Puthavathana, P., Auewarakul, P., Charoenying, P. C., Sangsiriwut, K., Pooruk, P., Boonnak, K., Khanyok, R., Thawachsupa, P., Kijphati, R., and Sawanpanyalert, P. (2005). Molecular characterization of the complete genome of human influenza H5N1 virus isolates from Thailand. *J. Gen. Virol.* **86:**423–433.

Rao, P., Yuan, W., and Krug, R. M. (2003). Crucial role of CA cleavage sites in the cap-snatching mechanism for initiating viral mRNA synthesis. *EMBO J.* **22:**1188–1198.

Reid, A. H., Taubenberger, J. K., and Fanning, T. G. (2001). The 1918 Spanish influenza: Integrating history and biology. *Microbes Infect.* **3:**81–87.

Rott, R., Orlich, M., and Scholtissek, C. (1979). Correlation of pathogenicity and gene constellation of influenza A viruses. III. Nonpathogenic recombinants derived from highly pathogenic parent strains. *J. Gen. Virol.* **44:**471–477.

Ryan, D. M., Ticehurst, J., Dempsey, M. H., and Penn, C. R. (1994). Inhibition of influenza virus replication in mice by GG167 (4-guanidino-2,4-dideoxy-2,3-dehydro-N-acetylneuraminic acid) is consistent with extracellular activity of viral neuraminidase (sialidase). *Antimicrob. Agents Chemother.* **38:**2270–2275.

Schultz-Cherry, S., and Hinshaw, V. S. (1996). Influenza virus neuraminidase activates latent transforming growth factor beta. *J. Virol.* **70:**8624–8629.

Seo, S. H., Hoffmann, E., and Webster, R. G. (2002). Lethal H5N1 influenza viruses escape host antiviral cytokine responses. *Nat. Med.* **8:**950–954.

Shih, S. R., and Krug, R. M. (1996). Novel exploitation of a nuclear function by influenza virus: The cellular SF2/ASF splicing factor controls the amount of the essential viral M2 ion channel protein in infected cells. *EMBO J.* **15**:5415–5427.

Shih, S. R., Nemeroff, M. E., and Krug, R. M. (1995). The choice of alternative 5′ splice sites in influenza virus M1 mRNA is regulated by the viral polymerase complex. *Proc. Natl. Acad. Sci. USA* **92**:6324–6328.

Shimuzu, K., Iguchi, A., Gomyou, R., and Onu, Y. (1999). Influenza virus inhibits cleavage of the HSP70 pre-MRNAs at the polyadenylation site. *Virology* **254**:213–219.

Shinya, K., Hamm, S., Hatta, M., Ito, H., Ito, T., and Kawaoka, Y. (2004). PB2 amino acid at position 627 affects replicative efficiency, but not cell tropism, of Hong Kong H5N1 influenza A viruses in mice. *Virology* **320**:258–266.

Skehel, J. J., Hay, A. J., and Armstrong, J. A. (1978). On the mechanism of inhibition of influenza virus replication by amantadine hydrochloride. *J. Gen. Virol.* **38**:97–110.

Staeheli, P., Horisberger, M. A., and Haller, O. (1984). Mx-dependent resistance to influenza viruses is induced by mouse interferons alpha and beta but not gamma. *Virology* **132**:456–461.

Stieneke-Grober, A., Vey, M., Angliker, H., Shaw, E., Thomas, G., Roberts, C., Klenk, H. D., and Garten, W. (1992). Influenza virus hemagglutinin with multibasic cleavage site is activated by furin, a subtilisin-like endoprotease. *EMBO J.* **11**:2407–2414.

Suarez, D. L., Perdue, M. L., Cox, N., Rowe, T., Bender, C., Huang, J., and Swayne, D. E. (1998). Comparisons of highly virulent H5N1 influenza A viruses isolated from humans and chickens from Hong Kong. *J. Virol.* **72**:6678–6688.

Subbarao, K., Klimov, A., Katz, J., Regnery, H., Lim, W., Hall, H., Perdue, M., Swayne, D., Bender, C., Huang, J., Hemphill, M., Rowe, T., Shaw, M., Xu, X., Fukuda, K., and Cox, N. (1998). Characterization of an avian influenza A (H5N1) virus isolated from a child with a fatal respiratory illness. *Science* **279**:393–396.

Sumpter, R., Jr., Loo, Y. M., Foy, E., Li, K., Yoneyama, M., Fujita, T., Lemon, S. M., and Gale, M., Jr. (2005). Regulating intracellular antiviral defense and permissiveness to hepatitis C virus RNA replication through a cellular RNA helicase, RIG-I. *J. Virol.* **79**:2689–2699.

Suzuki, H., Saito, R., Masuda, H., Oshitani, H., Sato, M., and Sato, I. (2003). Emergence of amantadine-resistant influenza A viruses: Epidemiological study. *J. Infect. Chemother.* **9**:195–200.

Talon, J., Horvath, C. M., Polley, R., Basler, C. F., Muster, T., Palese, P., and Garcia-Sastre, A. (2000). Activation of interferon regulatory factor 3 is inhibited by the influenza A virus NS1 protein. *J. Virol.* **74**:7989–7996.

To, K. F., Chan, P. K., Chan, K. F., Lee, W. K., Lam, W. Y., Wong, K. F., Tang, N. L., Tsang, D. N., Sung, R. Y., Buckley, T. A., Tam, J. S., and Cheng, A. F. (2001). Pathology of fatal human infection associated with avian influenza A H5N1 virus. *J. Med. Virol.* **63**:242–246.

Tran, T. H., Nguyen, T. L., Nguyen, T. D., Luong, T. S., Pham, P. M., Nguyen, V. C., Pham, T. S., Vo, C. D., Le, T. Q., Ngo, T. T., Dao, B. K., Le, P. P., Nguyen, T. T., Hoang, T. L., Cao, V. T., Le, T. G., Nguyen, D. T., Le, H. N., Nguyen, K. T., Le, H. S., Le, V. T., Christiane, D., Tran, T. T., Menno de, J., Schultsz, C., Cheng, P., Lim, W., Horby, P., and Farrar, J., World Health, Organization International Avian, Influenza Investigative, Team. (2004). Avian influenza A (H5N1) in 10 patients in Vietnam. *New Engl. J. Med.* **350**:1179–1188.

Ungchusak, K., Auewarakul, P., Dowell, S. F., Kitphati, R., Auwanit, W., Puthavathana, P., Uiprasertkul, M., Boonnak, K., Pittayawonganon, C., Cox, N. J., Zaki, S. R., Thawatsupha, P., Chittaganpitch, M., Khontong, R., Simmerman, J. M., and

Chunsutthiwat, S. (2005). Probable person-to-person transmission of avian influenza A (H5N1). *N. Engl. J. Med.* **352:**333–340.

Varghese, J. N., Laver, W. G., and Colman, P. M. (1983). Structure of the influenza virus glycoprotein antigen neuraminidase at 2.9 A resolution. *Nature* **303:**35–40.

von Itzstein, M., Wu, W. Y., Kok, G. B., Pegg, M. S., Dyason, J. C., Jin, B., Van Phan, T., Smythe, M. L., White, H. F., Oliver, S. W., Colman, P. M., Varghese, J. N., Ryan, D. M., Woods, J. M., Bethell, R. C., Hotham, V. J., Cameron, J. M., and Penn, C. R. (1993). Rational design of potent sialidase-based inhibitors of influenza virus replication. *Nature* **363:**418–423.

Woods, J. M., Bethell, R. C., Coates, J. A., Healy, N., Hiscox, S. A., Pearson, B. A., Ryan, D. M., Ticehurst, J., Tilling, J., and Walcott, S. M. (1993). 4-Guanidino-2,4-dideoxy-2,3-dehydro-N-acetylneuraminic acid is a highly effective inhibitor both of the sialidase (neuraminidase) and of growth of a wide range of influenza A and B viruses *in vitro*. *Antimicrob. Agents Chemother.* **37:**1473–1479.

Wright, P. F., and Webster, R. G. (2001). Orthomyxoviruses. *In* "Fields Virology, 4th edition" (D. M. Knipe and P. M. Howley, eds.), pp. 1533–1579. Lippincott, Williams, and Wilkins, Philadelphia.

Yoneyama, M., Kikuchi, M., Natsukawa, T., Shinobu, N., Imaizumi, T., Miyagishi, M., Taira, K., Akira, S., and Fujita, T. (2004). The RNA helicase RIG-I has an essential function in double-stranded RNA-induced innate antiviral responses. *Nat. Immunol.* **5:**730–737.

Yuan, W., and Krug, R. M. (2001). Influenza B virus NS1 protein inhibits conjugation of the interferon (IFN)-induced ubiquitin-like ISG15 protein. *EMBO J.* **20:**362–371.

Zhirnov, O. P., Konakova, T. E., Wolff, T., and Klenk, H. D. (2002). NS1 protein of influenza A virus downregulates apoptosis. *J. Virol.* **76:**1617–1625.

Zitzow, L. A., Rowe, T., Morken, T., Shieh, W. J., Zaki, S., and Katz, J. M. (2002). Pathogenesis of avian influenza A (H5N1) viruses in ferrets. *J. Virol.* **76:**4420–4429.

ADVANCES IN VIRUS RESEARCH, VOL 65

ALTERATION AND ANALYSES OF VIRAL ENTRY WITH LIBRARY-DERIVED PEPTIDES

Keith Bupp and Monica J. Roth

Department of Biochemistry, Robert Wood Johnson Medical School, University of Medicine and Dentistry of New Jersey, Pisacataway, New Jersey 08854

I. Introduction

The use of peptide libraries has revolutionized the ability to mutate and alter various properties of proteins. Initial studies employing peptide libraries utilized bacteriophage as a display platform (Cwirla et al., 1990; Devlin et al., 1990; Scott and Smith, 1990) or synthetic peptides on beads (Geysen et al., 1984; Lam et al., 1991). The initial applications screened simply for peptides binding to target molecules (streptavidin) (Devlin et al., 1990; Lam et al., 1991), or for peptide epitopes that bound to antibodies (Cwirla et al., 1990; Geysen et al., 1984; Scott and Smith, 1990). Since then, vast amounts of literature have been published employing peptide display libraries to alter and analyze various characteristics of proteins.

In this chapter, we discuss the use of peptide libraries to examine and alter the interactions of mammalian viruses with cellular receptors. Investigations using antibodies derived from library screens to study immunological aspects of viruses and/or alter virus–cell interactions will not be discussed. Screening peptide libraries is a useful method for targeting retroviral, adenoviral, and adeno-associated viral gene therapy vectors to specific cells. In the case of adenovirus (Ad) and adeno-associated virus (AAV), peptides have been obtained from phage display libraries and attached to the surface of the virus, resulting in new targeting properties. Libraries of AAV and retroviral peptide display

0065-3527/05 $35.00
DOI: 10.1016/S0065-3527(05)65005-1

particles have also been employed in the development of gene therapy vectors.

II. Examining Virus–Receptor Interactions Using Libraries

Peptide libraries have been used to define regions of virus surface proteins that may be involved in receptor binding and to identify cellular proteins that may serve as virus receptors. One way of investigating the interactions of viruses with cellular receptors is to screen libraries of peptides displayed on phage for their ability to bind the virus of interest. This can have potential therapeutic applications similar to those of antibodies to prevent viruses from infecting cells *in vivo* (Ferrer, 1999; Ramanujam *et al.*, 2002). In addition, peptides derived from library screens can also provide insights into virus–cell interactions due to their homology to either cellular receptors or to the viruses themselves.

A. Peptides Corresponding to Receptor Sequences

Sequences of peptides obtained by screening libraries for virus-binding proteins can be compared with known surface proteins, which could then be further investigated as candidate receptor proteins. But this approach depends upon the prior cloning of the receptor protein. Also, database searches using short peptide sequences tend to identify many possible receptor candidates. The list of possibilities must then somehow be narrowed down for further experimental analyses. Outlined later are two cases where the approach has been useful in determining the surface proteins that the viruses bind to.

Peptides binding to the human picornaviruses echovirus EV22 and coxsackievirus A9 were obtained by phage library screening (Pulli *et al.*, 1997). An LRSG motif enriched in the peptide screen on EV22 is found in several proteins on the surface of cells. The most abundant of the two peptides isolated from the screen had the sequence LRSGRG, which is identical to a sequence found in matrix metalloproteinase 9. Another surface protein containing the LRSG motif is the β_1 integrin subunit. Since the VP1 capsid subunit of EV22 has an RGD sequence, recognized in fibronectin by $\alpha_5\beta_1$ and $\alpha_V\beta_1$ integrins, it therefore followed that EV22 might use an integrin for entry. Antibodies to α_V and β_1 subunits blocked entry of EV22 antibodies to MMP-9, and the β_3 integrin subunit were less effective, and anti-α_5 was totally ineffective in preventing EV22 entry. These experiments support the model that $\alpha_V\beta_1$ integrin and MMP-9 play roles in EV22 entry.

The phage peptide library screen on coxsackievirus A9 (CAV9) (Pulli *et al.*, 1997) led to the enrichment of peptides containing a VWD consensus sequence. This sequence is found in at least ten cell-surface proteins, including integrin β_5. Neither the VWD-containing peptide nor anti-β_5 antibodies were able to block CAV9 infection. Results of blocking experiments using the same anti-integrin and anti-MMP-9 antibodies used for the EV22 experiments were similar to those with EV22. But CAV9 entry was less sensitive to blocking at higher dilutions of the antibodies. These blocking experiments support the conclusion that the peptides identified by library screening constitute a portion of the receptor-binding sites of CAV-9.

B. Peptides Corresponding to Viral Sequences

A reverse approach was taken to identify measles virus (MV) peptides that interacted with the signaling lymphocyte activation molecule (SLAM) receptor (Hu *et al.*, 2004). A fragment of SLAM (amino acids 27-135) known to bind to measles hemagglutinin (H) was the target of a phage peptide library screen. Two major groups of homologous peptides were identified. One group with the consensus *SXXXPLXTHX*, including 3 with *SGF DPLITHA*, was homologous to amino acids 429–438 of MV H protein (*SGFGPLITHG*). The other group with the consensus *MAEVLXXXXG* did not share homology to an MV protein. The peptides homologous to H protein, when displayed by phage, were able to inhibit binding of H to the SLAM peptide, full-length SLAM expressed on CHO cells and also infection of CHO-SLAM cells by MV. The nonhomologous peptide was unable to inhibit these interactions. Thus, the homologous peptides appear to mimic the virus-receptor interaction whereas the nonhomologous peptide binds to SLAM by an alternate mechanism.

Salone *et al.* used a similar library screening strategy to identify sequences in the penton base of adenovirus serotype 5 that interact with $\alpha_3\beta_1$ integrin (Salone *et al.*, 2003). The hexon, penton base, and fiber are the major structural proteins of the adenovirus capsid (Fig. 1) (Medina-Kauwe, 2003). The hexon is the most abundant protein, but is not involved in cellular entry. The five subunits of the penton base (Krasnykh *et al.*, 2000) each contain flexible loops with RGD sequences that interact with integrins. The trimeric fiber protein associates with the penton base through its amino-terminal tail domain. The central shaft domain of the fiber projects the C-terminal knob domain away from the virion to facilitate interactions with the cellular receptor. Ad binding to coxsackie-adenovirus receptor (CAR) is mediated by

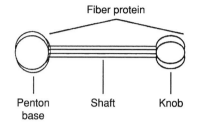

Fig 1. The trimeric fiber protein of adenovirus consists of a C-terminal knob domain, a shaft domain, and an N-terminal tail domain connecting the fiber protein to the penton base on the surface of the virus.

discontinuous sequences in the knob (Krasnykh *et al.*, 2000). The initial interaction between the knob and the receptor protein is followed by interaction of the Ad penton base with integrins during internalization (Wickham, 2000).

Interestingly, no RGD or LDV-containing peptides were identified from the random hexapeptide library screening on $\alpha_3\beta_1$ integrin. This was somewhat surprising in light of previous binding and peptide competition experiments, indicating that the RGD and LDV penton base sequences were involved in other Ad interactions with integrins (Mathias *et al.*, 1994; Wickham *et al.*, 1993). Instead, phage with sequences homologous to other portions of the penton base were recovered. The peptide NNAGFL was the most abundant peptide isolated from the screen and is homologous to the NNAIV sequence found at position 185 to 189 and NNSG found at position 311 to 314 of the penton base. Other peptides isolated from the screen had little or no homology to the penton base, and the authors suggest that these represented discontinuous $\alpha_3\beta_1$ binding sites.

A more direct approach to identifying viral peptides that interact with cells is to express fragments of the virus surface protein, rather than random peptides, in the context of phage and screen for those that bind to cells. In this way, fragments of the rotavirus attachment protein VP4 were screened for their ability to bind to MA104 cells (Jolly *et al.*, 2001). Peptides spanning several overlapping and nonoverlapping portions of the protein were obtained. Upon testing, some of these peptides inhibited viral infection better than others. Interestingly, several of the regions identified by the screen had not been previously implicated in virus–receptor interactions. Conversely, a number of amino-acid residues that had been implicated in receptor binding by previous mutational analyses were not covered by peptides selected from the library. This might be due to different conformations adapted by the peptides in the context of phage. Peptides located within the hemagglutinating domain of VP4 and another with homology to galactose-binding lectins were recovered from the screen. Further experiments are required to further elucidate the possible roles of these sequences in virus–receptor interactions.

III. Targeting Viruses for Gene Therapy

Peptides derived from libraries have also been used to target viral vectors to specific cells for gene therapy purposes. The use of small peptides should not disrupt the overall structure of viral surface proteins to the extent of large molecules such as antibodies, and should therefore more likely result in functionally retargeted vectors. While this has generally been the case, there are examples of peptides that fail to retarget (Grifman *et al.*, 2001; Wu *et al.*, 2000), and there are examples of large molecules that have been used successfully for retargeting (Katane *et al.*, 2002; Martin *et al.*, 2002). Several reviews have been published recently, covering various aspects of viral vector targeting including the use of larger targeting molecules (Buning *et al.*, 2003; Larochelle *et al.*, 2002; Sandrin *et al.*, 2003; Wickham, 2003). This review will focus, therefore, on the application of short peptides derived from libraries to vector targeting.

A. Targeting Adenovirus

1. Cross-Linking Peptides to Adenovirus

Cross-linking peptides derived from library screening to adenovirus for retargeting has met with some success. In one case, peptides were obtained by screening phage on either primary normal human bronchial epithelial cells (NHBE) or small airway epithelial cells (SAEC) (Romanczuk *et al.*, 1999). Several phage were obtained, and the dodecameric peptides could be aligned into four different consensus sequence groups. Phage sss.10 showed specific binding to both NHBE and SAEC primary cells but did not bind to several cell lines tested (HeLa, 9L, COS, 293, and CV-1). Wild-type phage had the reverse binding properties, binding better to the cell lines than the primary cells. The peptide derived from another phage, sss.17, was chosen to crosslink to adenovirus for further analysis since it gave the highest binding on NHBE cells. Sss.10 and sss.17 do not share homology.

Sss.17 was cross-linked to Ad2 lysine residues via a polyethyleneglycol (PEG) crosslinker and shown to increase gene transduction of well-differentiated human airway epithelial cells. The increase was not observed using a scrambled sss.17 peptide sequence. An increase in transduction of HeLa cells by sss.17-Ad2 compared with PEG-treated wild-type Ad2 was not statistically significant. Other cell lines were not tested for specificity. However, the interaction with HeLa cells was blocked by competing Ad fiber knob protein, but the interaction with the well-differentiated airway epithelial cells was not, indicating the specificity of the sss.17 peptide for the airway epithelial cells.

A similar approach was taken to target Ad5 to chronic lymphocytic leukemia (CLL) cells. A library of 20-mers displayed on phage was screened on B-CLL cells (Takahashi et al., 2003). No consensus sequences were observed among selected peptides. Eight peptides bound nonspecifically to CLL cells, B cells, T cells, and monocytes; thirteen preferentially bound normal B and B-CLL cells compared with T cells and monocytes, and eight were relatively specific for CLL cells. Peptide 4-5*, a peptide that bound all cells tested, and peptide 1-5, a CLL-specific binding peptide, were each cross-linked to Ad5 using an SMCC cross-linker, and gene transfer was measured. Ad5/4-5* and Ad5/1-5 mediated 2- to 3-fold and 4- to 17-fold increased gene transfer efficiencies, respectively, compared with the Ad5/SMCC control on CLL cells. The authors remark that increasing the activity of these peptides by affinity-maturation using subsequent library screens will be essential for their further development.

2. Integrating Peptide-Coding Sequences into the Adenovirus Fiber Protein Gene

Integrating peptide-coding DNA into the genes for adenovirus surface fiber protein has been more commonly employed than the cross-linking procedure for retargeting. It involves a less complicated method of preparation than cross-linking, making it more suitable for clinical preparations. There are several issues that need to be addressed in undertaking this method. Is the targeting peptide compatible with the structure of the adenovirus? Will the peptide be in an exposed location where it can interact with a cell surface receptor? Will the peptide adapt an appropriate conformation at the site of insertion? Will the use of wild-type receptors be eliminated?

   a. Inserting Peptides into the Shaft Domain   In the strategy developed by Gaden et al., library-derived peptide sequences were transferred from phage and inserted between a shortened shaft domain and

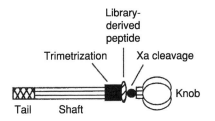

FIG 2. Phage-derived peptides were inserted between the shaft and the knob of the Ad fiber protein (Gaden *et al.*, 2004). The presence of the knob domain allowed increased virus growth, but also could be proteolytically removed by factor Xa cleavage to allow the virus to be targeted via the peptides rather than the cell-binding site of the knob domain. The neck region peptide of the lung surfactant protein was added as a trimerization signal that could maintain the trimeric shaft structure in the absence of the knob domain.

the knob domain of the fiber protein (Fig. 2) (Gaden *et al.*, 2004). The knob could be removed by factor Xa cleavage to reveal the targeting peptide. The phage were screened on the cystic fibrosis transmembrane conductance regulator (CFTR)-deficient human tracheal gland cell line CF-KM4. This cell line is not readily infected by Ad5 due to the absence of the CAR (coxsackievirus-adenovirus receptor) protein. The peptide QM10, which had little effect on virus binding, led to increased endocytosis, redirecting the Ad vector to a lysosomal/low pH compartment where it was proposed to result in increased gene delivery to the nucleus. This increased gene delivery was more efficient on several cell lines other than the CF-KM4 target of the screen, but it also enabled increased infection of the CF-KM4 cells compared with the nonliganded parental construct. The short shaft of both the liganded and nonliganded constructs led to significantly decreased titer compared with the long-shaft virus. Removal of the knob by proteolysis resulted in increased gene delivery efficiency via the QM10 peptide that was placed between the fiber and the knob. Thus, the knob domain was not required in the initial entry steps, where it normally plays an important role.

b. Inserting Peptides into the Knob Domain For adenoviruses, peptides derived from libraries have been most frequently inserted into the HI loop of the C-terminal knob domain of the fiber protein. This loop is not involved in CAR binding, since an HI loop-deleted protein still binds CAR (Roelvink *et al.*, 1999). Other deletion analyses suggested that some sequences within this loop were involved either directly or indirectly in fiber trimerization (Santis *et al.*, 1999). But trimerization was still observed when library-selected peptides targeting the human transferrin receptor (hTfR) were inserted there (Xia *et al.*, 2000). These nonamer peptides were obtained by screening a phage display library on the extracellular domain of hTfR protein. Out of 43 peptides isolated, 31 fell into three homology groups. Only a few had limited homology to transferrin sequences. Two peptides that were isolated multiple times were inserted into the HI loop. Gene delivery to hTfR-expressing cells by these targeted adenoviruses was inhibited by the presence of either transferrin or soluble hTfR extracellular domain in the medium, indicating the specificity of the targeting.

Using a slightly different strategy for phage library screening, the entire knob domain of Ad5 was expressed on the surface of phage λ via a phage D protein-knob fusion (Fontana *et al.*, 2003). This allowed the production of trimerized knob protein—the normal structure on the surface of the Ad5 particle. The phage were screened on CAR-negative NIH 3T3 cells. The CAR-binding site from amino acids 489–492 was deleted from the knob protein for screening. The 14 amino-acid random peptides flanked by cysteine residues were incorporated into the HI loop. Three homologous but nonidentical phage expressing peptides with the highest binding to 3T3 cells were transferred to adenovirus and shown to promote adenoviral gene delivery to 3T3 cells and CHO cells, which also do not express the CAR protein. Since gene transfer to the CAR-positive NMuLi cell line was not inhibited by soluble CAR, the peptides appeared to be promoting entry through a nonCAR pathway. Entry was blocked, however, by an RGD containing peptide, but not an RGE-containing peptide, indicating that the internalization of the peptide-modified virions still depended on integrins for entry. Two of the peptides also improved the efficiency of adenovirus-mediated gene delivery to mouse immature dendritic cells and human monocyte-derived immature dendritic cells.

Several studies have employed similar strategies to target vascular endothelial and vascular smooth muscle cells (Nicklin *et al.*, 2001b, 2004; Work *et al.*, 2004). Two 7-mer peptides were recovered from a phage-display screen on human umbilical vein endothelial cells (HUVECs) (Nicklin *et al.*, 2000). Originally, these peptides were tested in a two component system where they were encoded at the N-terminus of an antibody directed against the Ad knob domain. This targeted "adenobody" served as a bridge between the vector and the endothelial cells (Nicklin *et al.*, 2000). However, due to concerns about *in vivo* stability and the difficulties in developing a two component system into a drug preparation (Krasnykh *et al.*, 2000; Nicklin *et al.*, 2001b;

Wickham, 2000), the peptides were subsequently directly incorporated into the HI loop of the knob protein (Nicklin *et al.*, 2001b).

When incorporated into the HI loop, the SIGYPLP peptide was able to mediate increased specific transduction of primary vascular endothelial cells compared to primary vascular smooth muscle cells and the HepG2 liver cell line (Nicklin *et al.*, 2001b). But the other peptide (LSNFHSS) and the control Ad were most efficient on HepG2 cells, in contrast to its preference for HUVEC cells when attached to the adenobody. When the knob protein was mutated to decrease CAR binding, the SIGYPLP peptide was still able to mediate transduction of endothelial cells specifically, and was not inhibited by soluble knob protein, suggesting that a different receptor was being used. 12-mer peptides were also isolated by phage library screening on HUVECs (Nicklin *et al.*, 2004). When tested *in vivo*, the 12-mer targeted Ad vectors were distributed similarly compared with the controls, except for a statistically significant twofold increase in gene transduction in the abdominal aorta/vena cava by the MSLTTPPAVARP peptide.

Screening a 7-mer phage library on human saphenous vein smooth muscle cells (HSVSMCs) yielded 82 peptide sequences after four rounds (Work *et al.*, 2004). The two that were obtained more than once were inserted into the HI loop of Ad5 and shown to improve transduction of HSVSMCs compared with the control Ad5 by about twofold to fourfold. In contrast to the results with HSVSMCs, HSV endothelial cell transduction via EYHHYNK was decreased compared with a control. The question of whether an alternate receptor was being used was left unanswered since an anti-CAR antibody blocked neither the EYHHYNK-targeted SMC transduction nor the control Ad transduction.

B. Targeting Adeno-Associated Virus

1. Targeting Adeno-Associated Virus with Phage Library-Derived Peptides

Vasculature-targeting peptides have also been incorporated into the capsid of adeno-associated virus-2 (AAV-2). The insertion sites were chosen based on the homology of AAV-2 capsid to the canine parvovirus (CPV) capsid protein, the structure of which had been previously solved (Tsao *et al.*, 1991). The library-derived peptides were inserted after position 587 since molecular modeling suggested that this position should be hydrophilic and flexible (Girod *et al.*, 1999). Virus containing an insertion of an RGD-containing peptide at this position was shown to maintain a relatively high titer and was retargeted to an integrin receptor (Girod *et al.*, 1999).

The SIGYPLP peptide used to retarget Ad5 (discussed previously) was also inserted after position 587 of the AAV-2 capsid protein, and shown to enhance transduction of both HUVECs and human saphenous vein endothelial cells (HSVECs) (Nicklin *et al.*, 2001a). Transduction of HSVSMCs and HepG2 liver cells was reduced relative to the control wild-type AAV, indicating the specificity of the retargeting to ECs. The retargeted AAV did not bind to heparan sulphate proteoglycan (HSPG) nor was its infection of HUVECs blocked in the presence of HSPG. In contrast, the wild-type AAV demonstrated an HSPG interaction in both of these assays. Furthermore, transduction by the retargeted AAV was increased in the presence of bafilomycin A_2, which inhibits endosome acidification. Bafilomycin A_2 had the opposite effect on wild-type AAV. Thus, the retargeted AAV used an alternate entry pathway.

Another pair of EC-targeting peptides was recovered from a subtractive phage display library screening (White *et al.*, 2004). The library was first precleared on VSMCs, HepG2 cells, PBMCs and finally screened on HUVECs. The 2 out of 49 peptides that were isolated at the highest frequency, MTPFPTSNEANL (MTP) and MSLTTPPAVARP (MSL), were chosen for integration after residue 587. The resulting AAV-2 vectors gave transduction efficiencies similar to wild-type AAV on HUVECs but decreased efficiencies on HepG2 cells. In mice, 28 days after infusion, the MTP- and MSL-targeted vectors showed decreased levels of genomic particles in liver, spleen, lung, and aorta compared to wild-type, but increased levels in the vena cava. Transgene expression was also detected 28 days after infusion in cells at the luminal surface of the vena cava by MTP-AAV vector, but not wild-type AAV. Transgene expression was less in liver homogenates from MTP-AAV infused animals than wild-type infused controls. Heparin decreased transduction by wild-type and MSL AAV but increased MTP-AAV transduction.

The SMC-targeting peptide EYHHYNK used to retarget Ad5 (Work *et al.*, 2004) was also inserted after capsid protein position 587 of AAV-2. Enhanced transduction of HSVSMCs and human coronary artery smooth muscle cells (HCASMCs) by this construct was observed with a concomitant decrease in HSVEC transduction (Work *et al.*, 2004). Like the control, AAV, the peptide-targeted construct still bound to HSPG and infection was inhibited in the presence of heparin. But, unlike the control, the SMC-targeted construct was insensitive to proteosome inhibitors, suggesting possibly altered transport properties.

A peptide obtained from a phage screen on tumor angiogenic vasculature (NGRAHA) (Arap *et al.*, 1998; Pasqualini *et al.*, 2000) was inserted into AAV-2 at either position 449 or 588 of the capsid protein

(Grifman *et al.*, 2001). Position 449 is located within a region of AAV that binds neutralizing antibodies. Deletion of the putative loop (loop III) containing residue 449 had no effect on virus particle production but resulted in decreased transduction efficiency. The loop was predicted from a molecular model based on the CPV capsid structure. Insertion of a Myc epitope tag at 449 also demonstrated the surface accessibility of this region. The NGRAHA peptide receptor had previously been shown to be CD13 (Pasqualini *et al.*, 2000). When a peptide containing the NGRAHA sequence was incorporated at position 588 or in a deletion of loop IV (586–592), transduction correlated with the level of CD13 on the surface of the target cell—RD and KS1767 cells being the most susceptible. This was 10- to 20-fold the wild-type level of transduction. These NGR-containing viruses were able to bind to heparin. Other targeting peptides isolated from libraries were unable to retarget AAV when inserted at position 588 (Grifman *et al.*, 2001). Thus, simply transferring a peptide from a bacteriophage to an AAV particle does not always lead to functional retargeting.

2. *Adeno-Associated Virus Libraries*

The complex structural interactions required for AAV virus particle formation, target cell binding, and gene delivery place stringent demands on targeting peptides. Therefore, Perabo *et al.*, developed an alternative to the indirect phage screen (Perabo *et al.*, 2003). In this study, random heptapeptides were inserted directly into position 587 of capsid (Cap) proteins of AAV particles and screened for their ability to functionally target AAV to desired cell types (Fig. 3, Strategy B). To do this, plasmids expressing Rep and mutated Cap genes located between AAV ITR elements were cotransfected with a plasmid providing adenovirus helper functions into 293 cells. Virus was then harvested from these cells and tested on target cells. If a capsid protein containing a library peptide was compatible with proper protein folding, incorporation into an AAV particle, cell binding and entry, and gene delivery, then the Rep/Cap minigenome could be rescued by coinfecting adenovirus through subsequent rounds of infection. After several rounds of infection and rescue, certain peptide sequences began to dominate the culture as determined by sequencing viral DNA. Screening the library on a megakaryocytic cell line (M-07e) yielded two peptides containing the RGD sequence. Screening on a B-cell chronic lymphocytic leukemia (B-CLL)-derived cell line yielded different peptide sequences. Inhibition experiments on three different cell lines (M-07e, CO-115 and HeLa) using heparin indicated that three of the four mutants were not using the wild-type AAV receptor heparan sulfate proteoglycan as a receptor,

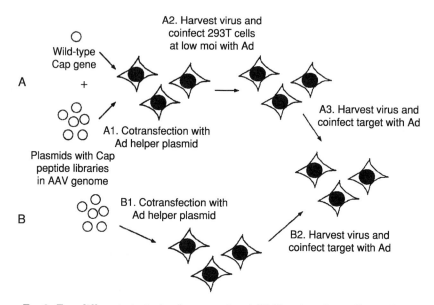

Fig 3. Two different strategies for screening AAV libraries of peptides within the Capsid (Cap) protein of AAV. In the upper strategy, A, the library screened on target cells was produced after two steps in order to minimize mismatching the mutated Cap protein and its corresponding gene in the same AAV particle (Muller *et al.*, 2003). In the lower strategy, B, target cells were screened directly with the library produced from cells cotransfected with the AAV plasmid library and Ad helper plasmid (Perabo *et al.*, 2003).

consistent with the disruption of the HSP binding site by the peptide insertion. An RGD containing peptide was able to inhibit both of the M-07e-targeted capsids and also one of the B-CLL-targeting capsids (MecA). The other B-CLL targeted capsid (MecB) was not inhibited by RGD, but was inhibited by heparin on the cell lines. In contrast to wildtype AAV, MecA and MecB were able to infect primary B-CLL cells. No inhibition by heparin was observed on the primary cells.

Müller *et al.* took a slightly different approach (Muller *et al.*, 2003) (Fig. 3, Strategy A). The first step in the protocol developed by Perabo *et al.* was a transfection of the library along with adenovirus helper genes followed by infection of target cells by resulting AAV particles, said infection depending on a functional targeting peptide sequence. A complication of this approach is that multiple library plasmids can be transfected into a single producer cell. This could result in AAV particles with a particular functional, peptide-containing Cap protein on their surface not encapsulating the corresponding *cap* gene. As a result, the functional gene could be lost. In contrast, the first step of Müller *et al.* entailed cotransfection of peptide library and adenovirus helper plasmids with a plasmid encoding wild-type Cap and lacking IST sequences. AAV from these cells were subsequently used to infect 293T cells at low m.o.i. This initial library transfer step, therefore, depended on a wild-type Cap protein rather than a functional library peptide. Functional screening began with virus from the second population of 293T AAV producer cells where the mismatching of *cap* genes and Cap proteins had been minimized due to the low m.o.i. infection. Multiple rounds of infection on target cells were carried out in the presence of Ad5 helper virus.

Using this modified protocol, Müller *et al.* screened two different libraries on primary human coronary artery endothelial cells resulting in the isolation of three predominant peptide consensus sequence motifs. NSVRDL^{G}/s and PRSVTVP were isolated after three rounds from one library, and NSVSSX^{S}/A was isolated from the second library after two rounds. As a control, none of these sequences was observed after a single round of screening on Kasumi acute myeloid leukemia cells. Infection and gene transduction of the primary endothelial cells were improved compared with the wild-type AAV-2. Improvement of transduction varied from 4- to 40-fold depending on the peptide. Infection of HeLa cells by the library-derived isolates was less efficient compared to wild-type. Tail-vein injection of the NSSRDLG isolate lead to fivefold increased reporter gene expression in the heart when compared to wild-type. Expression in the liver appeared to be lower than wild-type. Expression in the heart was greater than all other tissues examined. A peptide corresponding to one of the peptides isolated in the first round, NDVRAVS, inhibited transduction of primary endothelial cells by AAV expressing NDVRAVS-containing capsid approximately 10-fold indicating virus-cell interaction was mediated by the peptide.

C. Retrovirus Libraries

Difficulties in retargeting retroviruses by inserting previously characterized targeting peptides into the surface envelope (Env) protein have been encountered. Several different peptide library-screening approaches involving peptide display directly on retroviruses have therefore been used to examine and alter retroviral entry properties. They have been used to characterize the determinants of cell tropism, the R-peptide cleavage site of gibbon ape leukemia virus (GaLV) Env, and to retarget retroviruses either by selecting for protease-cleavable linkers connecting targeting ligands to Env or by altering cell-targeting sequences to target novel receptors.

1. Protease-Cleavable Linkers

It has previously been shown that the attachment of epidermal growth factor (EGF) to the N-terminus of MuLV Env proteins enabled the binding of viruses coated with these proteins to cells expressing the EGF receptor (EGFR) (Cosset et al., 1995). However, the interaction with the normal Env receptor was blocked, and the viruses were ultimately unable to deliver genes to the cells. A multistep strategy was therefore designed, whereby the viruses would bind to the EGFR-expressing cells, a protease sensitive linker between EGF and Env would be cleaved, and the resultant Env would be able to enter EGFR-expressing cells via its normal receptor (Nilson et al., 1996; Peng et al., 1997).

In order to obtain protease sensitive linker sequences, libraries of MuLV Env proteins with random peptides linking EGF to the N-terminus of Env were screened for their ability to infect EGFR-expressing HT1080 cells. The screen was carried out using ecotropic Moloney MuLV (Mo-MuLV) on HT1080 cells expressing a transfected ecotropic receptor (MCAT) (Buchholz et al., 1998), or using the amphotropic 4070A Env protein that could use the amphotropic receptor present on HT1080 cells after cleavage of the peptide linker sequence (Schneider et al., 2003). Plasmids with replication–competent retroviral genomes containing the substituted env genes were transfected into cells, and viruses were allowed to spread on appropriate target cells. The Mo-MuLV construct was first transfected into 293 cells that were then mixed with target HT1080 cells expressing MCAT. Mo-MuLV does not infect 293 human cells, and these were subsequently eliminated by treatment of the coculture with G418 to which the HT1080-MCAT cells were resistant. In the subsequent study using the amphotropic 4070A Env, the libraries were transfected directly into HT1080 cells. After several days in culture, viral RNA was reverse transcribed and amplified, then subcloned and sequenced to examine the evolution of selected linker sequences. Viruses with linker peptides that were efficiently cleaved by cellular proteases could propagate through the culture, and these sequences were obtained with higher frequency upon subcloning.

In the initial study with ecotropic virus (Buchholz et al., 1998), the library peptides had been totally randomized except for some bias to reduce the amount of termination codons. This result led to the somewhat surprising result that the protease(s) cleaving the selected peptides appeared to be Golgi-localized proteases such as furin since there were positively charged residues at positions corresponding to the consensus cleavage sites of these proteases. As a result, rather than having EGF removed upon cell binding, virus particles were actually produced from cells with EGF already removed.

In the second study employing the amphotropic Env, the random substituted linker was based on a matrix metalloprotease (MMP) substrate (Schneider *et al.*, 2003). Three of the six residues surrounding the cleavage site were randomized in two separate sets of screening experiments on human HT1080 cells. In the first set, the two residues at the cleavage site and an upstream proline were held constant, and the occurrence of positively charged residues was suppressed. In the second, the upstream proline and the G and L residues framing the cleavage site were randomized within the consensus sequence PQGLY^K/Q obtained in the first screen. A and Q were made the two alternative choices in the library in place of ^{K}/Q. The P, G, and L residues were the most frequently recovered at their respective positions in this screen, and A predominated at the "^{K}/Q" position.

EGF-Env proteins carrying the linker sequences PQGLYA and PQGLYQ were studied in further detail. EGF was cleaved from these proteins when they were expressed in cultures of MMP positive HT1080 cells but not MMP negative 293T cells. MMP-2 added to virus produced from the 293T cells was able to cleave the constructs appropriately. The constructs were also found to spread through a culture of HT1080 cells more rapidly than a previously characterized control construct bearing the MMP-sensitive linker PLGLWA.

2. Targeting Murine Leukemia Virus

Bahrami *et al.* (2003) used libraries of Akv MuLV to examine the structural constraints of amino acids R85 and D86 in the receptor binding domain of Env. These residues had previously been shown to be important in cell binding of Mo-MuLV and Friend MuLV, respectively (Davey *et al.*, 1999; MacKrell *et al.*, 1996). Libraries of Env proteins encoding combinations of residues at these positions were tested for their ability to spread through a population of NIH/3T3 cells that expressed Gag and Pol proteins (a "semi-packaging" cell line) from a transfected plasmid. The Env proteins were expressed from a packageable, retroviral cassette that expressed an upstream G418 resistance gene and a downstream substituted *env* gene that were separated by an IRES element (Fig. 4). This library of plasmids was first transfected into the BOSC 23 cell line, expressing Gag, Pol, and the Mo-MuLV Env, allowing the initial transfer of library *env* genes to the NIH/3T3 semi-packaging cells.

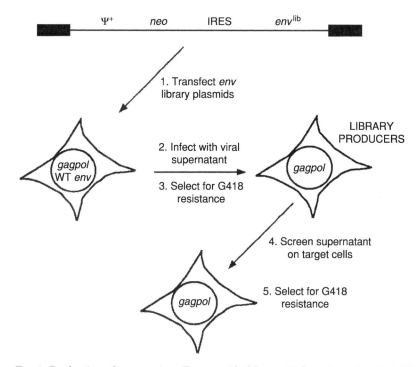

Fig 4. Production of a retrovirus Env peptide library (Bahrami *et al.*, 2003). The packageable (Ψ^+) retroviral cassette vector expresses the library of peptide substituted *env* genes downstream of an IRES (internal ribosome entry site) element. The upstream gene is the selectable marker for G418 resistance (*neo*). Black boxes are LTR (long terminal repeat) sequences. A library of these cassettes was transfected into a packaging cell line expressing Gag and Pol and wild-type Mo-MuLV Env. The virus produced from this cell line was used to infect semi-packaging cells expressing only Gag and Pol. After drug selection, viruses expressing library Env proteins on their surfaces and packaging env genes were tested for their ability to infect target cells. *Env* genes isolated from G418 resistant targets were further characterized.

The second population of cells was selected for G418 resistance to yield constitutive producer cells. Supernatant from these producers was then harvested and used to infect a target population of the semi-packaging cells. Successful infection in this round depended on functional Env protein variants. Neo^R clones derived from this infection were analyzed to determine the amino-acid substitutions at positions 85 and 86. The analysis revealed that arginine, hydrophilic, and hydrophobic residues could substitute for R85. Hydrophobic and positively charged residues were excluded from position D86. The functionality of these substituted Env proteins was confirmed by titering particles expressing them on their surface. Different clones with the same substitution yielded producer cells with different titers. For example, R85S + D86 yielded two clones that were 1.0×10^4 and 2.0×10^6 neo^R cfu per ml compared to the R85, D86 control that was 5.6×10^6 per ml. 1.0×10^4 was the lowest titer reported among the obtained mutants. These results were consistent with the earlier, more limited mutagenesis studies (Davey *et al.*, 1999; MacKrell *et al.*, 1996). Sequence alignment indicated that R85 of Akv Env corresponds to D84 of Mo-MuLV Env. The earlier study by MacKrell *et al.* showed that the charge-reversing Mo-MuLV D84K change was not permitted, but D84E was permitted, consistent with the lack of charge switching at this position observed in the Akv library screen. D84V resulted in a lower but measurable titer ($<1\%$) in Mo-MuLV. In Friend MuLV Env (Davey *et al.*, 1999), D86N gave 40% wild-type infection level, D86A gave 0.4%, D86V and D86W were both $<0.1\%$ of wild-type levels.

A similar approach was used to alter the tropism of the SL3-2 polytropic MuLV Env protein. In contrast to the closely related MCF-247 polytropic Env, SL3-2 does not infect human or mink cells. Both infect murine cells. Exchanging segments that varied between the two Env proteins indicated that the VR3 region (amino acids 199–213) just upstream of the proline rich region was responsible for the ability of MCF-247 to infect human cells. The involvement of this region in receptor interactions had been previously suggested (Battini *et al.*, 1995; Ott and Rein, 1992). An Env library with substitutions at each of the 5 positions in this region where MCF-247 and SL3-2 diverged was screened on human TE671 cells. In this case, there were only two substitutions at each position: the alternate amino acids from either MCF-247 or SL3-2. After selection, the residues recovered at positions 199, 200, and 202 were equally divided between the two possibilities. But at positions 212 and 213 the residues from MCF-247, G212, and I213, were favored by a 9:1 ratio indicating that positions 212 and 213 determined human-cell tropism.

A second library was then screened where positions 212 and 213 were randomized in the SL3-2 backbone. The most frequent combination of amino acids recovered from the screen on TE671 cells was MV (6 out of 29). The titers of this construct on TE671 and mink CCL-64 cells were comparable to the construct with G212/I213 (10^6–10^7 per ml) —higher than the titer of MCF-247 on these cells. The difference on TE671 cells was the most pronounced, at nearly five orders of magnitude higher than MCF-247. Interference assays on cells pre-infected with MCF-247 indicated that the GI, MV and LI substituted constructs were using the polytropic receptor on NIH/3T3, CCL-64, and TE671 cells.

3. Targeting Feline Leukemia Virus

In order to retarget retroviruses to specific cell types for gene therapy, it is essential to be able to have them target cell-specific receptor proteins that are not normally utilized by the wild-type Env protein. Feline leukemia virus subgroup A (FeLV-A) preferentially infects feline cells, FeLV-C infects cells of many species including humans, and they use different receptors for entry. It had been demonstrated that the tropism and receptor usage of FeLV-A could be switched to those of FeLV-C by substituting a 19 amino acid Vr-1 sequence near the N-terminus of FeLV-A by a 16-amino acid sequence from FeLV-C (Rigby et al., 1992). These experiments suggested the possibility that introduction of targeting sequences into this region might direct the FeLV-A Env to novel receptors. Since the insertion of potential targeting peptides obtained from other sources such as phage libraries might not lead to functional retargeting of the Env protein due to conformational constraints, an alternative was to select Env proteins from a library of retroviruses with random peptide substitutions in Vr1.

The Env-expressing construct employed in this library screening is shown at the top of Fig. 5. It is a bicistronic, packageable retroviral cassette. The upstream gene is the substituted *env* gene and the downstream gene is the *neo* drug-resistance marker. Ten random amino acids were substituted in Vr1 with a central R held constant since a positively charged residue was found in most naturally occurring FeLV-A and -C isolates. The protein otherwise is a hybrid of FeLV-A and FeLV-C. The backbone sequence is FeLV-A as is the KY sequence downstream of the random peptides. A 3 amino-acid insertion within this region of FeLV-A compared with FeLV-C was deleted. The K commonly found in FeLV-A was replaced by R which is found in FeLV-C isolates. Thus, the construct is neither FeLV-A nor FeLV-C and this has likely been important in directing library derived Env proteins to nonFeLV receptors.

The initial library screen (not shown) was on AH927 feline fibroblast cells (Bupp and Roth, 2002). Retroviruses expressing the library of Env proteins on their surfaces were produced by transient transfection of two plasmids into 293T cells. One encoded Gag and Pol, and the other was the Env library vector. Supernatant from this transfection was harvested, and used to inoculate AH927 cells which were then selected for G418 resistance. The only functional isolate recovered preferentially infected D17 canine osteosarcoma cells using a FeLV-C receptor. The

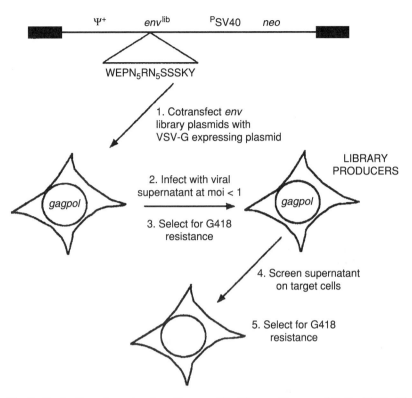

F<small>IG</small> 5. Production of a retrovirus Env peptide library (Bupp and Roth, 2003). The packageable (Ψ^+) retroviral cassette vector expressed the library of peptide substituted *env* genes downstream of retroviral LTR sequences (black boxes). Randomized residues are denoted by N. The W corresponds to W52 of the mature 61E FeLV-A protein (Donahue *et al.*, 1988). The selectable marker for G418 resistance (*neo*) is expressed from the SV40 promoter. A library of these cassettes was cotransfected with a VSV-G protein expressing plasmid into a cell line expressing Gag and Pol. Virus produced from this cell line was used to infect cells expressing Gag and Pol. After drug selection, viruses expressing library Env proteins on their surfaces and copackaging *env* genes were tested for their ability to infect target cells. *Env* genes isolated from G418 resistant targets were further characterized.

ability to switch the cellular tropism from preferentially infecting feline fibroblasts to being specific for canine osteosarcoma cells was an important advance.

Constitutive library producer cells were subsequently employed (Bupp and Roth, 2003), as shown in Fig. 5. These were created by transfecting 293T cells expressing Gag and Pol proteins with the library plasmids and a plasmid encoding the VSV-G protein. The next step was to harvest virus-containing supernatant and inoculate cells expressing Gag and Pol proteins at an m.o.i. <1. In comparison to the transient transfection system where multiple Env proteins could be produced from a single transiently transfected cell, only one or two Env proteins will be produced from a single constitutive producer cell, decreasing the likelihood that there will be a mismatch between the surface-expressed protein and the packaged *env* cassette.

These constitutive producer cell lines were used to screen libraries on AH927 cells, D17 cells, and 143B human osteosarcoma cells (Bupp and Roth, 2003; Bupp *et al.*, 2005). These cells were inoculated with supernatant from producer cells and subsequently selected for G418 resistance. The cassettes delivered to the target cells were then transferred to the TELCeB6 lacZ marker cell line to examine titers on various cells and to perform interference assays. The screen on D17 cells yielded multiple Env proteins that infected D17 cells specifically through a FeLV-C receptor. The screen on AH927 cells resulted in two isolates that did not use FeLV receptors or the amphotropic 4070A receptor. The screen on 143B cells yielded a single isolate that only infected human cells through a nonFeLV, non4070A receptor. Transduction efficiency on 143B cells was approximately 10^4 lacZ cfu per ml, and on 293T it was 10^5 per ml. Other cell lines tested were minimally infected. These results indicate the potential of this technique to retarget retroviruses to desired cell types.

Comparisons of the targeting peptides of these Env proteins failed to reveal little, if any, sequence conservation, even among proteins targeting the same cells using the same receptor (Bupp *et al.*, 2005). The exact role of these sequences in binding and entry remains to be elucidated. However, mutagenesis studies of the R that was held constant in the middle of the randomize peptide indicated different structural requirements for this residue in different targeted proteins. In a D17 targeted isolate, various hydrophilic and charged residues including D could substitute for R, albeit with lower resulting titers. But in an AH927 targeted isolate, only K could substitute. Identification of the receptors used by the AH927- and 143B-targeted Env proteins using well-developed techniques (Quigley *et al.*, 2000) will yield important insights into Env-receptor interactions.

4. Pseudotyping HIV-1 with GaLV Env

The ability to pseudotype HIV-1 lentiviral vectors with nonHIV viral surface proteins confers the targeting properties of the pseudotyping protein onto the vector (see (Larochelle *et al.*, 2002; Sandrin

et al., 2003)). However, pseudotyping by the native Gibbon ape leukemia virus (GaLV) Env protein yields defective particles with low levels
of Env incorporated into the virion without R peptide cleavage. The
defect can be corrected by deletion of the cytoplasmic C-terminal R
peptide, substitution of the C-terminus by MuLV sequences, or by
substituting three amino acids surrounding the R peptide cleavage
site of GaLV by MuLV sequences (Christodoulopoulos and Cannon,
2001; Stitz *et al.*, 2000).

A more detailed examination of the sequence requirements for GaLV
R peptide cleavage was carried out using libraries of GaLV Env proteins (Merten *et al.*, 2005). The three positions closest to the cleavage
site that differ from GaLV and MuLV were substituted by random
amino acids. The substituted Env variants were expressed from a
bicistronic, packageable HIV-1 vector, where the upstream HA-tagged
GaLV *env* gene was separated from the downstream *neo* gene by an
IRES element. The HA tag had no effect on infectivity. In order to
ensure that only one or two *env* genes would be present in a single
library producer cell for transfer, the retrovirus library was made
using a two-step procedure where 293T cells were transiently transfected with the library vector, an HIV-1 Gag-Pol packaging construct
and a plasmid expressing the VSV-G protein. The resulting supernatant was used to infect 293 cells, which were then selected for G418
resistance. The resultant cell line contained only the library vectors.
For library screening, these cells were transfected with the packaging
construct to produce the virus library and cocultivated, but separated
from HT1080 target cells by a permeable membrane in a transwell.

The four substituted variants isolated from the library after G418
selection of the HT1080 cells had approximately 10-fold lower titers
than the Env proteins with a deleted R peptide (ΔR) or with the three
GaLV cleavage site residues substituted by corresponding MuLV residues (RTM). Wild-type GaLV Env was nonfunctional. As with the ΔR
and RTM mutants, viral incorporation was higher for the library
derived Env proteins than the wild-type Env. One of the four variants
showed an increase in R peptide cleavage relative to RTM. The other
three were cleaved at a comparable level to RTM, but the cleavage site
appeared altered. Since it had previously been shown that only a low
number (7–14 trimers) of HIV Env proteins onto HIV particles was
required for entry (Chertova *et al.*, 2002), Merten *et al.* concluded that
the difference in titers between the library-derived Env proteins and
wild-type was due to increased R-peptide processing rather than
increased Env incorporation (Merten *et al.*, 2005). While this procedure did not result in increased infection efficiency, it demonstrates the

sequence variations possible at the R peptide cleavage site and, in one case, actually led to increased R cleavage relative to the RTM mutant.

IV. Conclusions

Peptide libraries have become useful tools for examining and altering the interactions between viruses and their cellular receptors. The recovery of peptides that bind to viruses and that share homology with cellular proteins has resulted in the identification of new viral receptor candidates. Peptides obtained by screening libraries on cells or virus receptors have yielded new data concerning viral sequences that may be involved in receptor binding. Further studies of these proposed interactions will be necessary to confirm the results.

Several different approaches employing peptide libraries to alter the properties of gene therapy vector surface proteins have been developed. Initially, library-derived peptides were either cross-linked to viral vectors or they were inserted into the coding regions of the surface proteins. However, the conformations of peptides upon transfer from phage coat proteins to viral gene therapy vectors are not always maintained, and therefore this strategy has met with some failure. More direct approaches involving the construction of peptide libraries within the surface proteins of the therapeutic viral vectors themselves have, therefore, been successfully developed. Although some experiments using the altered vectors on primary cells and animals have been performed, most studies have involved cell lines. Clinical application of these vectors for therapeutic purposes remains for future development.

Acknowledgment

This work was supported by award 1RO1 CA49932–11 from the National Institutes of Health to M. J. R.

References

Arap, W., Pasqualini, R., and Ruoslahti, E. (1998). Cancer treatment by targeted drug delivery to tumor vasculature in a mouse model. *Science* **279:**377–380.

Bahrami, S., Jespersen, T., Pedersen, F. S., and Duch, M. (2003). Mutational library analysis of selected amino acids in the receptor binding domain of envelope of Akv murine leukemia virus by conditionally replication competent bicistronic vectors. *Gene* **315:**51–61.

Battini, J. L., Danos, O., and Heard, J. M. (1995). Receptor-binding domain of murine leukemia virus envelope glycoproteins. *J. Virol.* **69:**713–719.

Buchholz, C. J., Peng, K. W., Morling, F. J., Zhang, J., Cosset, F. L., and Russell, S. J. (1998). *In vivo* selection of protease cleavage sites from retrovirus display libraries. *Nature Biotechnol.* **16:**951–954.

Buning, H., Ried, M., Perabo, L., Gerner, F. M., Huttner, N. A., Enssle, J., and Hallek, M. (2003). Receptor targeting of adeno-associated virus vectors. *Gene Ther.* **10:**1142–1151.

Bupp, K., and Roth, M. J. (2002). Altering retroviral tropism using a random-display envelope library. *Mol. Ther.* **5**(3):329–335.

Bupp, K., and Roth, M. J. (2003). Targeting a retroviral vector in the absence of a known cell-targeting ligand. *Hum. Gene Ther.* **14:**1557–1564.

Bupp, K., Sarangi, A., and Roth, M. J. (2005). Probing sequence variation in the receptor-targeting domain of feline leukemia virus envelope proteins with peptide display libraries. *J. Virol.* **79**(3):1463–1469.

Chertova, E., Bess, J. W., Crise, B. J., Sowder, R. C., Schaden, T. M., Hilburn, J. M., Hoxie, J. A., Benveniste, R. E., Lifson, J. D., Henderson, L. E., and Arthur, L. O. (2002). Envelope glycoprotein incorporation, not shedding of surface envelope glycoprotein (gp120/SU), is the primary determinant of SU content of purified human immunodeficiency virus type 1 and simian immunodeficiency virus. *J. Virol.* **76**(11):5315–5325.

Christodoulopoulos, I., and Cannon, P. M. (2001). Sequences in the cytoplasmic tail of the Gibbon ape leukemia virus envelope protein that prevent its incorporation into lentivirus vectors. *J. Virol.* **75:**4129–4138.

Cosset, F. L., Morling, F. J., Takeuchi, Y., Weiss, R. A., Collins, M. K. L., and Russell, S. J. (1995). Retroviral retargeting by envelopes expressing an N-terminal binding domain. *J. Virol.* **69:**6314–6322.

Cwirla, S. E., Peters, E. A., Barrett, R. W., and Dower, W. J. (1990). Peptides on phage: A vast library of peptides for identifying ligands. *Proc. Natl. Acad. Sci. USA* **87:**6378–6382.

Davey, R. A., Zuo, Y., and Cunningham, J. M. (1999). Identification of a receptor-binding pocket on the envelope protein of Friend murine leukemia virus. *J. Virol.* **73:**3758–3763.

Devlin, J. J., Panganiban, L. C., and Devlin, P. E. (1990). Random peptide libraries: A source of specific protein binding molecules. *Science* **249:**404–406.

Ferrer, M. (1999). Peptide ligands to human immunodeficiency virus type 1 gp120 identified from phage display libraries. *J. Virol.* **73**(7):5795–5802.

Fontana, L., Nuzzo, M., Urbanelli, L., and Monaci, P. (2003). General strategy for broadening adenovirus tropism. *J. Virol.* **77:**11094–11104.

Gaden, F., Franqueville, L., Magnusson, M. K., Hong, S. S., Merten, M. D., Lindholm, L., and Boulanger, P. (2004). Gene transduction and cell entry pathway of fiber-modified adenovirus type 5 vectors carrying novel endocytic peptide ligands selected on human tracheal glandular cells. *J. Virol.* **78**(13):7227–7247.

Geysen, H. M., Meloen, R. H., and Barteling, S. J. (1984). Use of peptide synthesis to probe viral antigens for epitopes to a resolution of a single amino acid. *Proc. Natl. Acad. Sci. USA* **81:**3998–4002.

Girod, A., Ried, M., Wobus, C., Lahm, H., Leike, K., Kleinschmidt, J., Deleage, G., and Hallek, M. (1999). Genetic capsid modificatins allow efficient re-targeting of adeno-associated virus type 2. *Nature Med.* **5**(9):1052–1056.

Grifman, M., Trepel, M., Speece, P., Gilbert, L. B., Arap, W., Pasqualini, R., and Weitzman, M. D. (2001). Incorporation of tumor-targeting peptides into recombinant adeno-associated virus capsids. *Mol. Ther.* **3**(6):964–975.

Hu, C., Zhang, P., Liu, X., Qi, Y., Zou, T., and Xu, Q. (2004). Characterization of a region involved in binding of measles virus H protein and its receptor SLAM (CD150). *Biochm. Biophys. Res. Commun.* **316:**698–704.

Jolly, C. L., Huang, J. A., and Holmes, I. H. (2001). Selection of rotavirus VP4 cell receptor binding domains for MA104 cells using a phage display library. *J. Virol. Meth.* **98:**41–51.

Katane, M., Takao, E., Kubo, Y., Fujita, R., and Amanuma, H. (2002). Factors affecting the direct targeting of murine leukemia virus vectors containing peptide ligands in the envelope protein. *EMBO Rep.* **3:**899–904.

Krasnykh, V. N., Douglas, J. T., and van Beusechem, V. W. (2000). Genetic targeting of adenoviral vectors. *Mol. Ther.* **1**(5)**:**391–405.

Lam, K. S., Salmon, S. E., Hersh, E. M., Hruby, V. J., Kazmierski, W. M., and Knapp, R. C. (1991). A new type of synthetic peptide library for identifying ligand-binding activity. *Nature* **354:**82–84.

Larochelle, A., Peng, K.-W., and Russell, S. J. (2002). Lentiviral vector targeting. *Curr. Top. Microbiol. Immunol.* **261:**143–163.

Mac Krell, A. J., Soong, N. W., Curtis, C. M., and Anderson, W. F. (1996). Identification of a Subdomain in the Moloney Murine Leukemia Virus Envelope Protein Involved in Receptor Binding. *J. Virol.* **70:**1768–1774.

Martin, F., Chowdhury, S., Neil, S., Phillipps, N., and Collins, M. K. (2002). Envelope-targeted retrovirus vectors transduce melanoma xenografts but not spleen or liver. *Mol. Ther.* **5:**269–274.

Mathias, P., Wickham, T., Moore, M., and Nemerow, G. (1994). Multiple adenovirus serotypes use av integrins for infection. *J. Virol.* **68**(10)**:**6811–6814.

Medina-Kauwe, L. K. (2003). Endocytosis of adenovirus and adenovirus capsid proteins. *Adv. Drug Deliv. Rev.* **55:**1485–1496.

Merten, C. A., Stitz, J., Braun, G., Poeschla, E. M., Cichutek, K., and Buchholz, C. J. (2005). Directed evolution of retrovirus envelope protein cytoplasmic tails guided by functional incorporation into lentivirus particles. *J. Virol.* **79:**834–840.

Müller, O. J., Kaul, F., Weitzman, M. D., Pasqualini, R., Arap, W., Kleinschmidt, J. A., and Trepel, M. (2003). Random peptide libraries displayed on adeno-associated virus to select for targeted gene therapy vectors. *Nat. Biotechnol.* **21:**1040–1046.

Nicklin, S. A., Buening, H., Dishart, K. L., de Alwis, M., Girod, A., Hacker, U., Thrasher, A. J., Ali, R. R., Hallek, M., and Baker, A. H. (2001a). Efficient and selective AAV-2-mediated gene transfer directed to human vascular endothelial cells. *Mol. Ther.* **4:**174–181.

Nicklin, S. A., Von Seggern, D. J., Work, L. M., Pek, D. C. K., Dominczak, A. F., Nemerow, G. R., and Baker, A. H. (2001b). Ablating adenovirus type 5 fiber-CAR binding and HI loop insertion of the SIGYPLP peptide generate an endothelial cell-selective adenovirus. *Mol. Ther.* **4:**534–542.

Nicklin, S. A., White, S. J., Nicol, C. G., von Seggern, D. J., and Baker, A. H. (2004). *In vitro* and *in vivo* characterization of endothelial cell selective adenoviral vectors. *J. Gene Med.* **6:**300–308.

Nicklin, S. A., White, S. J., Watkins, S. J., Hawkins, R. E., and Baker, A. H. (2000). Selective targeting of gene transfer to vascular endothelial cells by use of peptides isolated by phage display. *Circulation* **102:**231–237.

Nilson, B. H. K., Morling, F. J., Cosset, F.-L., and Russell, S. J. (1996). Targeting of retroviral vectors through protease-substrate interactions. *Gene Ther.* **3:**280–286.

Ott, D., and Rein, A. (1992). Basis for Receptor Specificity of Nonecotropic Murine Leukemia Virus Surface Glycoprotein gp70^{SU}. *J. Virol.* **66:**4632–4638.

Pasqualini, R., Koivunen, E., Kain, R., Lahdenrants, J., Sakamoto, M., Stryhn, A., Ashmun, R. A., Shapiro, L. H., Arap, W., and Ruoslahti, E. (2000). Aminopeptidase N is a receptor for tumor-homing peptides and a target for inhibiting angiogenesis. *Cancer Res.* **60:**722–727.

Peng, K.-W., Morling, F. J., Cosset, F.-L., Murphy, G., and Russell, S. J. (1997). A gene delivery system activatable by disease-associated matrix metalloproteinases. *Hum. Gene Ther.* **8:**729–738.

Perabo, L., Buning, H., Kofler, D. M., Ried, M. U., Girod, A., Wendtner, C. M., Enssle, J., and Hallek, M. (2003). *In vitro* selection of viral vectors with modified tropism: The adeno-associated virus display. *Mol. Ther.* **8:**151–157.

Pulli, T., Koivunen, E., and Hyypia, T. (1997). Cell-surface interactions of echovirus 22. *J. Biol. Chem.* **272:**21176–21180.

Quigley, J. G., Burns, C. C., Anderson, M. M., Lynch, E. D., Sabo, K. M., Overbaugh, J., and Abkowitz, J. L. (2000). Cloning of the cellular receptor for feline leukemia virus subgroup C (FeLV-C), a retrovirus that induces red cell aplasia. *Blood* **95:**1093–1099.

Ramanujam, P., Tan, W. S., Nathan, S., and Yusoff, K. (2002). Novel peptides that inhibit the propagation of Newcastle disease virus. *Arch. Virol.* **147:**981–993.

Rigby, M. A., Rojko, J. L., Stewart, M. A., Kociba, G. J., Cheney, C. M., Rezanka, L. J., Mathes, L. E., Hartke, J. R., Jarrett, O., and Neil, J. C. (1992). Partial dissociation of subgroup C phenotype and *in vivo* behaviour in feline leukaemia viruses with chimeric envelope genes. *J. Gen. Virol.* **73:**2839–2847.

Roelvink, P. W., Mi, L. G., Einfeld, D. A., Kovesdi, I., and Wickham, T. J. (1999). Identification of a conserved receptor-binding site on the fiber proteins of CAR-recognizing adenoviridae. *Science* **286:**1568–1571.

Romanczuk, H., Galer, C. E., Zabner, J., Barsomian, G., Wadsworth, S. C., and O'Riordan, C. R. (1999). Modification of an adenoviral vector with biologically selected peptides: A novel strategy for gene delivery to cells of choice. *Hum. Gene Ther.* **10:**2615–2626.

Salone, B., Martina, Y., Piersanti, S., Cundari, E., Cherubini, G., Franqueville, L., Failla, C. M., Boulanger, P., and Saggio, I. (2003). Integrin a3b1 is an alternative cellular receptor for adenovirus serotype 5. *J. Virol.* **77:**13448–13454.

Sandrin, V., Russell, S. J., and Cosset, F.-L. (2003). Targeting retroviral and lentiviral vectors. *Curr. Top. Microbiol. Immunol.* **281:**137–178.

Santis, G., Legrand, V., Hong, S. S., Davison, E., Kirby, I., Imler, J. L., Finberg, R. W., Bergelson, J. M., Mehtali, M., and Boulanger, P. (1999). Molecular determinants of adenovirus serotype 5 fibre binding to its cellular receptor CAR. *J. Gen. Virol.* **80:**1519–1527.

Schneider, R. M., Medvedovska, Y., Hartl, I., Voelker, B., Chadwick, M. P., Russell, S. J., Cichutek, K., and Buchholz, C. J. (2003). Directed evolution of retroviruses activatable by tumour-associated matrix metalloproteases. *Gene Ther.* **10:**1370–1380.

Scott, J. K., and Smith, G. P. (1990). Searching for peptide ligands with an epitope library. *Science* **249:**386–390.

Stitz, J., Buchholz, C. J., Engelstadter, M., Uckert, W., Bloemer, U., Schmitt, I., and Cichutek, K. (2000). Lentiviral vectors pseudotyped with envelope glycoproteins derived from Gibbon ape leukemia virus and murine leukemia virus 10A1. *Virology* **273:**16–20.

Takahashi, S., Mok, H., Parrott, M. B., Marini, F. C., III, Andreef, M., Brenner, M. K., and Barry, M. A. (2003). Selection of chronic lymphocytic leukemia binding peptides. *Cancer Res.* **63:**5213–5217.

Tsao, J., Chapman, M. S., Agbanje, M., Keller, W., Smith, K., Wu, H., Luo, M., Smith, T. J., Rossmann, M. G., Compans, R. W., and Parrish, C. R. (1991). The three-dimensional structure of canine parvovirus and its functional implications. *Science* **251**:1456–1464.

White, S. J., Nicklin, S. A., Buning, H., Brosnan, M. J., Leike, K., Papadakis, E. D., Hallek, M., and Baker, A. H. (2004). Targeted gene delivery to vascular tissue *in vivo* by tropism-modified adeno-associated virus vectors. *Circulation* **109**:513–519.

Wickham, T. J. (2000). Targeting adenovirus. *Gene Ther.* **7**:110–114.

Wickham, T. J. (2003). Ligand-directed targeting of genes to the site of disease. *Nature Med.* **9**:135–139.

Wickham, T. J., Mathias, P., Cheresh, D. A., and Nemerow, G. R. (1993). Integrins alpha v beta 3 and alpha v beta 5 promote adenovirus internalization but not virus attachment. *Cell* **73**:309–319.

Work, L. M., Nicklin, S. A., Brain, N. J. R., Dishart, K. L., Von Seggern, D. J., Hallek, M., Buning, H., and Baker, A. H. (2004). Development of efficient viral vectors selective for vascular smooth muscle cells. *Mol. Ther.* **9**:198–208.

Wu, B. W., Lu, J., Gallagher, T. K., Anderson, W. F., and Cannon, P. M. (2000). Identification of regions in the Moloney murine leukemia virus SU protein that tolerate the insertion of an integrin-binding peptide. *Virology* **269**:7–17.

Xia, H., Anderson, B., Mao, Q., and Davidson, B. L. (2000). Recombinant human adenovirus: targeting to the human transferrin receptor improves gene transfer to brain microcapillary endothelium. *J. Virol.* **74**:11359–11366.

FURTHER READING

Donohue, P. R., Hoover, E. A., Beltz, G. A., Riedel, N., Hirsch, V. M., Overbaugh, J., and Mullins, J. I. (1988). Strong sequence conservation among horizontally transmissible, minimally pathogenic feline leukemia viruses. *J. Virol.* **62**:722–731.

ADVANCES IN VIRUS RESEARCH, VOL 65

A DECADE OF ADVANCES IN IRIDOVIRUS RESEARCH

Trevor Williams,*,† Valérie Barbosa-Solomieu,‡ and
V. Gregory Chinchar§

*Departmento de Producción Agraria, Universidad Pública de Navarra
31006 Pamplona, Spain
†ECOSUR, Tapachula, Chiapas 30700, Mexico
‡Unité de Virologie Moléculaire et Laboratoire de Génétique des Poissons, INRA
78350 Jouy-en-Josas, France
§Department of Microbiology, University of Mississippi Medical Center, Jackson
Mississippi 39216

DOI: 10.1016/S0065-3527(05)65006-3

The serendipitous discovery of *Frog virus 3* (FV-3), the best characterized member and type species of the family *Iridoviridae*, resulted
from attempts by Allan Granoff and his coworkers to identify cell lines
that would support the *in vitro* growth of Lucke herpesvirus (Granoff,
1984; Granoff *et al.*, 1966). This amphibian herpesvirus was causally
connected with tumor development in frogs, but had proven intracta-
ble to propagation in cell culture. Reasoning that cultured amphibian
cells might prove useful for Lucke herpesvirus propagation, primary
cultures from normal and tumor-bearing kidneys were prepared from
Rana pipiens. Surprisingly, a fraction of the primary cultures displayed cytopathic effects (CPE) consistent with viral infection, and several, likely identical, viruses were isolated. FV-3, which was derived
from the kidney of a tumor-bearing frog, became the focus of early
studies since it was initially thought that the virus might play a role
in tumor development. While the association between FV-3 infection
and tumor development turned out to be incorrect, continued study of
FV-3 revealed a unique replication cycle and established iridoviruses
as members of a distinct taxonomic family.

Early studies highlighted an interesting dichotomy between FV-3
growth *in vitro* and *in vivo*. Whereas infections *in vitro* occurred in a
wide variety of mammalian, amphibian, and piscine cell lines and
resulted in marked CPE, infections *in vivo* appeared to be sporadic,
subclinical, and confined to only anurans, i.e., frogs and toads
(Chinchar, 2002; Chinchar and Mao, 2000; Willis *et al.*, 1985). The
observation that adult animals were relatively resistant to infection
was confirmed and extended by Tweedel and Granoff (1968) who
showed that adult frogs easily withstood inoculation with 10^6 plaque-
forming units of FV-3, but that injection of as few as 900 plaque-forming
units into frog embryos was lethal. Moreover, while localized outbreaks of FV-3-related disease were sometimes noted and virus could
be isolated from cultured kidney cells, infections did not become widespread or involve large numbers of animals (Clark *et al.*, 1968; Wolf
et al., 1968). The clinical impression that developed from these and
other studies suggested that FV-3 was neither very virulent nor a
major threat to amphibian populations. This benign view of vertebrate
iridoviruses persisted until the mid-1980s, when it became apparent
that iridoviruses were responsible for widespread clinical illness in a
variety of fish species (reviewed in Ahne *et al.*, 1997; Chinchar, 2000,
2002). Since that time, iridovirus disease has become increasingly

apparent as infections have been noted in various species of freshwater and marine fish, amphibians, and reptiles. However, while the number of reported disease outbreaks attributed to iridoviruses has increased markedly, it is not clear whether this is due to an actual increase in the prevalence of iridovirus disease, or in our collective ability to detect iridovirus infections. By 2004, iridoviruses have been linked to disease in wild and cultured frogs in Asia, North America, South America, Australia, and Europe (Cullen and Owens, 2002; Cunningham *et al.*, 1996; Wolf *et al.*, 1968; Zhang *et al.*, 2001; Zupanovic *et al.*, 1998b), salamanders in western North America (Bollinger *et al.*, 1999; Docherty *et al.*, 2003; Jancovich *et al.*, 1997), wild and cultured fish species in Asia, Australia, Europe, and North America (Ahne *et al.*, 1989, 1997; Langdon *et al.*, 1986; Nakajima *et al.*, 1998; Plumb *et al.*, 1996), and in reptile populations in Europe, North America, Africa, and Asia (Chen *et al.*, 1999; Drury *et al.*, 2002; Hyatt *et al.*, 2002; Mao *et al.*, 1997; Marschang *et al.*, 1999; Telford and Jacobson, 1993).

Most of the serious infections noted previously have been attributed to members of two genera within the family *Iridoviridae*: *Megalocytivirus* and *Ranavirus*. Megalocytiviruses (i.e., red seabream iridovirus [RSIV], rock bream iridovirus [RBIV], and infectious spleen and kidney necrosis virus [ISKNV]) infect a wide variety of marine fish in Southeast Asia, including several species that are widely used in aquaculture (Nakajima *et al.*, 1998). Outbreaks in these settings can be extensive, with mortalities approaching 100% (He *et al.*, 2000). The reasons for the explosive nature of recent outbreaks are not known, but may reflect one or more events, including the introduction of a novel pathogen into populations of susceptible fish species, ease of disease transmission among intensively cultured fish, or increased susceptibility to viral disease in immunologically stressed populations. Megalocytiviruses have been isolated from more than twenty species of marine fish and, although variously designated by the species which they infect or the disease they cause (e.g., infectious spleen and kidney necrosis virus), recent evidence suggests they are very similar (Do *et al.*, 2004, 2005). Analysis of individual viral genes (e.g., the major capsid protein [MCP] and ATPase genes) from a large number of viral isolates and whole genome analysis of three selected isolates (RSIV, RBIV, and ISKNV), indicates that these viruses share extensive sequence identity, yet differences are noted, and it is not yet clear whether these viruses represent strains/isolates of the same viral species or different species (Do *et al.*, 2004; He *et al.*, 2001; Sudthongkong *et al.*, 2002). Recent efforts have focused primarily on genetic analysis of various megalocytivirus isolates, identification of new isolates, and development of diagnostic techniques and a viral vaccine. Little is known about the molecular events in megalocytivirus-infected cells; similarities to the replication cycle of the better characterized ranaviruses have yet to be determined.

In contrast to megalocytiviruses, members of the genus *Ranavirus* appear to be more genetically diverse and, in addition, infect cold-blooded vertebrates of three different taxonomic classes: bony fish, amphibians, and reptiles (Chinchar, 2002). While many ranaviruses share high levels of sequence similarity within the MCP gene and other viral genes, differences in host range, restriction fragment length polymorphism (RFLP), and protein profiles, and gene content suggest that they represent different viral species, rather than strains or isolates of the same viral species (Chinchar *et al.*, 2005; Hyatt *et al.*, 2000). Since the identification of epizootic hematopoietic necrosis virus (EHNV) from redfin perch (*Perca fluviatilis*) by Langdon *et al.* (1986), ranaviruses have been detected in sheatfish and catfish in Europe (Ahne *et al.*, 1997), ica (Plumb *et al.*, 1996), ornamental fish imported into the United States from Southeast Asia (Hedrick and McDowell, 1995), frogs on all continents except Africa and Antarctica (most likely for lack of study in the former and lack of hosts in the latter), and also in salamanders, turtles, and snakes (Bollinger *et al.*, 1999; Hyatt *et al.*, 2002; Jancovich *et al.*, 1997; Marschang *et al.*, 1999). In a few cases it appears that viruses infecting animals of one taxonomic class, e.g., fish, also infect animals of a different taxonomic class, e.g., amphibians (Mao *et al.*, 1999a; Moody and Owens, 1994). Whether cross-class infections are common in nature is not known, although the isolation of strains of FV-3 from fish and frogs indicates that interclass transmission is possible. Such events may explain the sudden appearance of disease in previously healthy populations.

The third genus of vertebrate iridoviruses, *Lymphocystivirus*, has been known by the disease it causes since the early 1900s, and was the first iridovirus genome to be sequenced (Tidona and Darai, 1997). However, because of its inability to be grown easily in culture, its replication cycle has not been studied. Infection with lymphocystis disease virus (LCDV) leads to the development of wart-like lesions, generally on the external surface of infected fish (Weissenberg, 1965). However, unlike papilloma virus infections that involve cellular hyperplasia, LCDV results in the massive enlargement of individual epidermal cells. Infections are not believed to be fatal, but large numbers of wart-like lesions may impair the mobility and feeding of infected fish and indirectly contribute to morbidity. Collectively, vertebrate iridoviruses are a cause of localized die-offs in wild and farmed fish, reptile, and amphibian species, and are recognized as key factors in wildlife disease throughout the world (Daszak *et al.*, 1999, 2003).

II. Taxonomy

A. Definition of the Family Iridoviridae

Iridoviruses are large, icosahedral viruses with a linear, double-stranded DNA genome (Chinchar *et al.*, 2005) (Table I). Unlike other DNA viruses, viral particles can be either enveloped or nonenveloped, depending upon whether they are released from the cell by lysis, or bud from the plasma membrane. Although both enveloped and naked virions are infectious, the specific infectivity of the former is higher, suggesting that one or more viral envelope proteins play an important role in virion entry. In addition to the outer envelope found in viruses that bud from the plasma membrane, iridoviruses also possess a lipid membrane that lies between the viral DNA core and the capsid. Iridovirus infections result in the appearance within the cytoplasm of large, morphologically distinct viral assembly sites (AS). AS serve as a concentration point for viral protein and DNA and are the site of virion assembly. Mature virions accumulate within the cytoplasm in large paracrystalline arrays or bud from the plasma membrane. The appearance within the cytoplasm of arrays of icosahedral particles usually 120–200 nm in diameter is pathognomic for iridovirus infections. LCDV and white sturgeon iridovirus (WSIV) infections may involve much larger particles, exceeding 300 nm diameter (Adkison *et al.*, 1998; Madeley *et al.*, 1978; Paperna *et al.*, 2001).

In addition to their distinctive size and cytoplasmic location, iridoviruses are distinguished from other virus families by their genomic organization. The iridovirus genome is circularly permuted and terminally redundant, and the unique region ranges in size from ~100 kbp for ranaviruses and megalocytiviruses, to ~200 kbp for lymphocystiviruses and iridoviruses (Goorha and Murti, 1982). The terminal repetition adds another 10–30% to the size of the genome.

B. Features Distinguishing the Genera

Members of the family *Iridoviridae* infect only cold-blooded vertebrates (bony fish, amphibians, and reptiles) or invertebrates (primarily insects), but also crustaceans and mollusks (see Table II). Aside from

TABLE I

TAXONOMY OF IRIDOVIRUSES

Category	Designation	Distinguishing criteria
Family	*Iridoviridae*	Naked or enveloped icosahedral virions, 120–300 nm in diameter
		Genome is a single molecule of linear, dsDNA (102 – 212 kbp)*
		Genome is terminally redundant and circularly permuted
		Virions assemble and accumulate within the cytoplasm
Genus	*Ranavirus*	**Genome:** DNA is methylated[†]; ~105 kbp
		GC content: 49–55%
		Virion diameter: ~150 nm
		Host species: bony fish, amphibians, and reptiles worldwide
		Clinical illness: Systemic infection involving multiple internal organs; infections range from subclinical to fulminant with mortalities approaching 100%
	Lymphocystivirus	**Genome:** DNA is methylated; 103–186 kbp
		GC content: 27–29%
		Virion diameter: 200–300 nm; may possess a fibril-like fringe
		Host species: Marine and freshwater fish worldwide.
		Clinical illness: Wart-like, primarily external, lesions. Infected cells undergo massive enlargement
	Megalocytivirus	**Genome:** DNA is methylated; 105–118 kbp
		GC content: 53–55%
		Virion diameter: ~150 nm
		Host species: marine fish in SE Asia.
		Clinical illness: Systemic infection involving multiple internal organs; mortality approaching 100%
	Iridovirus	**Genome:** DNA is not methylated (no viral DNA methyltransferase); 212 kbp
		GC content: 27%
		Virion diameter: 120–130 nm
		Host species: Insects, crustaceans, possibly mollusks

(continues)

TABLE I (*continued*)

Category	Designation	Distinguishing criteria
		Clinical illness: Massive levels of virus replication result in a change in color of infected insects; covert infections may be common
	Chloriridovirus	**Genome:** DNA is not methylated; ∼135 kbp
		GC content: 54%
		Viirion diameter: ∼180 nm
		Host species: Mosquitoes and midges (Diptera)
		Clinical illness: Massive levels of virus replication result in a change in color of infected insects; covert infections may be common

*Genome size indicates unique region and does not include the terminal repeats that add an extra 10–30%.

†Singapore grouper iridovirus (SGIV) appears to lack a DNA methyltransferase. As a result, it is the only known vertebrate iridovirus that lacks a methylated genome.

this difference in host range, one nearly universal feature distinguishing the genera infecting vertebrates (*Megalocytivirus, Ranavirus, Lymphocystivirus*) from those infecting invertebrates (*Iridovirus, Chloriridovirus*) is the presence of a DNA methyltransferase in the former (Willis and Granoff, 1980; Willis *et al.*, 1984). The viral DNA methyltransferase targets cytosine residues within the sequence CpG, and as a result, ∼20% of cytosines within the viral genomic are methylated (Willis and Granoff, 1980). While the role of DNA methylation is not clear, methylation is required for virus replication, since drugs that inhibit DNA methylation, such as azacytidine, reduce viral yields by 100-fold and markedly impair viral DNA synthesis (Goorha *et al.*, 1984). The family *Iridoviridae* is currently organized into five genera based on particle size, host range, presence of a DNA methyltransferase, GC content, similarity within the MCP gene, clinical disease, and other criteria listed in Table I (Chinchar *et al.*, 2005). Cladistic analysis generally supports division of the *Iridoviridae* family into four genera (Wang *et al.*, 2003b). Differences between the trees generated therein are due to the tree building method used (maximum parsimony) and the choice of sequences that were analyzed.

The two invertebrate genera include invertebrate iridescent viruses (IIVs) that typically infect insects in moist or aquatic habitats (Table II). The genus *Chloriridovirus* comprises a single species, *Invertebrate*

TABLE II
SPECIES, TENTATIVE SPECIES, AND STRAINS REFERRED TO IN THIS CHAPTER

Genus *Species name (type species in bold)* Names of strains (abrreviations used)	Tentative species
Iridovirus	
Invertebrate iridescent virus 1 (IIV-1)	Anticarsia gemmatalis iridescent virus (AGIV)
Invertebrate iridescent virus 6 (IIV-6)	Invertebrate iridescent virus 2 (IIV-2)
	Invertebrate iridescent virus 9 (IIV-9)
	Invertebrate iridescent virus 16 (IIV-16)
	Invertebrate iridescent virus 21 (IIV-21)
	Invertebrate iridescent virus 22 (IIV-22)
	Invertebrate iridescent virus 23 (IIV-23)
	Invertebrate iridescent virus 24 (IIV-24)
	Invertebrate iridescent virus 29 (IIV-29)
	Invertebrate iridescent virus 30 (IIV-30)
	Invertebrate iridescent virus 31 (IIV-31)
Chloriridovirus	
Invertebrate iridescent virus 3 (IIV-3)	
Ranavirus	
Ambystoma tigrinum virus (ATV)	Rana esculenta iridovirus (REIR)
Regina ranavirus (RRV)	Testudo iridovirus (THIV)
Bohle iridovirus (BIV)	Singapore grouper iridovirus (SGIV)
Epizootic haematopoietic necrosis virus (EHNV)	
European catfish virus (ECV)	
European sheatfish virus (ESV)	
Frog virus 3 (FV-3)	
Box turtle virus 3	
Bufo bufo United Kingdom virus	
Bufo marinus Venezuelan iridovirus 1	
Lucké triturus virus 1	
Rana temporaria United Kingdom virus	
Redwood Park virus	
Stickleback virus	
Tadpole edema virus (TEV)	
Tadpole virus 2	
Tiger frog virus (TFV)	
Tortoise virus 5 (TV-5)	

(continues)

TABLE II (*continued*)

Genus Species name (type species in bold) Names of strains (abrreviations used)	Tentative species
Santee-Cooper ranavirus (SCRV)	
Largemouth bass virus (LMBV)	
Doctor fish virus (DFV)	
Guppy virus 6 (GV-6)	
Lymphocystivirus	
Lymphocystis disease virus 1 (LCDV-1)	Lymphocystis disease virus 2 (LCDV-2)
	Lymphocystis disease virus China (LCDV-C)
Megalocytivirus	
Infectious spleen and kidney	Rock bream iridovirus (RBIV)
***necrosis virus* (ISKNV)**	
Red Sea bream iridovirus (RSIV)	Olive flounder iridovirus (OFIV)
Sea bass iridovirus (SBIV)	
African lampeye iridovirus (ALIV)	
Grouper sleepy disease iridovirus (GSDIV)	
Dwarf gourami iridovirus (DGIV)	
Taiwan grouper iridovirus (TGIV)	
Unclassified	
White sturgeon iridovirus (WSIV)	

iridescent virus 3 (IIV-3) from the mosquito *Ochlerotatus* (*Aedes*) *taeniorhynchus*. The wild isolate has been referred to as "regular" mosquito iridescent virus in the literature to differentiate it from a turquoise-colored laboratory strain (Wagner and Paschke 1977). The genus status is based on large particle size (~180 nm diameter), host range (restricted to mosquito and possibly midge larvae), and studies on serological relationships, RFLP and DNA hybridization (reviewed by Williams, 1996). Surprisingly, given the global importance of mosquitoes as vectors of human disease, chloriridoviruses are among the least studied members of the family. Fortunately, the IIV-3 genome has recently been sequenced and is presently being analyzed (C. Alfonso, personal comment). The results of the analyses are eagerly awaited.

The *Iridovirus* genus comprises two species: *Invertebrate iridescent virus 1* (IIV-1), and the type species of the genus *Invertebrate iridescent virus 6* (IIV-6), often referred to as Chilo iridescent virus (CIV). There are an additional 11 tentative species recognized, for which insufficient characterization information is available to determine species status (Table II). Members of the genus are sometimes referred to as small iridescent viruses due to their particle size in thin section (120–130 nm diameter), which is smaller than the particle size of the chloriridovirus, IIV-3. There are several instances in which species and tentative species in the *Iridovirus* genus occur in the literature with host-derived names. This generates confusion, as there are clear examples of individual IIVs that infect different host species. For example, IIV-9 isolated from larvae of the moth *Wiseana cervinata* in New Zealand also naturally infects at least two other soil-dwelling insects in the same region (Williams and Cory, 1994). Originally, each of these isolates was given the host name and a type number, but this system was rationalized in the last ICTV report, and they are now united as the tentative species, invertebrate iridescent virus 9 (IIV-9). Levels of terminal redundancy in IIV genomes are typically 5–12%, depending on the method used (Delius *et al.*, 1984; Webby and Kalmakoff, 1999).

The species and tentative species within the *Iridovirus* genus can be assigned to one of three complexes based on serological findings: RFLP similarities, DNA hybridization, and partial sequence of the MCP (Webby and Kalmakoff, 1998; Williams and Cory, 1994). The largest complex comprises IIV-1 plus nine tentative species, the second comprises IIV-6 and strains thereof, and the third comprises IIV-31 from isopods and an isolate from the beetle *Popillia japonica*. The identification of these complexes has proved useful in the characterization of novel isolates (Just and Essbauer, 2001; Kinard *et al.*, 1995; Williams, 1994).

C. Features Delineating the Species

The delineation of iridovirus species is an ongoing process best described by several clear examples. Sequence analysis suggests that the genus *Megalocytivirus* may be composed either of a single virus species possessing a broad host range among marine fish, or a series of closely-related but molecularly distinct species from different hosts. For example, RBIV and ISKNV share 85 of 118 putative open reading frames (ORFs), but amino-acid identity varies among the shared genes from a high of 91–99% (53 ORFs), and 81–90% (25 ORFs), to a low of 60–80% (7 ORFs) (Do *et al.*, 2004). Recent genome sequence analysis of an Asian isolate of LCDV suggests that the genus *Lymphocystivirus* is composed of two species: a European/Atlantic Ocean species (LCDV-1) and a distinct Asian/Pacific Ocean species (LCDV-C) (Tidona and Darai, 1997; Zhang *et al.*, 2004a). LCDV-1 and LCDV-C are clearly members of the same genus, as indicated by their low GC content (27–29%) and high sequence identity within the MCP gene (88%). In contrast, members of the *Megalocytivirus* and *Ranavirus* genera show GC contents of ~55% and share only 49–53% amino-acid identity with the MCPs of the LCDV species. However, LCDV-1 and LCDV-C can be differentiated as distinct species for several reasons. First, there is little colinearity between the genomes of LCDV-1 and LCDV-C, suggesting that they have undergone a high degree of rearrangement. Moreover, even in one of the more similar regions (nts 15,055-25,423), the eight putative ORFs only show 68% identity at the nucleotide level. In addition, LCDV-C has the lowest coding density of any iridovirus and the highest level of tandem and overlapping repeat elements. For example, within one ~500 bp region there were eight 66 bp repeats. Thus, there are likely at least two species recognized in the genus *Lymphocystivirus*.

The genus *Ranavirus* comprises six virus species based on analysis of host range, sequence identity, and protein and RFLP profiles (Hyatt *et al.*, 2000; Mao *et al.*, 1997). FV-3, EHNV, *Bohle iridovirus* (BIV), *Ambystoma tigrinum virus* (ATV), and *European catfish virus* (ECV) comprise five closely-related viral species that share over 90% sequence identity within the highly conserved MCP and other genes, but clearly differ from each other in host range and RFLP profiles. *Santee-Cooper ranavirus* (SCRV) and the recently identified Singapore grouper iridovirus (SGIV, a tentative species) are the most divergent members of the genus and show approximately 80% and 70% sequence identity within the MCP gene, respectively (Mao *et al.*, 1999b; Ting *et al.*, 2004). Moreover, preliminary data indicate that SGIV is unique among the vertebrate iridoviruses in lacking a DNA methyltransferase, and thus a methylated genome, and in possessing a purine nucleoside phosphorylase (Song *et al.*, 2004; Ting *et al.*, 2004). The absence of a DNA methyltransferase is surprising, as the presence of a methylated genome was assumed to be a hallmark characteristic of vertebrate iridoviruses. Furthermore, the earlier suggestion that the DNA methyltransferase was part of a restriction/modification system (Goorha *et al.*, 1984; Willis *et al.*, 1989) suggests that the long-standing rationale for genome methylation, namely protection of the viral genome from virus-induced attack by host nucleases, may be incorrect.

Phylogenetic analysis, based on sequence comparison of the MCP genes of 15 iridovirus species, tentative species, and strains, supports the taxonomic divisions discussed previously (Fig. 1). MCP sequence information is not yet available for any member of the genus *Chloriridovirus*, although DNA polymerase analysis of IIV-3 provisionally supports the concept of a distinct genus (Stasiak *et al.*, 2003). The four

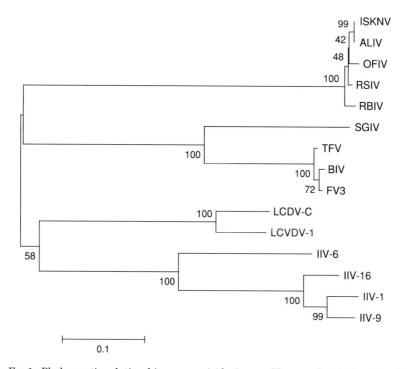

FIG 1. Phylogenetic relationships among iridoviruses. The complete inferred amino-acid sequence of the MCP of 15 iridoviruses representing four different genera were aligned using the CLUSTAL W program found within DNASTAR (Madison, WI). Based on that alignment, a phylogenetic tree was constructed using the Neighbor-Joining algorithm within Mega 2.1 (Kumar *et al.*, 2001). The tree was generated using the Poisson correction and was validated by 1000 bootstrap repetitions. Branch lengths are drawn to scale, and scale bar is shown. The numbers at each node indicate boot-strapped percentage values. The sequences used to construct the tree were obtained from GenBank: BIV (AY187046), FV-3 (U36913), TFV (AY033630); SGIV (AF364593), LCDV-1 (L63545); LCDV-C (AAS47819.1), IIV-1 (M33542); IIV-6 (AAK82135.1); IIV-9 (AF025774), IIV-16 (AF025775), ALIV (AB109368), ISKNV (AF370008), RBIV (AY533035), RSIV (AY310918), OFIV (AY661546).

genera for which extensive sequence information is available can be partitioned into four distinct taxonomic groups that share ~50% amino-acid similarity within the MCP. Within a given genus, sequence simi-larity is considerably higher, ranging from 64 to 100%. Moreover, the four extant genera differ markedly in their "within genus" variability. Megalocytiviruses show marked sequence similarity (97–100%) within the MCP, suggesting that they are isolates or strains of a single species.

The two species (LCDV-1, LCDV-C) of the genus *Lymphocystivirus* share 88% similarity between their MCP genes. Ranaviruses, on the other hand, show a cluster of viruses with high levels (>90%) of similarity (FV-3, ATV, BIV, TFV), which are distinguished from each other by host range and RFLP profiles, as well as two less-related members (SCRV and SGIV) with considerably greater sequence divergence. Finally, the one invertebrate iridovirus genus for which sequence data is available is comprised of a cluster of isolates with sequence similarities ranging from 64–93%. While most of these likely comprise separate viral species, those with the greatest similarity might also be considered strains of the same viral species. We recognize that phylogenetic trees generated by choosing a different gene (e.g., ATPase or DNA polymerase gene in place of the MCP gene) may yield trees with different topographies. While phylogenetic analysis of conserved viral genes such as the MCP supports the taxonomy described above, analysis of individual viral genomes highlights key similarities and differences among iridovirus genomes (Tidona *et al.*, 1998). The most accurate tree will likely be one based on whole genome sequence data (Herniou *et al.*, 2001), or a concatenated set of viral genes (Rokas *et al.*, 2003).

The two species recognized in the genus *Iridovirus*, IIV-1 and IIV-6, are quite dissimilar, and there is no doubt that they represent distinct species. Low levels of DNA annealing in solution, even under low stringency conditions, support the concept that certain members of the genus are quite distinct from one another (Webby and Kalmakoff, 1999). Perhaps the major challenge facing the classification of the IIVs is the ability to quantify heterogeneity within and between species. For the iridoviruses in general, and the IIVs in particular, species definitions are severely limited by the availability of data, particularly sequence information from a range of genes. This is probably due to the lack of interest in the IIVs as potential agents of biological control of insects, the low number of researchers with an active interest in IIVs, and difficulties in finding financial support for studies on these viruses.

For most of the tentative species presently recognized, there exists only a single isolate—an important obstacle to understanding within species variation. However, in the case of IIVs from the blackfly, *Simulium variegatum*, a number of isolates exist, each isolated from different infected individuals in Wales. Comparison of the RFLP profiles indicated that these isolates differed from one another in various degrees (Williams, 1995; Williams and Cory, 1993). Preliminary sequencing studies suggested that they could be divided into two groups: one group being IIV-22-like and another group being similar to IIV-2 (R. Webby, personal communication). Variation among IIV isolates from a single host species was also observed in fall armyworm larvae (*Spodoptera frugiperda*) collected in southern Mexico. RFLP analysis suggested that the *S. frugiperda* isolate may be a strain of IIV-6 (N. Hernández and T. Williams, unpublished data).

The need to quantify within species heterogeneity in IIV-6 has been highlighted in several studies. Williams and Cory (1994) noted that there were at least two strains of IIV-6 being used in laboratories in different parts of the world. The isolate that has been completely sequenced by Jakob *et al.* (2001) differs from the isolates used in New Zealand, Australia, and the USA. Recently reported isolates from crickets (Just and Essbauer, 2001; Kleespies *et al.*, 1999) appear to be strains of IIV-6 that share an average of 95.2% sequence identity in a selection of seven proteins, including the MCP (Jakob *et al.*, 2002). Moreover, IIV-21 and IIV-28 may have been contaminated by IIV-6 at some point, because they were originally reported as being distinct isolates, but now are almost identical to the IIV-6 isolate from Australia and New Zealand (Webby and Kalmakoff, 1998).

As indicated above, the task of defining virus species is not straightforward. The International Committee on Taxonomy of Viruses (ICTV) defines a virus species as "a polythetic class ... constituting a replicating lineage and occupying a particular ecological niche," but applying that definition to real virus populations is problematic (Van Regenmortel, 2000). For example, there are no accepted cut-off points for sequence similarity or RFLP identity that unambiguously define the boundaries between genera, species, and strains. Thus, virus species represent a class in which the presence or absence of no single character is sufficient to include (or exclude) membership in the group. Thus, ranavirus species are defined by a collection of sequence data, RFLP and protein profiles, host range, and physical characteristics. While these taxonomic distinctions may seem purely academic, they have potentially important commercial implications. For instance, if the ranavirus EHNV is strictly defined based on host species, RFLP profile, protein profile, and sequence analysis, then fish shipped from an EHNV-positive region, such as Australia, into an EHNV-free zone, such as North America, must be certified free of virus. However, if the species definition is more loosely applied, for example, if all ranaviruses with MCP sequence identity greater than 90% are considered to be members of the same virus species, then shipment of fish from Australia into the United States would not require certification, since the definition would include FV-3 and ATV that are already present in North America. Regardless of whether a given virus is classified as a distinct species, or as a strain of an existing species, there is a critical need for diagnostic tests to distinguish between relatively innocuous iridoviruses and those responsible for mortal disease. Development of these tests is discussed later.

D. Relationships with Other Families of Large DNA Viruses

Iridoviruses are members of a monophyletic clade of large, nucleo-cytoplasmic DNA viruses that includes the families *Poxviridae, Phycodnaviridae* and *Asfarviridae* (Iyer *et al.*, 2001). Members of the *Iridoviridae, Asfarviridae,* and *Poxviridae* replicate in the cytoplasm or, after starting replication in the nucleus, complete it in the cytoplasm. These viruses encode their own machinery for transcription and RNA modification including subunits of RNA polymerase and transcription factors, many of which appear to have been acquired from eukaryotic host cells (Tidona and Darai, 2000). Phycodnaviruses lack RNA polymerase subunit genes, and must therefore depend on host transcriptional mechanisms in the nucleus (Van Etten *et al.*, 2002). The clade shares a total of 31 ancestral genes including three genes for structural proteins: the capsid protein and two lipid membrane localized proteins. The structure of the ancestral virus appears likely to have been icosahedral, based on gene content. A recently described giant icosahedral virus (400 nm diameter) from amoebae may also be a member of this clade as indicated by analysis of ribonuclease reductase and topoisomerase II sequences (La Scola *et al.*, 2003).

In poxviruses, the ancestral capsid protein is conserved in one domain of the D13L protein of vaccinia virus, believed to form a scaffold for the construction of viral crescents during particle assembly (Sodeik *et al.*, 1994). The complex poxvirus particle appears to represent a derived state compared to the icosahedral ancestor. Based on the analysis of LCDV-1 and IIV-6 as the only fully sequenced iridoviruses at that time, Iyer *et al.* (2001) identified two ancestral genes that had been lost by the iridoviruses (ATP dependent DNA ligase and a capping enzyme required for RNA capping in the cytoplasm). The iridoviruses differed from the ancestral group by possessing seven additional genes including Rnase III, thymidylate synthase, and a cathepsin B–like cysteine protease. In contrast, ASFV and the phycodnaviruses only acquired a single gene (lambda type restriction exonuclease), and each lost five different genes from the ancestral compliment. The cladistic analysis may now require updating given the increase in the number of iridovirus genomes that have been sequenced completely.

The architectural similarities between iridoviruses, African swine fever virus (ASFV), and phycodnaviruses extend beyond the icosahedral symmetry of the capsid, a common inner-lipid membrane and an electron dense core. High-resolution models now reveal the presence of a core shell and twin inner membranes originating from collapsed cisterna derived from the endoplasmic reticulum in ASFV (Andrés *et al.*, 1998; Rouiller *et al.*, 1998). Vaccinia virus inner lipid envelopes are derived from an intracellular compartment between the endoplasmic reticulum and the Golgi apparatus, a process apparently facilitated by vimentin intermediate filaments concentrated in the viral factories (Risco *et al.*, 2002; Schmelz *et al.*, 1994). These observations suggest that the hypothesis of *de novo* synthesis of the iridovirus lipid component proposed during early electron microscope studies (Stoltz, 1971) should be re-examined. Phycodnaviruses have a less prominent inner lipid membrane, a right-handed (dextro) skewed capsid lattice comprising doughnut-shaped capsomers with a smooth outer surface, and lack the fibrilar structures seen in iridoviruses (Nandhagopal *et al.*, 2002; Yan *et al.*, 2000). Structural similarities between iridoviruses and the allantoid particles of ascoviruses are not immediately apparent although molecular evidence indicates clear relationships between these viruses.

Sequence comparisons of the virus-encoded δ DNA polymerase indicated putative evolutionary relationships among these viruses and ascoviruses of lepidopteran insects (Knopf, 1998; Stasiak *et al.*, 2000) (Fig. 2A). The relationship between iridoviruses and ascoviruses appears particularly close. The genome of the ascovirus DpAV-4 was found to be circular, contained large (1–3 kbp) interspersed repeats, and 76% of deoxycytidines were methylated. Examination of dinucleotide frequency ratios indicated that methylation occurred mostly at CpT and CpC dinucleotides. In contrast, the linear but circularly permuted genomes of IIV-3, IIV-6, and IIV-31 lack large regions of repetitive DNA and possess very low levels of methylation, less than 5% (Bigot *et al.*, 2000; Cheng *et al.*, 1999). The IIVs were most likely methylated at CpT residues, rather than the CpG residues targeted by vertebrate iridoviruses. Homologs to about 40% of the proteins encoded by *Spodoptera frugiperda ascovirus 1a* are found in IIV-6 with lower percentages seen among vertebrate iridoviruses, phycodnaviruses and ASFV (Stasiak *et al.*, 2003). Analysis of the major capsid protein revealed seven conserved domains that were used for alignment, resulting in a tree very similar to that generated by analysis of the DNA polymerase (Fig. 2B). Additional analyses of thymidine kinase, ATPase III and several other proteins yielded similar results

A **DNA Polymerase**

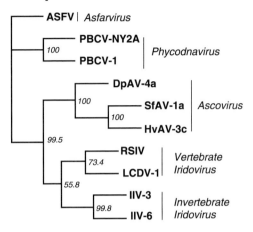

B **Major capsid protein**

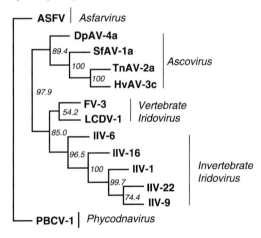

FIG 2. Consensus trees resulting from phylogenetic analyses of (A) δ DNA polymerase and (B) major capsid protein from one asfarvirus, *African swine fever virus* (ASFV); two phycodnaviruses, *Paramecium bursaria Chlorella virus 1* and *NY2A* (PBCV-1, PBCV-NY2A); four ascoviruses *Spodoptera frugiperda ascovirus 1a* (SfAV-1a), *Trichoplusia ni ascovirus 2a* (TnAV-2a) *Heliothis virescens ascovirus 3c* (HvAV-3c), *Diadromus pulchellus ascovirus 4a* (DpAV-4a); three vertebrate iridoviruses, *Frog virus 3* (FV-3), *Lymphocystis disease virus 1* (LCDV-1), Red sea bream iridovirus (RSIV); and six invertebrate iridescent viruses (IIV-1, IIV-3, IIV-6, IIV-9, IIV-16, IIV-22). Consensus trees based on parsimony procedure, numbers at each node indicate bootstrapped percentage values from 1000 repetitions. Modified with permission from Federici and Bigot (2003).

(Tidona *et al.*, 1998; Zhao and Cui, 2003). Multiple phylogenetic trees generated using different viral genes supported the conclusion that ascoviruses are more closely related to invertebrate iridoviruses than to vertebrate iridoviruses (Federici and Bigot, 2003; Stasiak *et al.*, 2003). Moreover, similarities in virus transmission support that view. Like insect iridoviruses, ascoviruses have very low *per os* infectivity, but are highly infectious by injection and depend on parasitoid wasps for transmission (Govindarajan and Federici, 1990). The recent observation of parasitoid mediated transmission of an IIV of Lepidoptera (López *et al.*, 2002) lends additional support to the hypothesis that ascoviruses evolved from invertebrate iridoviruses.

III. Genomic Studies

A. Genetic Content

Currently, nine iridovirus genomes have been sequenced in their entirety: IIV-6 (Jakob *et al.*, 2001), LCDV-1 (Tidona and Darai, 1997), LCDV-C (Zhang *et al.*, 2004a), TFV (He *et al.*, 2002), FV-3 (Tan *et al.*, 2004), ATV (Jancovich *et al.*, 2003), ISKNV (He *et al.*, 2001), RBIV (Do *et al.*, 2004), and SGIV (Song *et al.*, 2004) (Table III). Some of these viruses likely comprise distinct species within the same genera (e.g., LCDV and LCDV-C; ATV, SGIV, and FV-3), while others appear to be isolates of the same viral species (FV-3 and TFV; ISKNV, RBIV, and SGIV). Analysis of iridovirus genomes indicates that putative ORFs are closely packed, lack introns, and are generally nonoverlapping. Moreover, the few viral genes for which a 3′ terminus has been determined display terminal hairpins (i.e., an intragenic palindrome) that may play a role in transcriptional termination or in mRNA stability (Kaur *et al.*, 1995; Rohozinski and Goorha, 1992). Repeat regions and palindromes appear to be very common among iridoviruses. ATV contains 76 copies of a 14 bp palindrome while the closely-related TFV genome contains 52 copies and iridoviruses from other genera, LCDV, IIV-6, and ISKNV display three, one, and no copies of the same repeat, respectively (Jancovich *et al.*, 2003). The function of repeat regions is unknown. However, since they are often found within intergenic regions, they might play a role in viral gene transcription as initiation or termination signals. Alternatively, repeat regions through homologous pairing may be responsible for the high level of recombination seen among iridoviruses.

TABLE III

Comparison of Iridovirus Genomes

Genus	Virus	Genome size (kb)	Number of potential genes	GC content (%)	GenBank accession number
Ranavirus	FV-3	105,903	98	55	AY548484
	TFV	105,057	105	55	AF389451
	ATV	106,332	96	54	AY150217
	SGIV	140,131	162	49	AY521625
Lymphocystivirus	LCDV-1	102,653	110	29	L63515
	LCDV-C	186,247	176	27	AY380826
Megalocytivirus	ISKNV	111,362	105	55	AF371960
	RBIV	112,080	118	53	AY532606
Iridovirus	IIV-6	212,482	234	29	AF303741

Iridovirus genes can be classified into four categories: genes directly involved in virus replication, genes involved in immune evasion, genes homologous to those seen in other iridoviruses but not homologous to other genes in the database, and genes with no known homology (Table IV). Approximately 25–30% of the iridovirus genes encode polypeptides with homology to proteins of known, or presumed, replicative function, including DNA polymerase, the two largest subunits of DNA-dependent RNA polymerase II, the small and large subunits of ribonucleotide reductase, dUTPase, and DNA methyltransferase, or to known structural proteins (e.g., the major capsid protein) (Table IV). While it is reasonable to assume that viral homologs of cellular proteins of known function perform similar activities in virus replication, this hypothesis has not been verified in most situations. For example, although FV-3 encodes homologs of the two largest subunits of cellular DNA-dependent RNA polymerase II (Pol II), it has not been formally shown that they are involved in the transcription of late viral mRNA. Moreover, experience in other viral systems suggests that sequence homology does not always correlate with functional homology (Caposio *et al.*, 2004; Lembo *et al.*, 2004). Alternatively, some viral genes (e.g., those involved in the salvage pathway) may not be needed in all cell types and thus are dispensable for replication in those cells.

Among the 85 genes common to ISKNV and RBIV, 54 genes differ due to the insertion or deletion of amino acids within a given ORF or to alterations in the start or stop codon (Do *et al.*, 2004). Whether these

TABLE IV
CORE COMPLEMENT OF IRIDOVIRUS GENES

Category/gene product	Virus species and tentative species								
	FV-3	TFV	ATV	SGIV	LCDV-1	LCDV-C	IIV-6	ISKNV	RBIV
DNA/RNA Synthesis									
DNA polymerase	√	√	√	√	√	√	√	√	√
RNA polymerase II, α	√	√	√	√	√	√	√	√	√
RNA polymerase II, β	√	√	√	√	√	√	√	√	√
RAD2 repair enzyme	√	√	√	√	√	√		√	√
DNA methyltransferase	√	√	√		√	√		√	√
Rnase III	√	√	√	√	√	√	√	√	√
Helicase	√	√		√	√	√		√	√
Topoisomerase-like enzyme							√		
mRNA capping enzyme									√
Salvage Pathway									
Ribonucleotide reductase, α	√	√	√	√	√	√	√		
Ribonucleotide reductase, β	√	√	√	√	√	√	√	√	√
dUTPase	√	√	√	√			√		
Thymidylate synthase		√	√			√	√		
Thymidine kinase				√	√		√		
Thymidylate kinase	√		√			√	√		√
Purine nucleoside phosphorylase				√					
Immune evasion									
eIF-2α homolog	Δ*	√	√						
β-OH steroid oxidoreductase	√	√	Δ	√	√	√			
CARD-containing protein	√		√	√		√			
TNF receptor				√	√	√			
TRAF2-like protein									√
Bak-like protein				√					
Miscellaneous									
CTD-like phosphatase				√		√			
Reverse transcriptase-like						√			
ATP-dependent protease	√								
Inhibitor of apoptosis							√		
PCNA	√		√		√	√	√		
Major capsid protein	√	√	√	√	√	√	√	√	√

*Δ, gene present in a truncated form.

genetic changes are tolerated because they have little effect on protein function, or whether they reflect adaptation to a specific host species is not known. They are not, however, confined just to megalocytiviruses. Among the genomes of ATV, TFV, and FV-3, most viral genes show >95% sequence identity. However, inspection of those few genes that show <80% identity detects the presence of not only single amino-acid substitutions, most likely due to point mutation, but also the insertion or deletion of small blocks of amino acids. Interestingly, in some cases the indel occurs within a region of repeated amino-acid motifs. For example, at the C-terminus of ATV ORF 15L, there are 13 repeats of the sequence $[Q_nRPV]$ where $n = 4 - 8$, whereas within the corresponding ORFs of TFV and FV-3 there are six and nine copies, respectively.

While most viral genes with homology to known cellular genes are involved in DNA and RNA metabolism, a smaller number, similar to the poxvirus immunomodulatory genes, might play roles in evading host immune responses (Alcami and Koszinowski, 2000; Johnston and McFadden, 2003; Nash et al., 1999). For example, ranaviruses encode genes with homology to a β hydroxysteroid oxidoreductase (βSOR), the α subunit of eukaryotic initiation factor 2 (eIF-2α), and a caspase recruitment domain CARD-containing protein (vCOP) (Essbauer et al., 2001; Jancovich et al., 2003; Tan et al., 2004). βSOR may be involved in steroid metabolism and regulate immune responses (Alcami and Koszinowski, 2000). The viral homolog of eIF-2α (vIF2α) may function in maintaining viral protein synthesis in infected cells by serving as a decoy for activated protein kinase R (PKR), an interferon-induced, dsRNA-activated kinase that phosphorylates eIF-2α and thus blocks translational initiation (Beattie et al., 1991; Kawagishi-Kobayashi et al., 1997; Langland and Jacobs, 2002). Finally, vCOP may modulate caspase activation and thus control the induction of apoptosis or regulate the inflammatory response (Bouchier-Hayes and Martin, 2002). Additional immune evasion genes (e.g., TNF receptor-like, Bak-like, and TRAF2-like proteins, as well as an inhibitor of apoptosis [IAP] protein) have also been detected in some iridoviruses (Table III). However, for the most part, gene identities have been assigned based on BLAST analysis, so caution must be exercised in the assignment of functions to these putative ORFs.

In addition to the abovementioned proteins, for which a function is inferred by homology, more than half of the iridovirus ORFs encode putative proteins that lack significant homology to known cellular proteins in the current databases. Surprisingly, despite the lack of homology to known nonviral proteins, the vast majority of these are conserved among all iridoviruses, suggesting that they likely play key roles in viral replication, whether they function directly in virus replication by acting as transcription factors, assembly proteins, or structural components, or whether they downregulate immune responses by interfering with IFN, TNF, MHC, STAT, or other immune-related proteins, remains to be seen.

Viral genes involved with nucleic acid metabolism generally show marked sequence similarity compared to the corresponding genes of higher and lower vertebrates. For example, the identity/similarity between FV-3 Pol II α and the largest subunit of human Pol II is 28/45% over 707 amino acids. Similarities between cytokine genes of ectothermic vertebrates and mammals are, however, considerably lower, making it that much more difficult to decide if a given viral gene is a homolog of a cellular cytokine gene. For instance, the amino-acid sequence identity between fish TNF and interferon genes and those of mammals is only about 20% (Long *et al.*, 2004a; Zou *et al.*, 2003). If the same low level of identity exists between the TNF and interferon receptors of lower vertebrates and their mammalian counterparts, it would be extremely difficult to detect an iridovirus homolog of a cytokine receptor by comparison to its mammalian counterpart. However, the increasing availability of sequences of key immune-related genes in lower vertebrates, such as *Xenopus*, axolotl, zebrafish, etc., should greatly facilitate the identification of viral homologs. Since other large DNA viruses, including poxviruses, herpesviruses, and African swine fever virus, encode proteins that mediate viral evasion of host immunity (Johnston and McFadden, 2003; Tortorella *et al.*, 2000), it is likely that iridoviruses also contain such genes.

In IIV-6, five genes involved in DNA replication are clustered in shared orientation in one region of the genome: DNA polymerase, DNA topoisomerase II, a helicase, nucleoside triphosphatase I, and an exonuclease II (Muller *et al.*, 1999). Other notable putative genes include an NAD+ dependent DNA ligase, which contrasts with the ATP-dependent DNA ligase reported from PBCV-1 and ASFV, and a putative homolog of sillucin, a cysteine-rich peptide antibiotic. The genome contains six origins of replication (Jakob and Darai, 2002; Jakob *et al.*, 2001). The promoter regions of the MCP and DNA polymerase have recently been characterized (Nalcacioglu *et al.*, 2003). Two adjacent ORFs been detected in IIV-6 with truncated homology to the nuclear polymerizing (ADP-ribosyl) transferase from eukaryotic organisms (Berghammer *et al.*, 1999). Interestingly, the large subunit of the IIV-6 ribonucleotide reductase appears to contain an intein, a form of selfish genetic element that removes itself from the protein posttranslationally by an autocatalytic splicing mechanism. This is the first reported insect virus intein (Pietrokovski, 1998).

Baculovirus repeated ORFs (*bro* genes), known as ALI motifs, in entomopoxviruses due to the invariant alanine-leucine-isoleucine residues (Afonso *et al.*, 1999), are a multi-gene family of unknown function. The amino terminal residues contain a DNA binding domain and are conserved across many families of viruses, whereas the C-terminus is also conserved among members of the nucleocytoplasmic large DNA virus clade. Certain BRO proteins may influence host DNA replication or transcription by regulating host chromatin structure (Zemskov *et al.*, 2000). Three *bro*-like genes have been identified in the invertebrate iridovirus IIV-6 and one in IIV-31, compared to 2–11 *bro*-like genes in ascoviruses. Apart from baculoviruses, *bro*-like genes are also present in entomopoxviruses, phycodnaviruses and a number of bacteriophages and bacterial transposons (Bideshi *et al.*, 2003). Sequence analysis indicates that *bro*-like genes in IIVs are most closely related to those of an ascovirus (DpAV-4a), although two of those in IIV-6 appear to be derived from fusion of N- and C-termini of different origins. However, *bro* diversity appears largely independent of the evolutionary history of invertebrate viruses. The *bro*-like genes are not found in viruses of vertebrates.

B. Gene Order and Elucidation of Gene Function

Despite similarity in their complement of core genes involved in DNA/RNA metabolism and other conserved functions (Table IV), iridovirus genomes show a marked lack of colinearity. Dot plot analyses, in which the nucleic acid sequence of one virus genome is compared position-by-position to that of another, indicate that strains of an iridovirus species (FV-3 vs. TFV), show a high level of sequence colinearity, whereas comparisons between species (ATV vs. TFV), show clear evidence of rearrangement. Moreover, dot plot analyses of viruses from different genera (ATV vs. IIV-6, LCDV-1, and ISKNV) showed no colinearity, indicating considerable rearrangement from the ancestral precursor (Jancovich *et al.*, 2003). Reduced colinearity likely reflects a high rate of recombination during virus replication. As the only genetic constraint may be the need to maintain the integrity of essential open reading frames, rearrangements that shuffle the genome without inactivating those genes would not be selected against. With time, those gene constellations that promote efficient growth in one or more host species would likely be selected resulting in the establishment of viral lineages with markedly different gene order, host preference, or tissue tropism. Moreover, the high rate of recombination seen among various iridoviruses may reflect the presence of the terminal redundancies, the complex nature of concatameric DNA, or the presence of repeat elements throughout the genome.

Data from various iridovirus genome sequencing projects coupled with extensive sequencing studies involving cold-blooded vertebrates provides the basis for a systematic survey of the genetic complement of the family *Iridoviridae*. Earlier studies used temperature-sensitive and drug-resistant mutants to elucidate the function of a few replicative genes (Chinchar and Granoff, 1984, 1986). Contemporary approaches involving the use of small interfering RNA (siRNA) and antisense morpholinos (asMO) to knock down viral gene expression, and homologous recombination to knock out viral gene function, should allow us to link a specific viral gene to its function (Means *et al.*, 2003; Meister and Tuschl, 2004; Neuman *et al.*, 2004). Preliminary "proof of concept" experiments with FV-3 have successfully demonstrated that, as MO targeted against the MCP inhibited capsid protein synthesis and reduced viral yields (V. G. Chinchar, unpublished data). Continued progress using the above three methodologies should allow us to efficiently determine the function of key viral genes—surely one of the most exciting prospects in the study of these viruses.

IV. VIRAL REPLICATION STRATEGY

A. General Features of Iridovirus Replication

Most of what is known about iridovirus replication has been elucidated using FV-3 as a model (Fig. 3). Because of this, viral replication will be described from the viewpoint of FV-3, and other iridoviruses will only be mentioned where their replication differs from that of FV-3. The virus particle binds to a common, but currently unknown, cellular receptor, since cells of mammalian, piscine, and amphibian origin are readily infected at temperatures $<32\,°C$ (Chinchar, 2002; Goorha and Granoff, 1979; Granoff, 1984; Williams, 1996). As both naked and enveloped virions are infectious, there are likely at least two different host receptor proteins. Following binding to host cells, enveloped virus is thought to enter via receptor mediated endocytosis, whereas naked virions uncoat by fusion at the plasma membrane. Subsequently, viral DNA cores make their way to the nucleus, where they engage in the first phase of FV-3 replication.

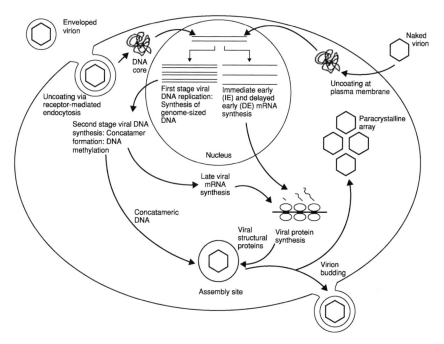

FIG 3. Iridovirus replication cycle. See text for details.

Viral gene expression begins in the nucleus, where two key events take place. In the first of these, early viral transcripts of two classes, immediate early (IE) and delayed early (DE), are synthesized using input virion DNA as a template. Early transcripts are thought to encode regulatory proteins and key catalytic enzymes such as the viral DNA polymerase and the viral homologs of cellular RNA polymerase II (Pol II). Following its translation in the cytoplasm, the viral DNA polymerase enters the nucleus where it synthesizes unit to twice unit size copies of the viral genome, representing first stage DNA synthesis (Goorha, 1982; Goorha et al., 1978). Newly synthesized viral DNA is subsequently transported into the cytoplasm where further rounds of DNA replication result in the formation of large concatameric structures, representing second stage DNA synthesis (Goorha and Dixit, 1984). In addition, viral DNA is methylated in the cytoplasm through the enzymatic activity of a virus-encoded cytosine DNA methyltransferase (Willis et al., 1984). Goorha et al. (1984) suggested that vertebrate iridoviruses contain a restriction-modification system similar to that seen in some bacteriophages. In this view, methylation protects the viral genome from degradation by a virus-encoded endonuclease. Virions demonstrate endonuclease activity, but no endonuclease that targets unmethylated CpG sequences has been molecularly identified. While the restriction-modification scenario is plausible, the identification of at least one vertebrate virus (SGIV) that lacks a DNA methyltransferase calls this model into question (Song *et al.*, 2004). Either this virus lacks both the DNA methyltransferase and the endonuclease, or methylation plays another role in the virus life cycle. An alternative role for viral DNA methylation is suggested by the observation that bacterial DNA containing unmethylated CpG residues induces innate immunity following its interaction with toll-like receptor 9 (TLR9) (Bauer *et al.*, 2001). Perhaps FV-3 methylates its genome to prevent TLR9-mediated induction of innate immunity.

Viral DNA, presumably in its concatameric form, is found within cytoplasmic viral assembly sites (AS). However, it is not known if viral DNA is directed to developing AS, or whether viral proteins coalesce around concatameric DNA and result in AS formation. Circumstantial evidence (see following) suggests that late (L) viral mRNA synthesis takes place either within the cytoplasm of infected cells or within AS, and is catalyzed by a virus-modified cellular polymerase, or a novel viral Pol II. In support of the latter, FV-3 and other iridoviruses encode multiple subunits of Pol II, including the two largest subunits (Tan *et al.*, 2004). Clearly, the synthesis of L viral messages in the cytoplasm requires either a novel viral polymerase, or some mechanism that transports cellular Pol II from the nucleus into the cytoplasm. While both mechanisms are possible, the existence of virus-encoded subunits favors the former. Late viral transcripts encode viral structural proteins that make their way into AS and take part in virion formation. The specific steps in virion formation are not known, but it is thought to be a concentration dependent process that involves the packaging of viral DNA by a "headful" mechanism using concatameric DNA as an intermediate (Goorha and Murti, 1982). FV-3 is believed to utilize a headful mechanism, in which the unit length genome along with an additional 20–30 kbp of genomic DNA is packed into a preformed viral capsid. As a consequence of this mode of packaging, the viral genome is circularly permuted and terminally redundant. While headful packaging ensures that each virion receives at least one copy of every gene, the terminal redundancy may have an evolutionary advantage. Since 10–30% of the viral genome is effectively diploid, it is possible that mutations within one of the two genes allow the emergence of viral proteins with new functions. Newly synthesized virions exit the AS, and accumulate either in large, paracrystalline arrays elsewhere in the cytoplasm, or bud from the plasma membrane and, in the process, acquire an envelope. Most FV-3 virions remain cell-associated, suggesting that accumulation within paracrystalline arrays is the predominant pathway. Both nonenveloped and enveloped forms of FV-3 are infectious, but the latter have a higher specific infectivity. Events in virus replication are summarized in Fig. 3.

B. RNA Synthesis: Virion-Associated and Virus-Induced Transcriptional Activators

FV-3 gene expression occurs in three coordinated phases leading to the synthesis of IE, DE, and L transcripts (Willis and Granoff, 1978; Willis et al., 1977). As described above, IE and DE transcription takes place in the nucleus and are catalyzed by host Pol II, whereas L transcription is a cytoplasmic event catalyzed by a virus-modified or virus-encoded Pol II (Goorha, 1981). Biochemical studies, using inhibitors such as cycloheximide (CHX, which block translation and allow only IE viral mRNAs to be synthesized) and flurophenylalanine (which blocks the appearance of L viral mRNAs), support the notion that one or more IE proteins are required to switch on DE transcription, and that one or more DE proteins are needed to activate L mRNA synthesis (Willis et al., 1977). In addition to DE proteins, full L gene expression is only seen in the presence of viral DNA replication (i.e., inhibitors of viral DNA polymerase such as phosphonoacetic acid or cytosine arabinoside, or temperature sensitive mutants that block DNA synthesis, result in markedly reduced levels of L transcripts and proteins) (Chinchar and Granoff, 1984, 1986). Events in the coordinated expression of viral genes are shown in Fig. 4. Compelling evidence supporting the role of host Pol II in the synthesis of IE transcripts comes from the observation that viral transcription is blocked in wild-type host cells by α-amanitin, an inhibitor of host Pol II, but not in cells that are α-amanitin resistant (Goorha, 1981). Data in support of the notion that IE and DE transcription takes place in the nucleus, whereas L mRNA synthesis occurs in the cytoplasm, is based on the observation that E, but not L, mRNAs possess 4–5 ^{m}6A residues/ 1000 nucleotides (Raghow and Granoff, 1980). Since the enzyme responsible for synthesis ^{m}6A methylation is thought to be a nuclear enzyme, it follows that only E mRNAs display this modification.

As indicated above, early viral transcription is catalyzed by host Pol II, whereas L viral mRNA synthesis is likely mediated by a virus-encoded enzyme. The largest subunit of vPol II is homologous to a wide range of eukaryotic Pol IIs and displays 28–30% sequence identity and

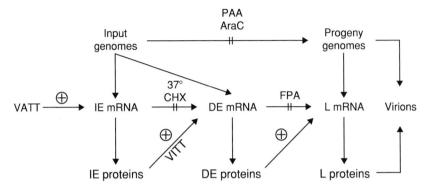

F𝐼G 4. RNA synthesis in FV-3 infected cells. VATT and VITT are the putative transcriptional activators of IE (VATT) and DE/L (VITT) mRNA synthesis. The points at which various drugs or elevated temperatures act to block viral RNA synthesis are shown.

44–46% sequence similarity to various fungal, plant, mammalian, and arthropod Pol IIs. Despite the marked homology, the viral protein differs from host Pol II by the absence of the carboxyl-terminal domain (CTD). The CTD is composed of multiple repeats (~50 in man and mouse) of the sequence YSPTSPS that are phosphorylated by various kinases. The CTD may be viewed as a docking platform for factors that mediate transcription (Barilla *et al.*, 2001). Truncation of the CTD results in a drop in transcriptional initiation, and recent evidence suggests that factors that mediate transcriptional termination and polyadenylation bind to the CTD (Kim *et al.*, 2004; Meininghaus and Eick, 1999). As vPOL II lacks a CTD, it is anticipated that it may utilize novel mechanisms to catalyze viral RNA synthesis.

In contrast to herpes simplex virus type 1, which also uses host Pol II for its transcription, FV-3 DNA is noninfectious. However, FV-3 DNA extracted from virus particles or present within heat-inactivated virions can be nongenetically reactivated by coinfection with UV-inactivated FV-3 (Willis *et al.*, 1979). It is thought that functional proteins from UV-inactivated FV-3 drive IE transcription using functional genomic DNA as a template. Subsequently, one or more of the newly synthesized IE proteins are believed to activate DE transcription and allow for the eventual synthesis of L genes and the formation of infectious virions. Both IE and DE transcription are catalyzed by host Pol II. Furthermore, one or more DE proteins are thought to be required for L mRNA synthesis. Whether the DE protein requirement for L mRNA synthesis reflects the need for a novel transcriptional transactivator, or whether it simply is due to the requirement for a virus-encoded RNA polymerase, is not known.

Willis and Granoff (1985) showed that a virion-associated transcriptional transactivator (VATT) is required for the synthesis of IE mRNA, whereas a newly synthesized virus-induced transcriptional transactivator (VITT) is needed to override the methylation-induced transcriptional block and synthesize DE mRNA (Willis *et al.*, 1989, 1990a,b). The molecular identities of VATT and VITT are unknown, as is their precise mechanism of action. VATT may alter the DNA template, modify cellular Pol II, or interact with cellular transcription factors and permit transcription of IE messages. Thompson *et al.* (1988) showed that a mutated IE promoter that was able to be methylated *in vitro* was still transcriptionally active, suggesting that IE genes do not contain methylatable sites in regions critical for transcription. They interpreted these results to suggest that the role of the VATT was not to override the inhibitory effect of methylation, but rather to permit IE transcription by a yet to be defined mechanism. In support of this model, a chloramphenicol acetyltransferase (CAT) reporter gene located downstream from an IE FV-3 promoter was only transcribed in FV-3 infected cells. CAT expression did not require full FV-3 gene expression as CAT mRNA synthesis was detected in cells infected in the presence of CHX, as well as in cells infected with UV-inactivated FV-3. CAT activity was, however, not seen in cells infected by heat-inactivated virus, suggesting that VATT was a heat-sensitive protein (Willis and Granoff, 1985). In contrast, a plasmid containing the CAT reporter gene downstream from a methylated adenovirus promoter (^{m}Ad-CAT) was only transcribed in productively FV-3-infected cells, suggesting that the role of the VITT is to override the inhibitory effect of a methylated promoter (Thompson *et al.*, 1986). As expected, ^{m}Ad-CAT was not expressed in cells infected with heat-inactivated or UV-inactivated FV-3, or in cells infected in the presence of CHX. Taken together, these results indicate that a virion-associated protein (VATT) is required to initiate IE transcription, while a newly synthesized viral protein (or virus-induced cellular protein, VITT), is required for DE transcription.

The patterns of transcription in IIV-6 and IIV-9 are similar to that described for FV-3. The invertebrate virus genomes are organized into distinct clusters of IE and DE genes, whereas the L genes are scattered throughout the genome (D'Costa *et al.*, 2001, 2004; McMillan and Kalmakoff, 1994). Of the 468 potential ORFs (234 of which are nonoverlapping) identified in the IIV-6 genome (Jakob *et al.*, 2001), just 137 transcripts have been identified and classified by temporal expression (D'Costa *et al.*, 2004).

C. Inhibition of Host Macromolecular Synthesis

FV-3 infection is accompanied by a rapid and profound inhibition of host macromolecular synthesis (Goorha and Granoff, 1974; Raghow and Granoff, 1979). It is thought that shut-off of cellular transcription and translation is a direct effect of virus infection, whereas inhibition of cellular DNA synthesis is a consequence of the inhibition of RNA and protein synthesis (Willis *et al.*, 1985). In productively infected cells, the shut-off of cellular protein synthesis is accompanied by a switch to the selective synthesis of viral proteins. The transition from host to viral protein synthesis is likely the result of several interacting events: the inhibition of cellular transcription, the degradation of host messages, the synthesis of abundant amounts of highly efficient viral transcripts, and the presence of a viral homolog of eukaryotic initiation factor 2 alpha (eIF-2α) that is thought to play a role in maintaining viral translation in the face of a shut-off of host protein synthesis (Chinchar and Dholakia, 1989; Chinchar and Yu, 1990, 1992). A translational activator of L protein synthesis has been suggested based on biochemical experiments, but the putative factor has not been identified based on molecular or genetic data (Raghow and Granoff, 1983). It should be noted that the VATT is heat-sensitive, whereas the factor responsible for host translational shutoff is a heat-stable protein, as shutoff takes place not only in productively-infected cells, but in cells infected with heat- and UV-inactivated FV-3. However, the identity of the inhibitor remains unknown.

Most, but not all, ranaviruses examined to date encode a ~260 amino-acid protein (vIF2α) that is homologous to eIF-2α at its amino terminus (Essbauer *et al.*, 2001; Jancovich *et al.*, 2003; Yu *et al.*, 1999). Interestingly, vaccinia virus, as well as other poxviruses, encodes a much shorter (88 amino acid) protein, designated K3L, that is also homologous to the amino terminus of eIF-2α (Kawagishi-Kobayashi *et al.*, 1997). Although sequence identity/similarity is limited, all three proteins share a highly conserved motif (KGYID[L/V]) thought to play a role in the interaction between PKR and eIF-2α. In the currently accepted model, K3L and vIF2α are thought to act as pseudo-substrates that bind activated PKR and prevent the phosphorylation, and thus inactivation, of eIF-2α. Surprisingly, ranavirus vIF2α shows no homology to eIF-2α or to any other protein in the database downstream of the KGYID motif. Whether the carboxyl-terminal two-thirds of vIF2α plays an additional function in virus replication is not known. Surprisingly, in the FV-3 isolate sequenced by Tan *et al.* (2004), the amino-terminal two-thirds of the putative vIF2α protein has been truncated. However, despite the loss of the domain thought to be involved in PKR binding, this FV-3 isolate readily shuts off host translation and selectively synthesizes viral proteins. Since vaccinia virus has at least two proteins that prevent PKR activation, K3L that binds PKR and blocks phosphorylation of eIF-2α and E3L that binds dsRNA and inhibits PKR activation (Langland and Jacobs, 2002), it is possible that FV-3 also possesses another protein that maintains viral protein synthesis in the face of host translational shut-off.

D. Apoptosis

Ranavirus infections are accompanied by marked cytopathic effect *in vitro*. Recent work indicates that cultured cells infected with ranaviruses undergo apoptosis, as indicated by Hoechst staining, DNA laddering, TUNEL assay, and the appearance of phosphatidylserine on the outer leaflet of the plasma membrane (Chinchar *et al.*, 2003; Essbauer and Ahne, 2002; Zhang *et al.*, 2001). In addition, apoptosis is blocked by a pan-caspase inhibitor (Z-VAD-FMK), strengthening the suggestion that caspase activation played a role in this process (Chinchar *et al.*, 2003). Apoptosis has been described in GF cells infected by RSIV, and it was suggested that caspase-3 and caspase-6 may be involved in the morphological changes seen in the middle and late stages of apoptosis (Imajoh *et al.*, 2004). While these studies make a strong case for apoptosis *in vitro*, it is unclear whether cell death seen in infected animals is due to necrosis or apoptosis. A recent study with LCDV suggests that there may be differences between *in vivo* and *in vitro* responses. Hu *et al.* (2004) employed Hoechst staining, DNA laddering, and caspase activation techniques to demonstrate that LCDV infection of FG-9307 cells resulted in apoptotic death. However, since infected animals typically show wart-like growths resulting from enlargement of individual cells, there appears to be a discrepancy between cell death *in vitro* and cell fate *in vivo*. In view of that, the authors suggest that virus-host interactions *in vivo* likely play a key role in determining the mode and the timing of the apopototic response. A TNF receptor-like homolog (ORF 167L) was recently identified in LCDV with possible involvement in the apoptosis cascade (Essbauer *et al.*, 2004).

Apoptosis inhibitors that block the activation of caspases (Seshagiri and Miller, 1997) have been detected in IIV-6 (Jakob and Darai, 2002).

These inhibitors of apoptosis proteins (IAP) are characterized by the presence of a C_3HC_4 Zinc/RING finger and BIR (baculovirus inhibitor repeat) domains. The three copies of this putative gene identified in IIV-6 vary in length and show 17.5–19.5% identity and 22.9–40.6% similarity in amino-acid sequence to the IAP protein of *Cydia pomonella granulovirus,* a demonstrated functional inhibitor of apoptosis (Birnbaum *et al.,* 1994). Three additional ORFs in IIV-6 showed lower levels of homology to established *iap* genes.

V. Vertebrate Iridoviruses: Pathology and Diagnosis

Pathology associated with vertebrate iridoviruses ranges from inapparent subclinical to fulminant fatal infections. Infection of marine and freshwater fish by LCDV is distinctive, since it is accompanied by the presence of wart-like lesions on the dermal surfaces of infected fish (Weissenberg, 1965). Histological examination reveals the presence of greatly enlarged individual cells, each encapsulated by a hyaline matrix. The duration of cell growth and virus proliferation is variable and probably temperature-dependent. Lymphocystis lesions regress spontaneously and are usually not life-threatening. However, when lesions affect the eyes (Dukes and Lawler, 1975) or other internal organs (Russell, 1974), infected fish may die. Mortality may also occur under culture conditions, particularly if fish are debilitated or suffer secondary bacterial infection. In contrast, fish infected with megalocytiviruses are lethargic, have abnormal coloration of the body, petechia of the gills, and lesions involving the spleen, gills, and digestive tract (Gibson-Kueh *et al.,* 2003; He *et al.,* 2000; Jung *et al.,* 1997). RSIV, for example, causes cell hypertrophy in the spleen, heart, kidney, liver, and gills and leads to necrosis of the renal and splenic hematopoietic tissues (Nakajima *et al.,* 1998; Oshima *et al.,* 1998; Qin *et al.,* 2002). Cellular hypertrophy is accompanied by the appearance of inclusion body-bearing cells (IBC). IBC contain a unique inclusion body that may be sharply delineated from the host cell cytoplasm by a limiting membrane. The inclusions contain viral AS as well as abundant ribosomes, rough endoplasmic reticulum, and mitochondria.

Ectothermic animals infected with ranaviruses show variable pathology. LMBV infections appear to involve only the swim bladder and result in a build up of a wax-like material in the lumen of the bladder (Hanson *et al.,* 2001). Infections with EHNV, FV-3, and ATV cause systemic infections involving the liver, kidney, and gastrointestinal tract (Ahne *et al.,* 1997; Bollinger *et al.,* 1999; Jancovich *et al.,*

1997; Wolf *et al.*, 1968). Typical symptoms of infection in fish include focal or generalized necrosis of the hematopoietic tissues and other organs, hemorrhages of internal organs, and peticheal hemorrhages of the skin. Altered swimming behavior has been observed in some cases. ATV infection of salamanders causes lesions involving internal organs and also leads to skin sloughing and the development of dermal polyps (Jancovich *et al.*, 1997). BIV is highly pathogenic for tadpoles and juveniles of several species (*Limnodynastes ornatus, Litoria caerulea, L. alboguttata, Cyclorana brevipes, Pseudophryne* sp.), as well as juvenile and adult cane toads, *Bufo marinus*. Experimental challenges in juvenile frogs have resulted in high mortality (Cullen and Owens, 2002). Metamorphs and adults generally do not show overt signs of infection, although adults occasionally suffer weakness, and hemorrhages may occur. Histopathological changes include acute necrosis of hematopoietic and lymphoid tissues, as well as leucocytes in liver, kidney, and respiratory organs. Significant hemosiderosis may occur in chronically infected frogs (Cullen and Owens, 2002). BIV experimentally transmitted to barramundi fish caused focal or diffuse necrosis of the hematopoietic tissues of the kidney, spleen and the liver, resulting in 100% mortality (Moody and Owens, 1994).

Experimental infection of adult, immunocompetent *Xenopus* indicates that these animals readily clear FV-3 infection and that viral antigens are detected only transiently in the kidney. In contrast, immunocompromised adult *Xenopus* succumb to disease and show viral antigens in the kidney, liver, and spleen. *Xenopus* tadpoles are readily infected with FV-3 and show high levels of mortality (Gantress *et al.*, 2003; J. Robert, unpublished data). Histologically, there appears to be a difference between ranavirus and megalocytivirus infections. The latter are characterized by the appearance of IBCs, whereas ranavirus infections show no such structures. In place of inclusion bodies, ranavirus infections are accompanied by the presence of one or more AS within the cytoplasm. As infection progresses, AS may become quite large, but unlike inclusion bodies, they are not set apart from the cytoplasm by a limiting membrane and lack ribosomes and mitochondria. Moreover, AS appear to be surrounded by elements of the cellular cytoskeleton (Murti and Goorha, 1989, 1990; Murti *et al.*, 1988). Whether these elements contribute to the formation and integrity of the AS, or whether they are simply excluded as the AS enlarges is not known. In this respect, microtubules, intermediate filaments, and vimentin structures are integral components of viral factories in ASFV-infected cells (Heath *et al.*, 2001).

Gross lesions induced by megalocytivirus and ranavirus infections are not unique, so that definitive identification of the viral agent requires additional tests. The presence of 120–300 nm nonenveloped icosahedral particles within the cytoplasm of infected cells establishes probable infection by an iridovirus, but does not indicate either the genus or species of the infecting agent. Classical techniques of virus isolation using cell culture are laborious and require subsequent molecular analyses (Drury *et al.*, 2002). To this end, a number of PCR-based and antigen-based tests have been developed (Gould *et al.*, 1995; Marsh *et al.*, 2002; Office International des Epizooties, 2000; Oshima *et al.*, 1996, 1998; Whittington *et al.*, 1997, 1999; Zupanovic *et al.*, 1998a,b). Definitive identification of the etiological agent depends on PCR primers that amplify a specific size fragment from a known viral target, or that flank a region containing a unique DNA sequence (Chinchar and Mao, 2000; Hyatt *et al.*, 2000). PCR detection tests have frequently targeted the MCP (Bollinger *et al.*, 1999; Mao *et al.*, 1997, 1999a,b; Marsh *et al.*, 2002; Murali *et al.*, 2002), DNA methyltransferase (Mao *et al.*, 1999b), DNA polymerase (Kurita *et al.*, 1998), ATPase (Sudthongkong *et al.*, 2002), and ribonucleotide reductase genes (Oshima *et al.*, 1996, 1998), or other ORFs (Ahne *et al.*, 1998; Cullen and Owens, 2002; Gould *et al.*, 1995; Tamai *et al.*, 1997), or characteristic restriction fragments of the viral genome (Kim *et al.*, 2002; Miyata *et al.*, 1997). Nested PCR targeted at a *Pst*I fragment of RSIV appeared to offer no advantages over conventional PCR (Wang *et al.*, 2003a).

Serological tests have been developed for detecting antiranavirus antibodies (Zupanovic *et al.*, 1998a) and iridoviral antigens (Qin *et al.*, 2002; Shi *et al.*, 2003), but need to be interpreted with caution because of cross-reactivity between members of the same viral genus (Hedrick *et al.*, 1992a). Interestingly, the monomeric form of the MCP may not be the most important protein antigenically in EHNV or SGIV. Only 3 out of 20 mAbs precipitated the 50 KDa MCP of EHNV, whereas the majority of mAbs recognized polypeptides of 15 and 18 KDa (Monini and Ruggeri, 2002). Similarly, mAbs to SGIV reacted strongly with proteins of 100 and 117 KDa, but showed no affinity to the MCP (Shi *et al.*, 2003).

A number of more complex diagnostic techniques have also been developed. *In situ* hybridization targeted at MCP gene sequences has been successfully employed to identify SGIV infected tissues in formalin fixed samples of Malabar grouper (Huang *et al.*, 2004). PCR + RFLP was successfully used to differentiate between ranaviruses (Marsh *et al.*, 2002). Most recently, amplified fragment length polymorphism (AFLP) and real-time quantitative PCR were applied to detect and differentiate between geographical isolates of LMBV (Goldberg *et al.*, 2003), and multiplex PCR has been employed for the diagnosis of different strains of RSIV using sets of primers targeted at three virus genes and a virus specific *Pst*I fragment (Jeong *et al.*, 2004). Loop-mediated isothermal amplification (LAMP) has also been shown to detect the presence of RSIV, using primers designed from the *Pst*I restriction fragment sequence (Caipang *et al.*, 2004). The LAMP procedure was ten times more sensitive than conventional PCR. A clear correlation between the number of copies of template DNA and the optical density of a magnesium pyrophosphate by-product of the reaction indicated that the LAMP technique could be used semi-quantitatively to estimate the concentration of RSIV in fish samples.

VI. Immune Responses to Vertebrate Iridoviruses

Our understanding of immune responses among lower vertebrates is incomplete, although experience with several species of bony fish and *Xenopus* indicate that fish and amphibians possess immune responses that are functionally and molecularly similar to those of higher vertebrates (DuPasquier, 2001; DuPasquier *et al.*, 1989; Miller *et al.*, 1998; Shen *et al.*, 2002). Lower vertebrates display many of the effectors associated with innate responses including antimicrobial peptides with antiviral activity (Chinchar *et al.*, 2001, 2004; Zhang *et al.*, 2004b), complement, Natural Killer cells (Hogan *et al.*, 1996), interferon (Long *et al.*, 2004a), and phagocytic cells. Amphibians and fish develop viral antibodies (Bengten *et al.*, 2000), and likely possess antigen-specific cytotoxic cells (Yoder, 2004). At the molecular level, genes for immunoglobulin, T cell receptor (TCR), and MHC class I and II have been identified (Antao *et al.*, 1999, 2001; Bengten *et al.*, 2000; Wilson *et al.*, 1998). Sequence analysis has revealed that lower vertebrates also possess antiviral cytokines such as TNF, interferon, and other immune-related molecules such as Fas Receptor and caspase 8, but the molecular identity between the genes from lower vertebrates and those from mammals is low (Long *et al.*, 2004a,b; Zou *et al.*, 2003). The low level of identity has frustrated efforts to identify CD4 and CD8 in some species (e.g., channel catfish) and interferon in amphibians. Moreover, several striking differences exist between lower and higher vertebrates. First, premetamorphic tadpoles lack MHC class I molecules and show impaired cytotoxic T cell (CTL) responses (DuPasquier *et al.*, 1989). Second, mammals possess five classes of immunoglobulin (IgA, IgD, IgE, IgG, and IgM), whereas channel catfish possess two (IgD and IgM), with the latter arranged as a tetramer, rather than the pentamer seen in mammals (Miller *et al.*, 1998). Third, mammalian NK cells possess two types of inhibitory receptors on their surface (Killer Cell Ig-like receptors [KIRs] and CD94/NKG2 in humans; Ly49 and CD94/NKG2 in mice) that recognize self MHC and prevent cell killing. In contrast, KIRs, Ly49-like molecules, or CD94/NKG2 have not yet been detected in fish. In their place, a large family of NITRs (novel immune-type receptors) has been identified that may serve as the functional equivalent of the inhibitory and activation receptors seen in mammals (Hawke *et al.*, 2001; Yoder *et al.*, 2004).

Little is known about the immune response in any lower vertebrate species against iridovirus infection. However, the ability to protect fish from megalocytivirus infection by vaccination suggests that immune responses play a role in protection from disease (Nakajima *et al.*, 1999). Interestingly, an iridovirus causing acute necrosis of the spleen and renal hematopoietic tissue in catfish (Pozet *et al.*, 1992) has been observed to suppress macrophage activity and lymphocyte proliferation in sheatfish cells *in vitro* (Siwicki *et al.*, 1999), although *in vivo* studies have yet to be reported. Furthermore, the widespread occurrence of subclinical infections with LMBV suggests that either innate or acquired responses control the virulence of iridovirus infections and prevent or moderate clinical disease (Hanson *et al.*, 2001). In *Xenopus*, Gantress *et al.* (2003) demonstrated that adult animals developed anti-FV-3 antibodies following intraperitoneal injection with FV-3. Furthermore, the ability of prior FV-3 infection to protect tadpoles and prevent clinical disease induced by a pathogenic ranavirus designated RGV-Z suggests that lower vertebrates can mount protective responses against iridoviruses (V. G. Chinchar and S. LaPatra, unpublished data). Whether protection involves both viral antibodies and antiviral cytotoxic T cells is presently unknown, but it is interesting to note that *Xenopus* tadpoles are much more sensitive to FV-3 infection than adults, and this may be due to the absence of MHC class I expression in premetamorphic animals (Gantress *et al.*, 2003). Somamoto *et al.* (2002) demonstrated the presence of cytotoxic T cell (CTL)-like effectors that lysed rhabdovirus-infected syngeneic crucian carp cells; however, whether anti-iridovirus CTLs are present in the various species of susceptible amphibians, reptiles, and fish remains to be determined. Hopefully, the *Xenopus*/FV-3 system will prove amenable to definitively identifying antiranavirus CTLs. If immunity to vertebrate iridovirus infection is similar to that seen in mammals, we might expect protection to be a multi-faceted phenomenon involving antimicrobial peptides, cytokines, and NK cells acting early as a first line of defense, followed by the appearance of antiviral CTLs and antiviral antibodies. While circumstantial evidence supports a role for all of these, in no one system have all components been identified and characterized.

VII. Ecology of Vertebrate Iridoviruses

A. Infections in Natural Populations

Our understanding of the ecology of iridovirus infections is fragmentary. Two examples may serve to illustrate the state of our knowledge. First is the case of *Ambystoma tigrinum virus*. ATV is the virus responsible for localized mass mortality of various subspecies of tiger salamanders throughout western North America. ATV die-offs have been seen in manmade water holes in Arizona, farm ponds in Saskatchewan, and natural water courses in Utah and other western states (Bollinger *et al.*, 1999; Docherty *et al.*, 2003; Jancovich *et al.*, 1997). Mortality is high, and although some populations recover, others stay depressed for several years (D. Schock, personal communication). Iridoviruses are relatively stable in aquatic environments, but it is unclear whether subsequent die-offs reflect the reintroduction of the virus, the activation of latent or subclincial infections in the surviving host population, transmission from asymptomatic sympatric alternative hosts (e.g., infected frogs or fish), or persistence of the virus in the environment. It is not clear if ATV infects only salamander species or also fish and other amphibian species (Jancovich *et al.*, 2001). If these latter species can be infected, they may serve as a reservoir for the virus when transmission opportunities in salamander populations are limited (D. Schock, unpublished data). Natural modes of transmission include water-borne exposure or cannibalism of infected animals or carcasses (Jancovich *et al.*, 2001). It is hypothesized that the recrudescence of active infection, perhaps induced by spawning or other stresses, leads to the release of virus and infection of susceptible hatchlings. Moreover, the transport by recreational fishermen of ATV-infected bait salamanders may serve as an unintended but effective method of dispersing ATV throughout western North America (Jancovich *et al.*, 2005).

Second is the case of largemouth bass virus. Following its initial detection in the Santee-Cooper Reservoir (South Carolina, USA) in the

1996, LMBV has been shown to be present through the southeastern United States, as far north as Michigan, and as far west as Texas (Grizzle and Brunner, 2003; Grizzle *et al.*, 2002; Plumb *et al.*, 1996). Although LMBV is responsible for the death of adult largemouth bass (*Micropterus salmoides*), infection is not invariably fatal, as large numbers of virus-positive, subclinically-infected fish have been detected in some lakes (Hanson *et al.*, 2001; Woodland *et al.*, 2002). The reasons why LMBV infection triggers die-offs in one case and leads to subclinical infections in another are not known. Whether resistance reflects pre-existing immunity, an effective response against a new infection, or environmental insults that lead to immune suppression is not known. However, different geographical isolates of LMBV were recently shown to exhibit major differences in pathogenicity and virulence in largemouth bass juveniles; the number of fish that died from LMBV infection and the titer of virus in the survivors differed markedly according to the origin of the isolate (Goldberg *et al.*, 2003). Genotypic differences in LMBV isolates appear to have marked implications for key phenotypic characteristics. Aside from largemouth bass, other species of centrachids are susceptible to infection, but do not develop clinical signs of disease. Whether these species serve as reservoirs for LMBV is not clear. Interestingly, doctor fish virus and guppy virus 6 isolated from tropical fish imported from Southeast Asia share sequence similarity with LMBV within the MCP gene, and all are classified as strains of SCRV (Mao *et al.*, 1997, 1999b). Whether this indicates that LMBV-like viruses have a worldwide distribution or whether it suggests that the virus was introduced into North America via imported tropical fish is not known.

B. Infections in Farmed Populations

The number of reports of iridovirus infection of fish has increased rapidly over the past decade, with abundant reports from China, Japan, Korea, and Taiwan. Matsuoka *et al.* (1996) reported that systemic iridovirus disease occurred in 19 species from 3 taxonomic orders of cultured marine fish in Japan. A comparison of iridovirus isolates from long-spined sea bream (*Argyrops spinifer*), amberjack (*Seriola dumerili*), striped jack (*Caranx delicatissimus*), and albacore (*Thunnus thynnus*) indicated each was pathogenic for red sea bream (*Pagrus major*) (Nakajima and Maeno, 1998). Other economically important marine species affected by iridoviruses in this region include sea bass, *Lateolabrax* sp. (Nakajima and Sorimachi, 1995), various species of grouper, *Epinephelus* spp. (Chou *et al.*, 1998; Chua *et al.*, 1994; Danayadol *et al.*, 1996; Murali *et al.*, 2002), red drum, *Sciaenops ocellata* (Weng *et al.*, 2002), large yellow croaker, *Larimichthys crocea* (Chen *et al.*, 2003), striped beakperch, *Oplegnathus fasciatus* (Jung and Oh, 2000) and turbot, *Scophthalmus maximus* (Shi *et al.*, 2004).

The white sturgeon iridovirus (WSIV) is a significant cause of mortality of juvenile white sturgeon (*Acipenser transmontanus*) in North America and among Russian sturgeon (*A. guldenstadi*) in northern Europe (Adkison *et al.*, 1998; Hedrick *et al.*, 1990; Watson *et al.*, 1998a). Other sturgeon species have been experimentally infected (Hedrick *et al.*, 1992b). WSIV has been detected in white sturgeon originating from wild-caught adults in California, Oregon, Washington, and Idaho (Hedrick *et al.*, 1990; LaPatra *et al.*, 1994, 1999). There is little antigenic relationship between WSIV, EHNV, or RSIV. The survival time of infected fish is negatively correlated with temperature (Watson *et al.*, 1998b). Pallid sturgeon (*Scaphirhynchus albus*) and shovelnose sturgeon (*S. platorynchus*) have suffered high levels of mortality in North and South Dakota, but the causative agent has yet to be characterized (MacConnell *et al.*, 2001).

It is unclear whether the mortality seen among cultured marine fish in Asia and among farmed frogs in China and the United States (V. G. Chinchar and S. LaPatra, unpublished data; Zhang *et al.*, 2001) is due to the introduction of virus into the affected species via an unknown animal reservoir, or represents an endemic pathogen of those species whose pathogenicity has been enhanced by environmental stress. Fish may succumb to endemic diseases, given the stress induced by intensive farming practices, as reported for WSIV infected sturgeon (Georgiadis *et al.*, 2001). Additionally, pathogens are presented with ample opportunities for transmission in high-density, immunologically compromised farmed animal populations. There is growing evidence that exposure to pesticides, increased UV irradiation, and other environmental insults can result in immune suppression that may render animals susceptible to parasite and pathogen loads that would not trigger disease in a healthy organism (Bly *et al.*, 1992; Carey *et al.*, 1999; Cheng *et al.*, 2002; Kiesecker, 2002; Quiniou *et al.*, 1998; Taylor *et al.*, 1999). Such effects may have major consequences at the population level, and much current attention is focused on the plight of amphibian populations globally (Blaustein *et al.*, 2003; Kiesecker *et al.*, 2001).

VIII. Structure of Invertebrate Iridescent Viruses

A highly detailed model of iridovirus particle structure has been defined by cryo-electron microscopy and three-dimensional image reconstruction (Yan *et al.*, 2000). Particles of IIV-6 were observed in closely packed quasi-crystalline hexagonal arrays with an interparticle distance of 4–6 nm. Particle diameter was calculated to be 162–165 nm along the two and threefold axes of symmetry, respectively. The outer capsid is composed of trimeric capsomers, each approximately 8 nm diameter and 7.5 nm high, arranged in a pseudo-hexagonal array. A thin fiber projects radially from the surface of each capsomer and is probably important in regulating interparticle distance (Fig. 5A). This would explain the results of Mercer and Day (1965), who reported remarkably stable interparticle distances (5–8 nm) of IIV-2 under radically different conditions of hydration, type of solvent, and the use of polymers and antibodies. Surface fibrils have been observed in a number of other iridoviruses from both vertebrates and invertebrates (reviewed by Williams, 1996).

At the base, the capsomers are interconnected, forming a contiguous icosahedral shell ~2.5 nm thick (Fig. 5B). The MCP exists externally as a noncovalent trimer and internally as a trimer linked by disulfide bonds (Cerutti and Devauchelle, 1985, 1990). The trimeric structures of the inner capsid shell are rotated ~60° relative to the outer capsomers (Fig. 5C). Surface labeling indicated that the surface structure of IIV-1 was fairly complex, with 10 or 11 polypeptides ^{125}I-labelled under nonreducing and reducing conditions, respectively (Watson and Seligy, 1997). A lipid bilayer, 4 nm thick, surrounds the DNA core, and is intimately associated with an additional inner shell beneath the fused layer of the capsid. The effect of lipid solvents and detergents on IIV infectivity appears to depend on the assay system used to measure virus titer; *in vitro* assays generally indicate higher sensitivity to solvents than *in vivo* assays (Martínez *et al.*, 2003). The capsomers are arranged into trisymmetron and pentasymmetron facets. Each particle consists of 20 trisymmetrons, composed of 55 capsomers, and 12 pentasymmetrons composed of 30 capsomers and one hexavalent capsomer of uncertain composition (Fig. 5D). Pentavalent capsomers are located at the vertices of the particle in the centre of each pentasymmetron. This gives a total of 1460 capsomers plus 12 pentavalent capsomers per particle, in agreement with previous calculations (Manyakov, 1977; Wrigley, 1970).

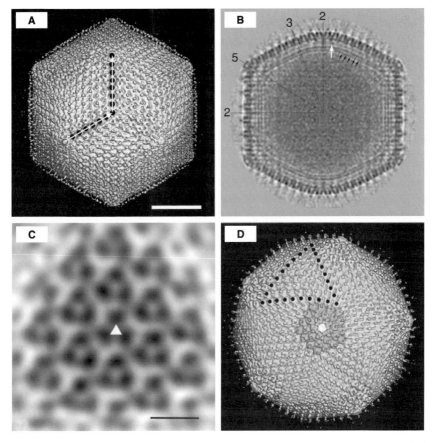

Fɪɢ 5. (A) Three dimensional reconstruction of IIV-6 particle viewed down a threefold axis of symmetry by cryo-electron microscopy. Circles indicate the position of capsomers along the h and k lattice used to calculate the triangulation number (T). The proximal ends of surface fibers are visible, whereas the distal ends have become lost during the reconstruction process because of their flexibility (white bar = 50 nm applies to images A, B, and D). (B) Central section of the reconstruction density map viewed along twofold axis indicating twofold, threefold, and fivefold axes of symmetry. A lipid bilayer is seen beneath the capsid shell (black arrows) with numerous connections to an additional shell (white arrow) beneath the outer shell. (C) Close-up view of the trimers comprising each capsomer shown as a planar section through a reconstruction density map viewed along threefold axis (triangle) (black bar = 10 nm). (D) Facets of capsid with five trisymmetrons (highlighted in green) arranged around one pentasymmetron (in pink). The central capsomer of the pentasymmetron (uncolored) is pentavalent. The edge of each trisymmetron comprises ten capsomers (black dots). Images reproduced with permission from Yan *et al.* (2000). (See Color Insert.)

The triangulation number, T, which describes the relationship between the fivefold axes of rotational symmetry of the particle, is calculated as $T = h^2 + hk + k^2$, where h and k indicate the number of steps (capsomers) required in two directions to reach the next pentasymmetron, i.e. the distance between apices. As the values of h and k are both 7, $T = 147$ (Yan *et al.*, 2000) (Fig. 5D).

IX. ECOLOGY OF INVERTEBRATE IRIDESCENT VIRUSES

Since the last comprehensive reviews of iridoviruses of invertebrates (Williams, 1996, 1998), principal advances in our understanding of IIV ecology relate to recognizing the importance of inapparent infections for virus survival and the consequences of covert infection on the reproductive capacity of the host.

IIVs infect many species of insects and several other invertebrate taxa, notably terrestrial isopods. The massive proliferation of IIV particles and their paracrystalline arrangement in the cytoplasm of host cells result in the distinctive iridescent hues of infected insects from which the family derives its name. This type of infection has been labeled patent infection to differentiate it from covert infections that may be common but less visible (Williams, 1993). The color of patently diseased insects ranges from lavender, blue, and turquoise for the small IIVs (genus *Iridovirus*) to yellow-green, orange, and red in infected aquatic Diptera. The shift in the color spectrum is indicative of larger particle size and greater interparticle spacing among viruses infecting Diptera (Hemsley *et al.*, 1994). There is no known adaptive advantage arising from iridescence. Natural infections are mainly observed in the immature stages (larvae, nymphs) of insects, although adults may also be patently infected in bees, isopods, crickets, and flour beetles.

A. Patent IIV Infections Are Usually Rare

In patently infected animals, the majority of host tissues are infected by IIV particles with particularly abundant replication in the body fat and epidermis (Federici, 1980; Hall and Anthony, 1971; Kelly, 1985). Individuals with patent infections that survive to pupate may suffer marked deformations of the pupa, particularly of the wing buds (Carter, 1974; Smith *et al.*, 1961; Stadelbacher *et al.*, 1978). Cellular pathology has been reviewed elsewhere (Williams, 1996).

Despite their visibility, the prevalence of patent IIV infection in host populations is usually very low. The majority of reports mention finding a few infected individuals, rarely exceeding 1% of the population, although abundant infections and even epizootics have occasionally been observed (Williams, 1998). However, the records of patent infections are probably not representative of their occurrence in nature, as frequent infections are far more likely to be reported than rare ones. Indeed, the low prevalence of infections in natural populations and difficulties in transmitting the virus to susceptible hosts are the principal factors that have deterred researchers interested in the biological control of insects from studying IIVs. These viruses are distributed globally across temperate and tropical regions and are most frequently observed in larval mosquito populations (Williams, 1998).

B. Seasonal Trends in Infections

An association of IIV infection with moisture is observed seasonally among soil-dwelling arthropods. The prevalence of patently infected terrestrial isopods (variously known as woodlice, sowbugs, and slaters) is highest in the coolest and wettest months of the year (December-March) in both northern Europe and the southern United States (Federici, 1980; Wijnhoven and Berg, 1999). Seasonal increases in the prevalence of IIV disease are concurrent with reductions in the spatial structure of the isopod population. In the drier months, the distribution of isopods is highly aggregated, within which patch densities are high and inter-patch distances are large, reducing the probability of dispersal. In the wetter months the situation is reversed—patches break down, isopods disperse more, and the prevalence of disease increases (Grosholz, 1993).

In southern Mexico, epizootics of patent infection were observed on an annual basis in multiple species of sympatric blackflies (Hernández et al., 2000). The prevalence of patent infection was also seasonally affected, with 41–100% of blackfly larvae infected at the end of the dry season when river levels were at a minimum and larval populations were high. This was followed by a marked decrease in patent infections, concurrent with the start of the rainy season, that caused a decline in larval population densities. The same patterns were seen yearly thereafter. However, attempts at quantifying covert infections by insect bioassay failed, as the virus did not replicate in *Galleria mellonella* (Hernández et al., 2000). Seasonal fluctuations in the prevalence of IIV infections of chironomid larvae have also been reported in the southern USA with infections only seen in March-May in Florida (Fukuda et al., 2002) or February-April in Louisiana (Chapman et al., 1968).

C. Covert Infections Can Be Common

Covert infections are not obvious to the naked eye, and infected hosts appear normal and may develop fully and reproduce as adults. Covert infections appear to be a low virulence infection rather than an early stage of a lethal disease that subsequently progresses to patent infection. An example of the latter is seen in terrestrial isopods from field populations that appear normal when collected, but develop patent infections shortly afterwards when observed under laboratory conditions, indicating that they were already infected when taken from the field (Federici, 1980). In contrast, covertly infected hosts do not spontaneously develop patent infections when incubated in the laboratory, at least in the case of blackflies (Williams, 1993, 1995).

The presence of abundant covert infections in blackfly populations (*Simulium variegatum*) in Wales, UK was originally detected using insect bioassay and PCR techniques (Williams, 1993). The prevalence of infection in the blackfly population was seasonally affected, with peaks of up to 37% infection during Spring and Autumn (Williams, 1995). However, the prevalence of infections was not correlated with host density. Patently infected individuals were only rarely observed during the study, but interestingly, they were observed immediately following the peak prevalence of covert infection. Sympatric blackfly species were not observed to suffer IIV infections.

Covert infections have been diagnosed by electron microscopic observation of low numbers of virus particles in the fatbody and hemocytes of mayfly larvae (Tonka and Weiser, 2000). Infected cells displayed abundant closed tubular structures of unknown function within viral assembly sites (Fig. 6). These closed tubular structures, although not commonly seen, are a result of aberrant assembly of capsid components (Buchatsky and Raikova, 1978; Darcy and Devauchelle, 1987; Devauchelle, 1977; Stoltz, 1973).

Laboratory studies on infection of *Aedes aegypti* by IIV-6 detected the prevalence of patent infection in the range 0.1–1.3%, whereas covert infections were approximately ten times more common (Marina *et al.*, 1999, 2003a,b,c). The nature of covert infection is poorly understood. The titer of virus particles in covertly infected hosts was shown to increase over time in both mosquitoes (Marina *et al.*, 1999) and *G. mellonella* larvae (Constantino *et al.*, 2001), but tissue tropisms and the density of particles in covertly infected insects remain unclear. IIV particles from covertly infected hosts can be manually transmitted to healthy hosts resulting in patent disease, indicating that the virulence of the pathogen is not a fixed characteristic and that genetically

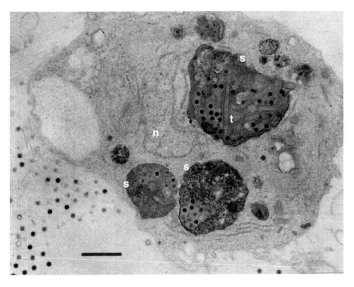

FIG 6. IIV particles in a hemocyte of a covertly infected mayfly larva showing abundance of tubular structures (t) and a low number of virus particles associated with viral assembly sites (s). In contrast, patent infection involves massive proliferation of IIV particles in infected insect cells. Cell nucleus (n), bar = 1 μm. Reproduced with permission from Tonka and Weiser (2000).

identical isolates can cause both types of infection. Doubtless, physiological stressor experiments and studies on covertly infected hosts inoculated with coinfecting pathogens or parasites have the potential to clarify the nature of covert infection and the mechanisms that trigger patent disease, as reported for Lepidopteran baculoviruses (Fuxa *et al.*, 1999; Hughes *et al.*, 1997).

The techniques used to detect covert infection include direct electron microscopic observation of infected tissues (Tonka and Weiser, 2000), PCR using primers targeted at the major capsid protein gene, endpoint dilution in cell culture and insect bioassay in larvae of the wax moth, *G. mellonella* (Constantino *et al.*, 2001). Immunological and DNA hybridization techniques have also been employed to determine the presence of infections in host range studies discussed below. The sensitivity of cell culture depends to a large extent on the cell line(s) employed. Of 12 different cell lines tested, the most sensitive were those of *Drosophila* (DR1, DR2) and the least sensitive were those of *Aedes aegypti, Ae. albopictus,* and *Plutella xylostella* (PX2-HNV3) (Constantino *et al.*, 2001). The *Ae. albopictus* cell line was previously described as nonpermissive to IIV-6 (Cerutti and Devauchelle, 1980), but semipermissive to IIV-1 (Tajbakhsh *et al.*, 1990). Lepidopteran lines such as the commonly used *Spodoptera frugiperda* Sf9 and Sf21 were 10-fold less sensitive than the *Drosophila* lines, and those from other noctuid species were of intermediate sensitivity. Between 15 and 64 particles of IIV-6 represented an infectious unit in Sf9 cells (Constantino *et al.* 2001). This compares to a sensitivity of approximately ten particles of IIV-3 in *Ae. aegypti* cells (Webb *et al.*, 1975). Czuba *et al.* (1994) reported that the sensitivity of plaque assay of IIV-1 in Sf9 cells varied according to whether virus concentration was estimated by optical density or direct counting, as many particles in their preparations did not contain a genome.

Insect bioassay is a simple technique in which a test insect is homogenized, semi-purified, and injected into *G. mellonella*. Patent infections of *G. mellonella* larvae were used to indicate that virus particles were present in the test insect. This technique is highly sensitive (Constantino *et al.*, 2001). Most, but not all, IIVs replicate well in *G. mellonella*; IIV-3, IIV-16, IIV-24, and an isolate from blackflies (Hernández *et al.*, 2001) cannot be detected by this method.

Nested and conventional PCR has been used successfully to amplify conserved sequences in the MCP gene from covertly infected insects (Constantino *et al.*, 2001; Williams, 1993). The sensitivity of the amplification technique was calculated to be 1000 particles/insect, much of which reflects the loss or dilution of genome copies that occurs during DNA purification. The use of carrier DNA, such as salmon sperm DNA, proved useful when attempting to amplify virus sequences from very small insects or insect tissues contaminated by insect gut material (Williams, 1993).

D. Covert Infections Are Detrimental to Host Fitness

There is increasing awareness of the importance of sublethal diseases on the dynamics of insect populations (Boots *et al.*, 2003). Sublethal infections by baculoviruses, entomopoxviruses, and small RNA viruses are frequently detrimental to the development, reproduction, and survival of their insect hosts (Rothman and Myers, 1996). However, not all studies on the sublethal effects of virus infections demonstrate the presence of the pathogen in the survivors. This oversight leaves open the possibility that survivors represent a resistant subset of the population with different demographic characteristics, or that a tradeoff exists between the cost of mounting an immune response and future investment in reproduction or longevity. With the advent of RT-PCR, the presence of the pathogen can now be confirmed with considerable certainty (see Burden *et al.*, 2003 for an example with baculovirus). To date, RT-PCR has not been employed to detect mRNA generated in covert IIV infections.

The influence of covert IIV infections on host fitness has been examined using the mosquito *Ae. aegypti* and IIV-6 as a model system (Marina *et al.*, 1999, 2003a,b,c). Mosquito larvae were briefly immersed in an IIV-6 suspension, subjected to repeated washing steps and then reared to adulthood, whereupon demographic parameters were determined. Insect bioassay was used to classify adult mosquitoes as (i) covertly infected, (ii) inoculated survivors that did not become infected or, (iii) controls. The findings have been remarkably consistent. Exposure to inoculum results in an extended immature development time. More importantly from the population viewpoint, covertly infected mosquitoes suffered a decrease in fecundity ranging from 22% on a single gonotrophic cycle to 36–39% on multiple (lifetime) gonotrophic cycles. Progeny production, reflected in the net reproductive rate (R_0) being the average number of female offspring produced by each mosquito, was reduced correspondingly by 22–50%. No effects on egg fertility have been detected, perhaps because the prevalence of infertility increases with age and covertly infected females die at younger ages than healthy conspecifics. Covert infection is also associated with reduced body size (wing length of females) and reduced adult longevity. The pattern of mortality also differs in covertly infected insects with an early peak in the death rate (d_x) and very few long-lived individuals (Marina *et al.*, 2003b). Reduced body size is likely to impact reproductive success negatively in natural mosquito populations, especially for males that have to compete for mates. A marked reduction in fecundity and reduced longevity have been reported in an aphid (*Toxoptera citricida*) fed crude suspensions of IIV-6. However, the replication of IIV particles in aphids has yet to be demonstrated explicitly (Hunter *et al.*, 2001b).

Why is covert infection costly to the host? The effects of sublethal infection may reflect the cost of mounting an immune response or because virus-induced cell death requires the diversion of host resources to repair damaged cells and tissues. Indeed, clear relationships between immune response (melanization) and larval developmental rate, adult body size, and fecundity have been reported in *Ae. aegypti* (Koella and Boëte, 2002; Schwartz and Koella, 2004). In contrast, the possibility that sublethal effects were a consequence of a toxic virus protein associated with the inner lipid membrane (Cerutti and Devauchelle, 1980; Ohba *et al.*, 1990) received little support from experimental studies with heat- or UV-inactivated virus. Heat- or UV-inactivated IIV-6 did not affect mosquito reproduction or longevity and caused only minor changes in adult body size compared to control insects. This may have been due to cellular apoptosis or reduced feeding of inoculated insects (Marina *et al.*, 2003c). Indeed, UV-inactivated ASFV and FV-3 trigger apoptosis in the absence of gene expression or replication (Carrascosa *et al.*, 2002; Chinchar *et al.*, 2003).

E. Host Range

Accurate appraisal of host range evidently requires diagnosis of infection on grounds other than the development of an iridescent patent infection alone. Historical studies have established that IIVs differ markedly in their host range, from those that naturally infect just one or very few species, to those, such as IIV-6, with broad host ranges. However, the range of species in which IIVs replicate depends very much on the route of infection. Laboratory studies in which the inoculum is injected indicate that many IIVs can productively infect species of agricultural and medical importance, and can often infect different taxonomic orders or even classes (Fukuda, 1971; Henderson *et al.*, 2001; Ohba, 1975; Ohba and Aizawa, 1979).

Replication *in vitro* is even more catholic. IIV-6 replicates in many insect cell lines and can even infect in reptile cells (McIntosh and Kimura, 1974). However, IIV-6 does not replicate *in vivo* in frogs (Ohba and Aizawa, 1982). Following intraperitoneal injection into *Rana limnocharis*, the titer of IIV-6 declined 100-fold over a period of four days. The *in vitro* host range of IIV-6 was recently extended to include cells from a whitefly (Funk *et al.*, 2001), a leafhopper, a lacewing (Hunter *et al.*, 2001a), and a root weevil (Hunter and Lapointe, 2003). Replication is abolished at temperatures of ~30°C (Tesh and Andreadis, 1992), so mammalian cells held at 37°C are not productively infected by IIV-6.

Laboratory studies involving the ingestion of high doses of IIV-6 also indicate a broad host range. An IIV-6-like isolate from commercial colonies of crickets (*Gryllus campestris, Acheta domesticus*) could be transmitted by feeding very high concentrations (2.2×10^{11} particles/ml) to other orthopteran species. Patent infection was subsequently observed in five of the eight species of locusts, crickets, and cock-roaches tested (Kleespies *et al.*, 1999). Dipping of cricket nymphs in concentrated virus suspensions also resulted in infection. The possibility of covert infection was not examined. Also, brown citrus aphids

(*Toxoptera citricida*) that consumed a crude suspension of IIV-6 were believed to have been infected, although viral replication or cytopathological effects were not demonstrated (Hunter *et al.*, 2001b).

We take this opportunity to call for caution in the interpretation of host range with reference to two recent studies. First, evidence for the *in vitro* replication of an IIV-6 isolate from a New Zealand virus collection in a whitefly cell line from *Bemisia tabaci* biotype B (Funk *et al.*, 2001) was followed shortly afterwards by a report of IIV-6 infecting natural populations of *B. tabaci* in Florida (Hunter *et al.*, 2001a). However, although the isolation procedure involved applying whitefly homogenates to the cell line, no virus particles were observed in whiteflies from which the homogenates were prepared. Moreover, the MCP gene sequence of the New Zealand isolate and the supposed whitefly isolate differ from that of the published IIV-6 sequence by ~5% (Jakob *et al.*, 2001; Stohwasser *et al.*, 1993), suggesting that the virus isolated from the whitefly may have been a contaminant from the initial cell line study. Second, a strain of IIV-6 has been reported from diseased reptiles (Just *et al.*, 2001). Various tissues were used to inoculate viper heart cells (VH2) incubated at 28°C. An IIV was isolated from VH2 cells showing cytopathic effects, leading to the conclusion that the reptiles were infected by IIV. However, SDS-PAGE, restriction endonuclease and sequence analysis of a PCR amplified fragment of the MCP gene revealed that the isolate from VH2 cells was identical in every way to an IIV-6 like strain isolated from patently infected crickets in the same laboratory at about the same time (Just and Essbauer, 2001). Since IIVs are capable of persisting almost indefinitely as low-level infections in cell lines (Mitsuhashi, 1967), the inferred infections of whiteflies and reptiles by IIV-6 will require confirmation.

Several studies have attempted to determine host range based on the presence of both symptomatic and asymptomatic infection. The indirect fluorescent antibody technique demonstrated the presence of virus antigen in head and abdominal squashes of mosquitoes, sandflies, and a triatomid bug following injection of IIV-22 from *Simulium variegatum*. The quantity of antigen was proportional to the interval between inoculation and testing. Low densities of IIV particles were also observed in the cells of one mosquito species by electron microscopy (Tesh and Andreadis, 1992).

Ward and Kalmakoff (1991) reported that dot-blot DNA hybridization could be used to detect viral replication in several species of Lepidoptera and Coleoptera prior to, or in the absence of, iridescence of host tissues, following injection of IIV-9 from *Wiseana* sp. (Lepidoptera). Similarly, Henderson *et al.* (2001) used dot-blot assays to demonstrate changes in the concentration of viral DNA signaling the replication of IIV-6 in injected cotton boll weevils (*Anthonomus grandis*).

While the host range of IIVs in nature is likely far more restricted than laboratory studies would lead us to believe, certain IIV species are capable of infecting a limited number of sympatric host species. Examples include IIV-9 isolated from soil dwelling Lepidoptera in New Zealand and others mentioned by Williams (1998).

During recent transmission studies, an IIV isolated from *Spodoptera frugiperda* (Lepidoptera) was observed to infect and kill the immature stages of hymenopteran endo- and ectoparasitoids that developed in or on infected hosts (López *et al.*, 2002). The interval between parasitism and virus infection was critical to parasitoid survival; parasitoids could only develop when hosts were infected shortly prior to parasitoid emergence and pupation. These studies were performed in the laboratory, but, as many species of Lepidoptera are host to parasitoids and IIV infections, it seems likely that IIVs may frequently infect and kill the developing stages of hymenopteran parasitoids in nature.

It has been suggested that at least six different genera of North American woodlice (Isopoda), and an isopod infecting nematode, are host to IIV-31 (Cole and Morris, 1980; Grosholz, 1993; Poinar *et al.*, 1980; Schultz *et al.*, 1982). However, to date, genetic similarities among isolates from different species indicating cross species transmission have only been demonstrated for certain species of *Porcellio* and *Armadillidium* (Cole and Morris, 1980; Federici, 1980). Similarly, eight different genera of terrestrial isopods were reported as hosts to IIV in northern Europe, based on iridescence of tissues (Wijnhoven and Berg, 1999) and electron microscope observations (Poinar *et al.*, 1985). In contrast, several reports of IIVs infecting natural populations of mosquitoes or blackflies suggest that patent infections are limited to one or a few sympatric species despite an abundance of alternative hosts (Anderson, 1970; Chapman *et al.*, 1971; Popelkova, 1982; Williams, 1995). It is intriguing that IIVs have never been reported from natural populations of *Anopheles* mosquitoes, despite successful infection in the laboratory. Given that many IIVs are capable of replicating in a range of invertebrate hosts but do not appear to infect more than a few sympatric species in nature, the key factors that determine the functional host range of these viruses are likely to be behavioral, or related to specific associations that favor transmission, such as parasitism, described later.

F. Transmission and Route of Infection

The route of infection remains unclear in many host-IIV systems. As *per os* infection almost invariably requires ingestion of very large doses of particles, cannibalism or predation of infected individuals is considered a likely mechanism of transmission in the mosquito *Ochlerotatus* (*Aedes*) *taeniorhynchus* (Linley and Nielsen, 1968a,b), terrestrial isopods (Federici, 1980; Grosholz, 1992), tipulid larvae (Carter 1973a,b), and the late instars of *Spodoptera frugiperda* (Chapman *et al.*, 1999; O. Hernandez and T. Williams, unpublished data).

As IIV-3 virions were degraded in large numbers in the gut of mosquito larvae, the peritrophic membrane appeared to be an effective barrier preventing infection of gut cells (Stoltz and Summers, 1971). By physically damaging the peritrophic membrane using an inoculum of IIV-3 containing silicon carbide "whiskers," a threefold increase in patent infection of mosquito larvae was observed (Undeen and Fukuda, 1994). Mixtures of IIV-6 and ground sand also resulted in an increase of patent and covert infection of *Ae. aegypti* (Marina *et al.*, 1999), but mixtures of IIV-6 with an optical brightener, know to degrade the chitin structure of the peritrophic membrane, or small silicon carbide grains, did not result in a significant increase in infection of mosquitoes, leading to the suggesting that IIVs depend on mechanisms other than ingestion to achieve infection, possibly wounding (Marina *et al.*, 2003a).

Intra- and interspecific aggression between individuals resulting in wounding has been identified as an important factor influencing IIV transmission in natural populations of isopods and laboratory populations of mosquitoes. Doubling the density of *Porcellio scaber* populations resulted in an insignificant change in the prevalence of infection by IIV-31, whereas introducing an equal density of *Porcellio laevis* was accompanied by a threefold increase in infection (Grosholz, 1992). Similarly, a 10-fold increase in the density of *Ae. aegypti* larvae resulted in a small but significant increase in nonspecific mortality, presumably due to increased aggression at high densities, and a concurrent threefold increase in the overall prevalence of infection (covert + patent infections combined) (Marina *et al.*, 2005). The transmission coefficient (v), representing the probability of becoming infected, decreased over time at both host densities. This was interpreted as evidence of heterogeneity in the susceptibility of individuals, with the most susceptible insects acquiring an infection quickly and the remaining individuals experiencing a reduced probability of infection, or possibly a developmental effect, with an increased probability of transmission in young larvae, and a reduced probability of infection in older larvae. Such nonlinearities in the transmission of insect pathogens are well recognized for Lepidopteran baculoviruses and have now been extended to include nonoccluded viruses in aquatic insects.

The very high pathogenicity seen following injection of IIV led to the suggestion that parasites and parasitoids may be able to transmit these viruses as they penetrate the host hemocoel. Enhanced transmission of an IIV from midge larvae (*Culicoides variipennis sonorensis*) by infected juvenile stages of a mermithid nematode was demonstrated in laboratory studies (Mullens *et al.*, 1999). Over 90% of infected midge larvae collected from the field in California were also infected by nematodes. In the laboratory, 40–100% of midge larvae became patently infected when incubated in a virus suspension containing juvenile nematodes, whereas almost no infections were observed in larvae incubated with virus suspension alone. In all cases, the virus killed the host larvae before the nematode could emerge. This is the first demonstration that nematodes can initiate infection during the act of host penetration.

Parasitoid mediated transmission was recently demonstrated in laboratory and field studies on *S. frugiperda* and a homologous IIV isolate from southern Mexico (López *et al.*, 2002). The ichneumonid endoparasitoid *Eiphosoma vitticolle* stung significantly more virus-infected than healthy larvae, apparently due to a lack of defense reactions in sluggish, virus-infected hosts. After stinging an infected larva, 100% of the female wasps transmitted the infection to healthy hosts during subsequent acts of stinging. Caged field experiments supported this result: virus transmission to healthy larvae only occurred in cages containing healthy and infected hosts and parasitoids. In contrast, a braconid ectoparasitoid was incapable of virus transmission, presumably because the parasitoid's ovipositor did not penetrate the hemocoel during oviposition. A report of IIV-like particles in the Varroa mite (Camazine and Liu, 1998), a parasite of honey bees, might represent an additional candidate for parasite-mediated transmission, as the original host to the mite *Apis cerana* suffers frequent IIV infections in northern India (Bailey *et al.*, 1976).

Evidence for vertical transmission has only been reported for IIV-3 infected mosquitoes. Adult females, infected as late instar larvae, appear capable of transmitting the virus to between 19 and 46% of their offspring (Woodard and Chapman, 1968). Progeny larvae develop normally until the third or fourth instar, when the signs of disease become manifest and the larvae become lethargic or die, and are consumed by conspecifics, resulting in horizontal transmission. The pattern of vertical transmission was highly aggregated—patently infected progeny appeared in the broods of just 8% of females. However, all of the larvae from those broods were infected (Linley and Nielsen, 1968a), although others have asserted that vertical transmission may not be so efficient (Fukuda and Clark, 1975). This system merits reexamination with modern molecular techniques.

G. Moisture is Crucial to IIV Survival in the Environment

The ability of IIVs to persist outside of a host is not well understood. The stability of these viruses in water has been attributed to virus particle structure, particularly the internal lipid membrane (Kelly, 1985). The infectivity of IIV-2 fell by 50% after 32 days at 4°C (Day and Gilbert, 1967). Similarly, a 10-fold reduction in titer was seen in IIV-6 after 50 days at either at 4°C or 25°C. Loss of IIV-6 infectivity was faster and more variable when tubes of virus suspension were placed in a small pond with a average water temperature of 27°C. The stability of IIV-6 was sensitive to extremes of pH (<4.0 or >9.0), but not to the presence of metal ions (Marina et al., 2000).

Laboratory and field observations indicate that IIVs are particularly sensitive to desiccation. An IIV from a midge retained infectivity for six weeks in water at unspecified laboratory temperatures, whereas infectivity was not observed in infected midge cadavers that had been left dry for eight weeks (Mullens et al., 1999). IIV-6 incorporated into a boll weevil bait gradually lost infectivity over 14 days and was inactivated more rapidly when sprayed on cotton plants (McLaughlin et al., 1972). Following the death of the host, IIV-31 in the cadavers of isopods remained infective for just five days at ambient laboratory temperatures (Grosholz, 1993).

Soil represents an important environmental reservoir for most entomopathogenic viruses. The infectivity of IIV-3 in mosquito larvae fell markedly after two days in fresh or brackish water at 27°C, but when inoculated onto damp soil, the virus failed to cause patent infections in mosquito larvae after 24 hours (Linley and Nielsen, 1968b). Virus extraction and insect bioassay techniques were recently applied to determine the effect of soil moisture and soil sterility on the persistence of IIV-6 at 25°C in the laboratory (Reyes et al., 2004). Virus extraction was problematic due to the high affinity for clay minerals; an albumen solution was used as desorbant. Loss of activity in dry soil (6.4% moisture) was very rapid and was not studied beyond 24 hours. However, soil moisture did not affect the rate of inactivation of virus in damp (17% moisture) or wet soil (37% moisture). In contrast, soil sterilization significantly improved the persistence of IIV-6 activity, both in damp and wet soil. These figures represent half-lives of 4.9 days in nonsterile soil, 6.3 days in sterilized soil (data pooled for 17% and 37% moisture treatments), compared to 12.9 days for aqueous virus suspensions incubated in the laboratory.

Exposure to solar UV light resulted in a very rapid inactivation of IIV-6 in water, with infectivity dropping by approximately nine logarithms after 24 hours' exposure to sunlight (A. Hernández, unpublished data). The presence of soil sediment greatly improved the persistence of the virus. Extra-host persistence in soil and aquatic habitats is likely to represent an important aspect of the ecology of these viruses.

Interestingly, a peptide from a diapausing chrysomelid beetle was identified with high homology to a putative peptide of unknown function identified in the genome of IIV-6 (Tanaka *et al.*, 2003). The diapause specific peptide was similar to antifungal peptides from plants and the conotoxin-like peptides from baculoviruses. When purified, the peptide inhibited growth of a dermatophyte pathogen, but not of the insect pathogen *Beauveria bassiana*. The possibility that IIV-6 produces this peptide to avoid opportunistic exploitation of IIV-infected insect cadavers by soil-dwelling fungi, thereby improving persistence and the probability of transmission, merits examination.

X. IRIDOVIRUSES IN NONINSECT MARINE AND FRESHWATER INVERTEBRATES

Information on iridoviruses infecting noninsect aquatic invertebrates is scarce and consists almost entirely of descriptions of particle morphology, DNA content, cytoplasmic assembly, and cytopathology and histopathology. Putative iridovirus infections have been reported in a marine annelid worm, *Nereis diversicolor* (Devauchelle, 1977; Devauchelle and Dorchon, 1973), the marine crustaceans *Balanus eburneus* (Leibovitz and Koulish, 1989), *Sacculina carcini* (Russell *et al.*, 2000), *Protrachypene precipua* (Lightner and Redman, 1993), and *Macropipus depurator* (Montanie *et al.*, 1993), the freshwater snail, *Lymnaea truncatula* (Barthe *et al.*, 1984; Rondelaud and Barthe, 1992; Ruellan, 1992), and the marine cephalopod, *Octopus vulgaris* (Rungger *et al.*, 1971).

A number of iridovirus infections have been reported in oysters. Gill necrosis virus disease is an epizootic disease characterized by ulceration of the gills, labial palps, and the mantle concurrent with the slimming of the oysters followed by death (Arvy and Franc, 1968;

Alderman and Gras, 1969). This disease is believed to be involved in the disappearance of the Portuguese oyster from major culture areas, including the Atlantic coast of France (Comps, 1970; Comps and Duthoit, 1976). Hemocytic infectious virus disease appears to have affected Portuguese oysters since the 1970s, although clinical signs were not reported. Virus infection is systemic and targets hemocytes for destruction (Comps, 1983). This disease also causes mortality in Pacific oysters in France (Comps, 1980; Comps and Bonami, 1977; Comps *et al.*, 1976). Recently, iridovirus-like particles were observed in adult Japanese oysters, *Crassostrea gigas,* suffering gill erosion and abnormal mortalities in Mexico (Cáceres-Martínez and Vásquez-Yeomans, 2003).

Larval stages of Pacific oysters infected by oyster velar virus disease (OVVD) (Elston, 1979; Elston and Wilkinson, 1985) have been described in hatcheries in Washington State, USA. Loss of ciliated velar epithelial cells, which may appear as blebs on the periphery of the velum as they are sloughed, form the characteristic "blisters" associated with the disease. Larvae lose the ability to swim. OVVD can cause near 100% mortality in affected hatchery tanks.

The absence of DNA sequence information for noninsect invertebrate iridoviruses represents a serious hurdle for the development of molecular diagnostic tools. Recent studies aiming at establishing a PCR protocol for the detection of oyster iridoviruses focused on conserved sequences from vertebrate and insect iridoviruses (Barbosa-Solomieu, 2004). Three genes encoding essential proteins (ATPase, MCP, and the small subunit of ribonucleotide reductase) were selected for assays. Preliminary findings indicated that primers designed from MCP and ATPase genes may be useful for the detection of mollusk iridoviruses. Studies are currently underway to sequence the PCR products and determine their relationship to sequences registered in the databases.

XI. Conclusions

This review has mainly focused on the advances in iridovirus research over the past decade. Here we summarize salient features of the abovementioned studies, and discuss their importance to future work.

Iridoviruses are emerging as major mortality factors in natural populations of amphibians and reptiles, and in commercially important marine and freshwater fish. It is not clear if this is a result of the introduction of exotic pathogens into new host species, or a consequence of alterations in the transmission or virulence of endemic pathogens in populations of physiologically stressed or immunocompromised hosts.

As our knowledge of iridovirus biology increases, so does our understanding of the taxonomic structure of the family. A new genus (*Megalocytivirus*) has recently been created to accommodate a number of viruses isolated from diseased fish in Southeast Asia, and an established one (*Chloriridovirus*) is about to be validated following the sequencing of the genome of the mosquito pathogen IIV-3.

The question of what defines an iridovirus species remains highly pertinent. A steadily increasing number of sequencing studies, together with a growing appreciation of the degree to which phenotypic characteristics can vary among strains of a virus, represent important advances in defining iridovirus species. However, we are still a long way from a quantitative understanding of intraspecies and interspecies variability and how it impacts the differentiation of one species from another.

Genomic sequencing projects have clarified evolutionary relationships between iridoviruses and other families of large, nucleocytoplasmic DNA viruses. Iridoviruses appear situated at the center of a clade of DNA viruses that include the poxviruses, phycodnaviruses, ASFV, ascoviruses, and possibly a new giant virus of amoebae (mimivirus). Further studies may shed light on the origins of these closely linked families.

We are beginning to understand the interactions between iridoviruses and the immune system of ectothermic vertebrates. Further studies on iridovirus pathobiology and the functioning of lower vertebrate immune systems will doubtless reveal intriguing viral strategies for overcoming host defenses and coevolutionary host responses that modulate the virulence of iridovirus infections and protect against clinical disease.

Not all iridovirus infections result in mortal disease. Indeed, the recognition that many iridovirus infections are chronic, especially among insects, should stimulate studies on the importance of such illnesses on host fitness. Sublethal effects have been demonstrated in infected insects, and theoretical studies suggest such effects have a major influence on the dynamics of host populations. However, the consequences of chronic disease on the demography (fecundity, fertility, death rate, etc.) of infected amphibian and fish populations are notable by their absence.

Our ability to detect iridovirus infections has advanced markedly with the adoption of highly sensitive LAMP and multiplex PCR techniques for rapid diagnosis of infected individuals. However, the identification of novel isolates is still hampered by a lack of adequate sequence information for key iridovirus genes and the difficulty of propagating some viruses (e.g., LCDV) in cell culture.

Following three decades of neglect, the study of virion structure has returned to the spotlight with the construction of a high-resolution model that offers the most precise representation of an iridovirus particle to date. Additional studies are in progress to further improve the resolution of the model.

An entire group of putative iridoviruses infecting marine and freshwater invertebrates has been largely ignored, despite being suspected of being responsible for serious diseases and mass mortalities in economically important species, particularly oysters. Molecular studies are required to ascertain the relationships between these noninsect invertebrate isolates and those from the established genera.

Elucidation of the complete genetic sequence of FV-3 and other iridoviruses opens the way for a detailed biochemical and genetic study of iridovirus replication. Understanding the role of key replicative genes such as the viral homolog of Pol II and putative immune control proteins such as the viral homolog of eIF-2α should now be possible using knock-down technology (e.g., siRNA and antisense morpholinos) and knock-out (e.g., homologous recombination) technology. This work, coupled with studies of antiviral immunity, should lead to the development of better vaccines and control vectors. Moreover, since DNA methylation often results in gene silencing and plays a key role in development and carcinogenesis, determining how the highly methylated genomes of the vertebrate iridoviruses are transcribed will shed light on a fundamental process in higher eukaryotic systems.

After an extended period of quiescence, interest in the iridoviruses as pathogens of ectothermic animals has flourished over the past decade. This has been stimulated by a growing appreciation of their impact as agents of morbidity and mortality in natural and commercially important animal populations. Following marked advances in our understanding of the taxonomic relationships within the *Iridoviridae* and our ability to define the genetic identity of virus species, future possibilities for research on iridoviruses appear very encouraging.

ACKNOWLEDGMENTS

We thank Misha Sokolov for translating Russian texts and Thomas Tonka, Jaroslav Weiser, Brian Federici, Xiaodong Yan and Timothy Baker for providing figures. This work was supported by the Spanish Ministerio de Ciencia y Tecnología (AGL2002-04320-C02-01 to TW) and the U.S. National Science Foundation (award 2002-35204-12211 to VGC).

REFERENCES

Adkison, M. A., Cambre, M., and Hedrick, R. P. (1998). Identification of an iridovirus in Russian sturgeon *Acipenser guldenstadi* from Northern Europe. *Bull. Eur. Assoc. Fish Pathol.* **18:**29–32.

Afonso, C. L., Tulman, E. R., Lu, Z., Oma, E., Kutish, G. F., and Rock, D. L. (1999). The genome of *Melanoplus sanguinipes* entomopoxvirus. *J. Virol.* **73:**533–552.

Ahne, W., Schitfekdt, H. J., and Thomsen, I. (1989). Fish viruses: Isolation of an icosahedral cytoplasmic deoxyribovirus from sheatfish (*Silurus glanis*). *J. Vet. Med. B.* **36:**333–336.

Ahne, W., Bremont, M., Hedrick, R. P., Hyatt, A. D., and Whittington, R. J. (1997). Iridoviruses associated with epizootic haematopoietic necrosis (EHN) in aquaculture. *World J. Microbiol. Biotechnol.* **13:**367–373.

Ahne, W., Bearzotti, M., Bremont, M., and Essbauer, S. (1998). Comparison of European systemic piscine and amphibian iridoviruses with epizootic haematopoietic necrosis virus and frog virus 3. *J. Vet. Med. B* **45:**373–383.

Alcami, A., and Koszinowski, U. H. (2000). Viral mechanisms of immune evasion. *Mol. Med. Today* **6:**365–372.

Alderman, D. J., and Gras, P. (1969). Gill disease of Portuguese oysters. *Nature* **224:**616–617.

Anderson, J. F. (1970). An iridescent virus infecting the mosquito *Aedes stimulans.* *J. Invertebr. Pathol.* **15:**219–224.

Andrés, G., García-Escudero, R., Simón-Mateo, C., and Viñuela, E. (1998). African swine fever virus is enveloped by a two-membraned collapsed cisterna derived from the endoplasmic reticulum. *J. Virol.* **72:**8988–9001.

Antao, A. B., Chinchar, V. G., McConnell, T. J., Miller, N. W., Clem, L. W., and Wilson, M. R. (1999). MHC class I genes of the channel catfish: Sequence analysis and expression. *Immunogenetics* **49:**303–311.

Antao, A. B., Wilson, M., Wang, J., Bengten, E., Miller, N. W., Clem, L. W., and Chinchar, V. G. (2001). Genomic organization and differential expression of channel catfish MHC class I genes. *Dev. Comp. Immunol.* **25:**579–595.

Arvy, L., and Franc, A. (1968). Sur un protiste nouveau, agent de destruction des branchies et des palpes de l'Huître Portugaise. *C. R. Acad. Sci. Paris Ser. D.* **267:**103–105.

Bailey, L., Ball, B. V., and Woods, R. D. (1976). An iridovirus from bees. *J. Gen. Virol.* **31:**459–461.

Barbosa-Solomieu, V. (2004). Detección de agentes virales en ostión japonés (*Crassostrea gigas*). Unpublished PhD thesis, CIBNOR, Mexico. p. 247.

Barilla, D., Lee, B. A., and Proudfoot, N. J. (2001). Cleavage/polyadenylation factor 1A associates with the carboxyl-terminal domain of RNA polymerase II in *Saccharomyces cerevisiea. Proc. Natl. Acad. Sci. USA* **98:**445–450.

Barthe, D., Rondelaud, D., Faucher, Y., and Vago, C. (1984). Infection virale chez le mollusque pulmoné *Lymnaea truncatula* Müller. *C. R. Acad. Sci. Paris Ser. D.* **298:**513–515.

Bauer, S., Kirschning, C. J., Hacker, H., Redecke, V., Hausmann, S., Akira, S., Wagner, H., and Lipford, G. B. (2001). Human TLR9 confers responsiveness to bacterial DNA via species-specific CpG motif recognition. *Proc. Natl. Acad. Sci. USA* **98**:9237–9242.

Beattie, E., Tartaglia, J., and Paoletti, E. (1991). Vaccinia virus-encoded eIF-2 alpha homolog abrogates the antiviral effect of interferon. *Virology* **183**:419–422.

Bengten, E., Wilson, M., Miller, N., Clem, L. W., Pilstrom, L., and Warr, G. W. (2000). Immunoglobulin isotypes: Structure, function, and genetics. *Curr. Topics Microbiol. Immunol.* **248**:189–219.

Berghammer, H., Ebner, M., Marksteiner, R., and Auer, B. (1999). pADPRT-2: A novel mammalian polymerizing(ADP-ribosyl)transferase gene related to truncated pADPRT homologues in plants and *Caenorhabditis elegans*. *FEBS Lett.* **449**:259–263.

Bideshi, D. K., Renault, S., Stasiak, K., Federici, B. A., and Bigot, Y. (2003). Phylogenetic analysis and possible function of *bro*-like genes, a multigene family widespread among large double-stranded DNA viruses of invertebrates and bacteria. *J. Gen. Virol.* **84**:2531–2544.

Bigot, Y., Stasiak, K., Rouleux-Bonnin, F., and Federici, B. A. (2000). Characterization of repetitive DNA regions and methylated DNA in ascovirus genomes. *J. Gen. Virol.* **81**:3073–3082.

Birnbaum, M. J., Clem, R. J., and Miller, L. K. (1994). An apoptosis-inhibiting gene from a nuclear polyhedrosis virus encoding a polypeptide with cys/his sequence motifs. *J. Virol.* **68**:2521–2528.

Blaustein, A. R., Romansic, J. M., Kiesecker, J. M., and Hatch, A. C. (2003). Ultraviolet radiation, toxic chemicals and amphibian population declines. *Diversity Distrib.* **9**:123–140.

Bly, J. E., Lawson, L. A., Dale, D. J., Szalai, A. J., Durborow, R. M., and Clem, L. W. (1992). Winter saprolegniosis in channel catfish. *Dis. Aquat. Org.* **13**:155–164.

Bollinger, T. K., Mao, J., Schock, D., Brigham, R. M., and Chinchar, V. G. (1999). Pathology, isolation and molecular characterization of an iridovirus from tiger salamanders in Saskatchewan. *J. Wildlife Dis.* **35**:413–429.

Boots, M., Greenman, J., Ross, D., Norman, R., Hails, R., and Sait, S. (2003). The population dynamical implications of covert infections in host-microparasite interactions. *J. Anim. Ecol.* **72**:1064–1072.

Bouchier- Hayes, L., and Martin, S. J. (2002). CARD games in apoptosis and immunity. *EMBO Rep.* **3**:616–621.

Buchatsky, L. P., and Raikova, A. P. (1978). Study of *Aedes caspius caspius* mosquito larvae affested with iridescence virus. *Voprosy Virusologii* **3**:366–369(in Russian).

Burden, J. P., Nixon, C. P., Hodgkinson, A. E., Possee, R. D., Sait, S. M., King, L. A., and Hails, R. S. (2003). Covert infections as a mechanism for long-term persistence of baculoviruses. *Ecol. Lett.* **6**:524–531.

Cáceres-Martínez, J., and Vásquez-Yeomans, R. (2003). Erosión branquial en el ostión japonés *Crassostrea gigas* y su relación con episodios de mortalidad masiva en el Noroeste de México. *Bol. PRONALSA*. March 2003, 15–19.

Caipang, C. M. A., Haraguchi, I., Ohira, T., Hirono, I., and Aoki, T. (2004). Rapid detection of a fish iridovirus using loop-mediated isothermal amplification (LAMP). *J. Virol. Meth.* **121**:155–161.

Camazine, S., and Liu, T. P. (1998). A putative iridovirus from the honey bee mite *Varroa jacobsoni* Oudemans. *J. Invertebr. Pathol.* **71**:177–178.

Caposio, P., Riera, L., Hahn, G., Landolfo, S., and Gribaudo, G. (2004). Evidence that the human cytomegalovirus 46 kDa UL72 protein is not an active dUTPase but a late protein dispensable for replication in fibroblasts. *Virology* **325**:264–276.

Carey, C., Cohen, N., and Rollins-Smith, L. (1999). Amphibian declines: An immunological perspective. *Devel. Comp. Immunol.* **23:**459–472.

Carrascosa, A. L., Bustos, M. J., Nogal, M. L., Gonzalez de Buitrago, G., and Revilla, Y. (2002). Apoptosis induced in an early step of African swine fever virus entry into vero cells does not require virus replication. *Virology* **294:**372–382.

Carter, J. B. (1973a). The mode of transmission of *Tipula* iridescent virus. II Source of infection. *J. Invert. Pathol.* **21:**123–130.

Carter, J. B. (1973b). The mode of transmission of *Tipula* iridescent virus. II Route of infection. *J. Invertebr. Pathol.* **21:**136–143.

Carter, J. B. (1974). *Tipula* iridescent virus infection in the developmental stages of *Tipula oleracea. J. Invertebr. Pathol.* **24:**271–281.

Cerutti, M., and Devauchelle, G. (1980). Inhibition of macromolecular synthesis in cells infected with an invertebrate virus (iridovirus type 6 or CIV). *Arch. Virol.* **63:**297–303.

Cerutti, M., and Devauchelle, G. (1985). Characterisation and localisation of CIV polypeptides. *Virology* **145:**123–131.

Cerutti, M., and Devauchelle, G. (1990). Protein composition of Chilo iridescent virus. *In* "Molecular Biology of Iridoviruses" (G. Darai, ed.), pp. 81–112. Kluwer, Boston.

Chapman, H. C., Petersen, J. J., Woodard, D. B., and Clark, T. B. (1968). New records of parasites of Ceratopogonidae. *Mosq. News* **28:**122–123.

Chapman, H. C., Clark, T. B., Anthony, D. W., and Glenn, F. E., Jr. (1971). An iridescent virus from larvae of *Corethrella brakeleyi* (Dipera: Chaoboridae) in Louisiana. *J. Invertebr. Pathol.* **18:**284–286.

Chapman, J. W., Williams, T., Escribano, A., Caballero, P., Cave, R. D., and Goulson, D. (1999). Age-related cannibalism and horizontal transmission of a nuclear polyhedrosis virus in larval *Spodoptera frugiperda. Ecol. Entomol.* **24:**268–275.

Chen, Z., Zheng, J., and Jiang, Y. (1999). A new iridovirus from soft-shelled turtle. *Virus Res.* **63:**147–151.

Chen, X. H., Lin, K. B., and Wang, X. W. (2003). Outbreaks of an iridovirus disease in maricultured large yellow croaker, *Larimichtys crocea* (Richardson) in China. *J. Fish Dis.* **26:**615–619.

Cheng, X. W., Carner, G. R., and Brown, T. M. (1999). Circular configuration of the genome of ascoviruses. *J. Gen. Virol.* **80:**1537–1540.

Cheng, W., Liu, C. H., and Chen, J. C. (2002). Effect of nitrite on interaction between the giant freshwater prawn *Macrobrachium rosenbergii* and its pathogen *Lactococcus garvieae. Dis. Aquat. Org.* **50:**189–197.

Chinchar, V. G. (2000). Ecology of viruses of cold-blooded vertebrates. *In* "Virus Ecology" (C. J. Hurst, ed.), pp. 413–445. Academic Press, New York.

Chinchar, V. G. (2002). Ranaviruses (family Iridoviridae): Emerging cold-blooded killers. *Arch. Virol.* **147:**447–470.

Chinchar, V. G., and Dholakia, J. N. (1989). Frog virus 3-induced translational shut-off: Activation of an eIF-2 kinase in virus-infected cells. *Virus Res.* **14:**207–224.

Chinchar, V. G., and Granoff, A. (1984). Isolation and characterization of a frog virus 3 variant resistant to phosphonoacetate: Genetic evidence for a virus-specific DNA polymerase. *Virology* **138:**357–361.

Chinchar, V. G., and Granoff, A. (1986). Temperature-sensitive mutants of frog virus 3: Biochemical and genetic characterization. *J. Virol.* **58:**192–202.

Chinchar, V. G., and Mao, J. (2000). Molecular diagnosis of iridovirus infections in cold-blooded animals. *Sem. Avian Exotic Pet Med.* **9:**27–35.

Chinchar, V. G., and Yu, W. (1990). Frog virus 3-mediated translational shut-off: Frog virus 3 messages are translationally more efficient than host and heterologous viral messages under conditions of increased translational stress. *Virus Res.* **16**:163–174.

Chinchar, V. G., and Yu, W. (1992). Metabolism of host and viral mRNAs in frog virus 3-infected cells. *Virology* **186**:435–443.

Chinchar, V. G., Wang, J., Murti, G., Carey, C., and Rollins-Smith, L. (2001). Inactivation of frog virus 3 and channel catfish virus by esculentin-2P and ranatuerin-2P, two antimicrobial peptides isolated from frog skin. *Virology* **288**:351–357.

Chinchar, V. G., Bryan, L., Wang, J., Long, S., and Chinchar, G. D. (2003). Induction of apoptosis in frog virus 3-infected cells. *Virology* **306**:303–312.

Chinchar, V. G., Bryan, L., Silphadaung, U., Noga, E., Wade, D., and Rollins-Smith, L. (2004). Inactivation of viruses infecting ectothermic animals by amphibian and piscine antimicrobial peptides. *Virology* **323**:268–275.

Chinchar, V. G., Essbauer, S., He, J. G., Hyatt, A, Miyazaki, T., Seligy, V., and Williams, T. (2005). Iridoviridae. *In* "Virus Taxonomy: 8th Report of the International Committee on the Taxonomy of Viruses" (C. M. Fauquet, M. A. Mayo, J. Maniloff, U. Desselberger, and L. A. Ball, eds.), pp. 163–175. Elsevier, London.

Chou, H. Y., Hsu, C. C., and Peng, T. Y. (1998). Isolation and characterization of a pathogenic iridovirus from cultured grouper (*Epinephelus* sp.) in Taiwan. *Fish Pathol.* **33**:201–206.

Chua, H. C., Ng, M. L., Woo, J. J., and Wee, J. Y. (1994). Investigation of outbreaks of a novel disease, 'Sleepy Grouper Disease', affecting the brown-spotted grouper, *Epinephelus tauvina* Forskal. *J. Fish Dis.* **17**:417–427.

Clark, H. F., Brennan, J. C., Zeigel, R. F., and Karzon, D. T. (1968). Isolation and characterization of viruses from the kidneys of *Rana pipiens* with renal adenocarcinoma before and after passage in the red eft (*Triturus viridescens*). *J. Virol.* **2**:629–640.

Cole, A., and Morris, T. J. (1980). A new iridovirus of two species of terrestrial isopods, *Armadillidium vulgare* and *Porcellio scaber*. *Intervirology* **14**:21–30.

Comps, M. (1970). La maladie des branchies chez les huîtres du genre *Crassostrea*, caractéristique et évolution des altérations, processus de cicatrisation. *Rev. Trav. Inst. Pêches Marit.* **34**:23–44.

Comps, M. (1980). Les infections virales associées aux épizooties des huîtres du genre *Crassostrea*. *Rapp P. V. Reun. Cons. Int. Explor. Mer.* **182**:137–139.

Comps, M. (1983). Recherches histologiques et cytologiques sur les infections intracellulaires des mollusques bivalves marins. Unpublished Ph.D. thesis. Université des Sciences et Techniques du Languédoc. France.

Comps, M., and Bonami, J. R. (1977). Infection virale associée a des mortalites chez l'huitre *Crassostrea gigas* Th. *C. R. Acad. Sci. Paris Ser. D.* **282**:1991–1993.

Comps, M., and Duthoit, J. L. (1976). Infection virale associée à la "maladie des branchies" de l'huître portugaise *Crassostrea angulata* Lmk *C. R. Acad. Sci. Paris Ser. D.* **283**:1595–1596.

Comps, M., Bonami, J. R., Vago, C., and Campillo, A. (1976). Une virose de l'huître portugaise (*Crassostrea angulata*). *C. R. Acad. Sci. Paris Ser. D.* **292**:1991–1993.

Constantino, M., Christian, P., Marina, C. F., and Williams, T. (2001). A comparison of techniques for detecting *Invertebrate iridescent virus 6*. *J. Virol. Meth.* **98**:109–118.

Cullen, B. R., and Owens, L. (2002). Experimental challenge and clinical cases of bohle iridovirus (BIV) in native Australian anurans. *Dis. Aquat. Organ.* **49**:83–92.

Cunningham, A. A., Langton, T. E. S., Bennett, P. M., Lewin, J. F., Drury, S. E. V., Gough, R. E., and Mac Gregor, S. K. (1996). Pathogical and microbiological findings from incidents of unusual mortality of the common frog *Rana temporaria. Phil. Trans. R. Soc. Lond. B* **351:**1539–1557.

Czuba, M., Tajbakhsh, S., Walker, T., Dove, M. J., Johnson, B. F., and Seligy, V. L. (1994). Plaque assay and replication of *Tipula* iridescent virus in *Spodoptera frugiperda* ovarian cells. *Res. Virol.* **145:**319–330.

Danayadol, Y., Direkbusarakom, S., Boonyaratpalin, S., Miyazaki, T., and Miyata, M. (1996). An outbreak of iridovirus-like infection in brown-spotted grouper (*Epinephelus malabaracus*) cultured in Thailand. *Aquat. Anim. Health Res. Inst. Newsletter* **5:**6.

Darcy, F., and Devauchelle, G. (1987). Iridoviridae. *In* "Animal Virus Structure" (M. V. Nermut and A. C. Steven, eds.), pp. 407–420. Elsevier, Amsterdam.

Daszak, P., Berger, L., Cunningham, A. A., Hyatt, A. D., Green, E., and Speare, R. (1999). Emerging infectious diseases and amphibian population declines. *Emerg. Inf. Dis.* **5:**735–748.

Daszak, P., Cunningham, A. A., and Hyatt, A. D. (2003). Infectious disease and amphibian population declines. *Diversity Distrib.* **9:**141–150.

Day, M. F., and Gilbert, N. (1967). The number of particles of *Sericesthis* iridescent virus required to produce infections of *Galleria* larvae. *Aust. J. Biol. Sci.* **20:**691–693.

D'Costa, S. M., Yao, H., and Bilimoria, S. L. (2001). Transcription and temporal cascade in Chilo iridescent virus infected cells. *Arch. Virol.* **146:**2165–2178.

D'Costa, S. M., Yao, H. J., and Bilimoria, S. L. (2004). Transcriptional mapping in Chilo iridescent virus infections. *Arch. Virol.* **149:**723–742.

Delius, H., Darai, G., and Flügel, R. M. (1984). DNA analysis of insect iridescent virus 6: Evidence for circular permutation and terminal redundancy. *J. Virol.* **49:**609–614.

Devauchelle, G. (1977). Ultrastructural characterization of an iridovirus from the marine worm *Nereis diversicolor* (O. F. Müller). *Virology* **81:**237–246.

Devauchelle, G., and Dorchon, M. (1973). Sur la présence d'un virus de type iridovirus dans les cellules mâles de *Nereis diversicolor* (O.F. Müller). *C. R. Acad. Sci. Paris Ser. D.* **277:**463–466.

Do, J. W., Moon, C. H., Kim, H. J., Ko, M. S., Kim, S. B., Son, J. H., Kim, J. S., An, E. J., Kim, M. K., Lee, S. K., Han, M. S., Cha, S. J., Park, M. S., Park, M. A., Kim, Y. C., Kim, J. W., and Park, J. W. (2004). Complete genomic DNA sequence of rock bream iridovirus. *Virology* **325:**351–363.

Do, J. W., Cha, S. J., Kim, J. S., An, E. J., Park, M. S., Kim, J. W., Lim, Y. C., Park, M. A., and Park, J. W. (2005). Sequence variation in the gene encoding the major capsid protein of Korean fish iridovirus. *Arch. Virol.* **150:**351–359.

Docherty, D. E., Meteyer, C. U., Wang, J., Mao, J., Case, S. T., and Chinchar, V. G. (2003). Diagnostic and molecular evaluation of three iridovirus-associated salamander mortality events. *J. Wildlife Dis.* **39:**556–566.

Drury, S. E. N., Gough, R. E., and Calvert, I. (2002). Detection and isolation of an iridovirus from chameleons (*Chamaeleo quadricornis* and *Chamaeleo hoehnelli*) in the United Kingdom. *Vet. Record.* **150:**451–452.

Dukes, T. W., and Lawler, A. R. (1975). The ocular lesions of naturally occurring lymphocystis in fish. *Can. J. Comp. Med.* **39:**406–410.

DuPasquier, L. (2001). The immune system of invertebrates and vertebrates. *Comp. Biochem. Physiol. B* **129:**1–15.

DuPasquier, L., Schwager, J., and Flajnik, M. F. (1989). The immune system of *Xenopus. Annu. Rev. Immunol.* **7:**251–275.

Elston, R. (1979). Virus-like particles associated with lesions in larval Pacific oysters (*Crassostrea gigas*). *J. Invertebr. Pathol.* **33:**71–74.

Elston, R. A., and Wilkinson, M. T. (1985). Pathology, management and diagnosis of oyster velar virus disease (OVVD). *Aquaculture* **48:**189–210.

Essbauer, S., and Ahne, W. (2002). The epizootic haematopoietic necrosis virus (Iridoviridae) induces apoptosis *in vitro*. *J. Vet. Med. B* **49:**25–30.

Essbauer, S., Bremont, M., and Ahne, W. (2001). Comparison of the eIF-2 alpha homologous proteins of seven ranaviruses (Iridoviridae). *Virus Genes* **23:**347–359.

Essbauer, S., Fischer, U., Bergmann, S., and Ahne, W. (2004). Investigations on the ORF 167L of lymphocystis disease virus (Iridoviridae). *Virus Genes* **28:**19–39.

Federici, B. A. (1980). Isolation of an iridovirus from two terrestrial isopods, the pill bug, *Armadillidium vulgare* and the sow bug, *Porcellio dilatatus*. *J. Invertebr. Pathol.* **36:**373–381.

Federici, B. A., and Bigot, Y. (2003). Origin and evolution of polydnaviruses by symbiogenesis of insect DNA viruses in endoparasitic wasps. *J. Insect. Physiol.* **49:**419–432.

Fukuda, T. (1971). *Per os* transmission of *Chilo* iridescent virus to mosquitoes,. *J. Invertebr. Pathol.* **18:**152–153.

Fukuda, T., and Clark, T. B. (1975). Transmission of the mosquito iridescent virus (RMIV) by adult mosquitoes of *Aedes taeniorhynchus* to their progeny. *J. Invertebr. Pathol.* **25:**275–276.

Fukuda, T., Kline, D. L., and Day, J. K. (2002). An iridescent virus and a microsporidium in the biting midge *Culicoides barbosai* from Florida. *J. Am. Mosq. Contr. Assoc.* **18:**128–130.

Funk, C. J., Hunter, W. B., and Achor, D. S. (2001). Replication of insect iridescent virus 6 in a whitefly cell line. *J. Invertebr. Pathol.* **77:**144–146.

Fuxa, J. R., Sun, J., Weidner, E. H., and LaMotte, L. R. (1999). Stressor and rearing diseases of *Trichoplusia ni*: Evidence of vertical transmission of NPV and CPV. *J. Invertebr. Pathol.* **74:**149–155.

Gantress, J., Bell, A., Maniero, G., Cohen, N., and Robert, J. (2003). *Xenopus*, a model to study immune responses to iridovirus. *Virology* **311:**254–262.

Georgiadis, M. P., Hedrick, R. P., Carpenter, T. E., and Gardner, I. A. (2001). Factors influencing transmission, onset and severity of outbreaks due to white sturgeon iridovirus in a commercial hatchery. *Aquaculture* **194:**21–35.

Gibson-Kueh, S., Netto, P., Ngoh-Lim, G. H., Chang, S. F., Ho, L. L., Qin, Q. W., Chua, F. H. C., Ng, M. L., and Ferguson, H. W. (2003). The pathology of systemic iridoviral disease in fish. *J. Comp. Path.* **129:**111–119.

Goldberg, T. L., Coleman, D. A., Grant, E. C., Inendino, K. R., and Philipp, D. P. (2003). Strain variation in an emerging iridovirus of warm-water fishes. *J. Virol.* **77:**8812–8818.

Goorha, R. (1981). Frog virus 3 requires RNA polymerase II for its replication. *J. Virol.* **37:**496–499.

Goorha, R. (1982). Frog virus 3 DNA replication occurs in two stages. *J. Virol.* **43:**519–528.

Goorha, R., and Dixit, P. (1984). A temperature-sensitive mutant of frog virus 3 is defective in second stage DNA replication. *Virology* **136:**186–195.

Goorha, R., and Granoff, A. (1974). Macromolecular synthesis in cells infected by frog virus 3: I. Virus-specific protein synthesis and its regulation. *Virology* **60:**237–250.

Goorha, R., and Granoff, A. (1979). Icosahedral cytoplasmic deoxyriboviruses. *In* "Comprehensive Virology" (H. Fraenkel-Conrat and R. R. Wagner, eds.), pp. 347–399. Plenum Press, New York.

Goorha, R., and Murti, K. G. (1982). The genome of frog virus 3, an animal DNA virus, is circularly permuted and terminally redundant. *Proc. Natl. Acad. Sci. USA* **79:**248–262.

Goorha, R., Murti, G., Granoff, A., and Tirey, R. (1978). Macromolecular synthesis in cells infected by frog virus 3: VIII The nucleus is a site of frog virus 3 DNA and RNA synthesis. *Virology* **84:**32–50.

Goorha, R., Granoff, A., Willis, D. B., and Murti, K. G. (1984). The role of DNA methylation in virus replication: Inhibition of frog virus 3 replication by 5-azacytidine. *Virology* **138:**94–102.

Gould, A. R., Hyatt, A. D., Hengstberger, S. H., Whittington, R. J., and Coupar, B. E. H. (1995). A polymerase chain reaction (PCR) to detect epizootic haematopoietic necrosis virus and bohle iridovirus. *Dis. Aquat. Org.* **22:**211–215.

Govindarajan, R., and Federici, B. A. (1990). Ascovirus infectivity and effects of infection on the growth and development of noctuid larvae. *J. Invertebr. Pathol.* **56:**291–299.

Granoff, A. (1984). Frog virus 3: A DNA virus with an unusual life-style. *Prog. Med. Virol.* **30:**187–198.

Granoff, A., Came, P. E., and Breeze, D. C. (1966). Viruses and renal carcinoma of *Rana pipiens*: I. The isolation and properties of virus from normal and tumor tissues. *Virology* **29:**133–148.

Grizzle, J. M., and Brunner, C. J. (2003). Review of largemouth bass virus. *Fisheries* **28:**10–13.

Grizzle, J. M., Altinok, I., Fraser, W. A., and Francis-Floyd, R. (2002). First isolation of largemouth bass virus. *Dis. Aquat. Org.* **50:**233–235.

Grosholz, E. D. (1992). Interactions of intraspecific, interspecific and apparent competition with host-pathogen population dynamics. *Ecology* **73:**507–514.

Grosholz, E. D. (1993). The influence of habitat heterogeneity on host-pathogen population dynamics. *Oecologia* **96:**347–353.

Hall, D. W., and Anthony, D. W. (1971). Pathology of a mosquito iridescent virus (MIV) infecting *Aedes taeniorhychus*. *J. Invertebr. Pathol.* **18:**61–69.

Hanson, L. A., Petrie-Hanson, L., Means, K. O., Chinchar, V. G., and Rudis, M. (2001). Persistence of largemouth bass virus infection in a northern Mississippi reservoir following a die-off. *J. Aquat Anim. Health* **13:**27–34.

Hawke, N. A., Yoder, J. A., Haire, R. N., Mueller, M. G., Litman, R. T., Miracle, A. L., Stuge, T., Miller, N., and Litman, G. W. (2001). Extraordinary variation in a diversified family of immune-type receptor genes. *Proc. Natl. Acad. Sci. USA* **98:**13832–13837.

He, J. G., Wang, S. P., Zeng, K., Huang, Z. J., and Chan, S. M. (2000). Systemic disease caused by an iridovirus-like agent in cultured mandarinfish, *Siniperca chuatsi* (Basilewsky), in China. *J. Fish Dis.* **23:**219–222.

He, J. G., Deng, M., Weng, S. P., Li, Z., Zhou, S. Y., Long, Q. X., Wang, X. Z., and Chan, S. M. (2001). Complete genome analysis of the mandarin fish infectious spleen and kidney necrosis iridovirus. *Virology* **291:**126–139.

He, J. G., Lu, L., Deng, M., He, H. H., Weng, S. P., Wang, X. H., Zhou, S. Y., Long, Q. X., Wang, X. Z., and Chan, S. M. (2002). Sequence analysis of the complete genome of an iridovirus isolated from the tiger frog. *Virology* **292:**185–197.

Heath, C. M., Windsor, M., and Wileman, T. (2001). Aggresomes resemble sites specialized for virus assembly. *J. Cell Biol.* **153:**449–455.

Hedrick, R. P., and McDowell, T. S. (1995). Properties of iridoviruses from ornamental fish. *Vet. Res.* **26:**423–427.

Hedrick, R. P., Groff, J. M., McDowell, T. S., and Wingfield, W. H. (1990). An iridovirus infection of the integument of the white sturgeon *Acipenser transmontanus*. *Dis. Aquat. Org.* **8**:39–44.

Hedrick, R. P., McDowell, T. S., Ahne, W., Torhy, C., and de Kinkelin, P. (1992a). Properties of three iridovirus-like agents associated with systemic infections of fish. *Dis. Aquat. Org.* **13**:203–209.

Hedrick, R. P., McDowell, T. S., Groff, J. M., Yun, S., and Wingfield, W. H. (1992b). Isolation and properties of an iridovirus-like agent from white sturgeon *Acipenser transmontanus*. *Dis. Aquat. Org.* **12**:75–81.

Hemsley, A. R., Collinson, M. E., Kovach, W. L., Vincent, B., and Williams, T. (1994). The role of self-assembly in biological systems: Evidence from iridescent colloidal sporopollenin in *Selaginella* megaspore walls. *Phil. Trans. Roy. Soc. Lond. B* **345**:163–173.

Henderson, C. W., Johnson, C. L., Lodhi, S. A., and Bilimoria, S. L. (2001). Replication of *Chilo* iridescent virus in the cotton boll weevil, *Anthonomus grandis*, and development of an infectivity assay. *Arch. Virol.* **146**:767–775.

Hernández, O., Maldonado, G., and Williams, T. (2000). An epizootic of patent iridescent virus disease in multiple species of blackflies in Chiapas, Mexico. *Med. Vet. Entomol.* **14**:458–462.

Herniou, E. A., Luque, T., Chen, X., Vlak, J. M., Winstanley, D., Cory, J. S., and O' Reilly, D. R. (2001). Use of whole genome sequence data to infer baculovirus phylogeny. *J. Virol.* **75**:8117–8126.

Hogan, R. J., Stuge, T. B., Clem, L. W., Miller, N. W., and Chinchar, V. G. (1996). Antiviral cytotoxic cells in the channel catfish (*Ictalurus punctatus*). *Dev. Comp. Immunol.* **20**:115–127.

Hu, G. B., Cong, R. S., Fan, T. J., and Mei, X. G. (2004). Induction of apoptosis in a flounder gill cell line by lymphocystis disease virus infection. *J. Fish Dis.* **27**:657–662.

Huang, C., Zhang, X., Gin, K. Y. H., and Qin, Q. W. (2004). *In situ* hybridization of a marine fish virus, Singapore grouper iridovirus with a nucleic acid probe of major capsid protein. *J. Virol. Meth.* **117**:123–128.

Hughes, D. S., Possee, R. D., and King, L. A. (1997). Detection of transcriptional factors in insect cells harbouring a persistent Mamestra brassicae nuclear polyhedrosis virus infection. *J. Gen. Virol.* **78**:1801–1805.

Hunter, W. B., and Lapointe, S. L. (2003). Iridovirus infection of cell cultures from the *Diaprepes* root weevil *Diaprepes abbreviatus*. *J. Insect Sci.* **3**:37.

Hunter, W. B., Patte, C. P., Sinisterra, X. H., Achor, D. S., Funk, C. J., and Polston, J. E. (2001a). Discovering new insect viruses: Whitefly iridovirus (Homoptera: Aleyrodidae: *Bemisia tabaci*). *J. Invertebr. Pathol.* **78**:220–225.

Hunter, W. B., Sinisterra, X. H., McKenzie, C. L., and Shatters, R. G., Jr. (2001b). Iridovirus infection and vertical transmission in citrus aphids. *Proc. Florida State Hort. Soc.* **114**:70–72.

Hyatt, A. D., Gould, A. R., Zupanovic, Z., Cunningham, A. A., Hengstberger, S., Whittington, R. J., Kattenbelt, J., and Coupar, B. E. (2000). Comparative studies of piscine and amphibian iridoviruses. *Arch. Virol.* **145**:301–331.

Hyatt, A. D., Williamson, M., Coupar, B. E., Middleton, D., Hengstberger, S. G., Gould, A. R., Selleck, P., Wise, T. G., Kattenbelt, J., Cunningham, A. A., and Lee, J. (2002). First identification of a ranavirus from green pythons (*Chondropython viridis*). *J. Wildlife Dis.* **38**:239–252.

Imajoh, M., Sugiura, H., and Oshima, S. (2004). Morphological changes contribute to apoptotic cell death and are affected by caspase-3 and caspase-6 inhibitors during red sea bream iridovirus permissive replication. *Virology* **322**:220–230.

Iyer, L. M., Aravind, L., and Koonin, E. V. (2001). Common origin of four diverse families of large eukaryotic DNA viruses. *J. Virol.* **75:**11720–11734.

Jakob, N. J., and Darai, G. (2002). Molecular anatomy of *Chilo* iridescent virus genome and the evolution of viral genes. *Virus Genes* **25:**299–316.

Jakob, N. J., Muller, K., Bahr, U., and Darai, G. (2001). Analysis of the first complete DNA sequence of an invertebrate iridovirus: Coding strategy of the genome of *Chilo* iridescent virus. *Virology* **286:**182–196.

Jakob, N. J., Kleespies, R. G., Tidona, C. A., Muller, K., Gelderblom, H. R., and Darai, G. (2002). Comparative analysis of the genome and host range characteristics of two insect iridoviruses: Chilo iridescent virus and a cricket iridovirus isolate. *J. Gen. Virol.* **83:**463–470.

Jancovich, J. K., Davidson, E. W., Morado, J. F., Jacobs, B. L., and Collins, J. P. (1997). Isolation of a lethal virus from the endangered tiger salamander *Ambystoma tigrinum stebbinsi. Dis. Aquat. Org.* **31:**161–167.

Jancovich, J. K., Davidson, E. W., Seiler, A., Jacobs, B. L., and Collins, J. P. (2001). Transmission of the *Ambystoma tigrinum* virus to alternative hosts. *Dis. Aquat. Org.* **46:**159–163.

Jancovich, J. K., Mao, J., Chinchar, V. G., Wyatt, C., Case, S. T., Kumar, S., Valente, G., Subramanian, S., Davidson, E. W., Collins, J. P., and Jacobs, B. L. (2003). Genomic sequence of a ranavirus (family Iridoviridae) associated with salamander mortalities in North America. *Virology* **316:**90–103.

Jancovich, J. K., Davidson, E. W., Parameswaran, N., Mao, J., Chinchar, V. G., Collins, J. P., Jacobs, B. L., and Storfer, A. (2005). Evidence for emergence of an amphibian iridoviral disease because of human-enhanced spread. *Mol. Ecol.* **14:**213–224.

Jeong, J. B., Park, K. H., Kim, K. Y., Hong, S., Kim, K. H., Chung, J. K., Komisar, J. L., and Jeong, H. D. (2004). Multiplex PCR for the diagnosis of red sea bream iridoviruses isolated in Korea. *Aquaculture* **235:**139–152.

Johnston, J. B., and McFadden, G. (2003). Poxvirus immunomodulatory strategies: Current perspectives. *J. Virology* **77:**6093–6100.

Jung, S. J., and Oh, M. J. (2000). Iridovirus-like infection associated with high mortalities of striped beakperch, *Oplegnathus fasciatus* (Temminck and Schlegel), in southern coastal areas of the Korean peninsula. *J. Fish Dis.* **23:**223–226.

Jung, S., Miyazaki, T., Miyata, M., Danayadol, Y., and Tanaka, S. (1997). Pathogenicity of iridovirus from Japan and Thailand for the red sea bream *Pagrus major* in Japan, and histopathology of experimentally infected fish. *Fish. Sci.* **63:**735–740.

Just, F. T., and Essbauer, S. S. (2001). Characterization of an iridescent virus isolated from *Gryllus bimaculatus* (Orthoptera: Gryllidae). *J. Invertebr. Pathol.* **77:**51–61.

Just, F., Essbauer, S., Ahne, W., and Blahak, S. (2001). Occurrence of an invertebrate iridescent-like virus (Iridoviridae) in reptiles. *J. Vet. Med. B* **48:**685–694.

Kaur, K., Rohozinski, J., and Goorha, R. (1995). Identification and characterization of the frog virus 3 DNA methyltransferase. *J. Gen. Virol.* **76:**1937–1943.

Kawagishi-Kobayashi, M., Siverman, J. B., Ung, T. L., and Dever, T. E. (1997). Regulation of the protein kinase PKR by the vaccinia virus pseudosubstrate inhibitor K3L is dependent on residues conserved between the K3L protein and the PKR substrate eIF2 alpha. *Mol. Cell. Biol.* **17:**4146–4158.

Kelly, D. C. (1985). Insect iridescent viruses. *Curr. Topics Microbiol. Immunol.* **116:**23–35.

Kiesecker, J. M. (2002). Synergism between trematode infection and pesticide exposure: A link to amphibian limb deformities in nature? *Proc. Natl. Acad. Sci. USA* **99:**9900–9904.

Kiesecker, J. M., Blaustein, A. R., and Belden, L. K. (2001). Complex causes of amphibian population declines. *Nature* **410**:681–684.

Kim, Y. J., Jung, S. J., Choi, T. J., Kim, H. R., Rajendran, K. V., and Oh, M. J. (2002). PCR amplification and sequence analysis of irido-like virus infecting fish in Korea. *J. Fish. Dis.* **25**:121–124.

Kim, M., Krogan, N. J., Vasiljeva, L., Rando, O. J., Nede, A. E., Greenblatt, J. F., and Buratowski, S. (2004). The yeast Rat1 exonuclease promotes transcriptional termination by RNA polymerase II. *Nature* **432**:517–522.

Kinard, G. R., Barnett, O. W., and Carner, G. R. (1995). Characterization of an iridescent virus isolated from the velvetbean caterpillar, *Anticarsia gemmatalis*. *J. Invertebr. Pathol.* **66**:258–263.

Kleespies, R. G., Tidona, C. A., and Darai, G. (1999). Characterization of a new iridovirus isolated from crickets and investigations on the host range. *J. Invertebr. Pathol.* **73**:84–90.

Knopf, C. W. (1998). Evolution of viral DNA-dependent DNA polymerases. *Virus Genes* **16**:47–58.

Koella, J. C., and Boëte, C. (2002). A genetic correlation between age at pupation and melanization immune response of the yellow fever mosquito *Aedes aegypti*. *Evolution* **56**:1074–1079.

Kumar, S., Tamura, K., Jakobsen, I. B., and Nei, M. (2001). MEGA 2: Molecular evolution genetics analysis software. *Bioinformat.* **17**:1244–1245.

Kurita, J., Nakajima, K., Hirono, I., and Aoki, T. (1998). Polymerase chain reaction (PCR) amplification of DNA of red sea bream iridovirus (RSIV). *Fish Pathol.* **33**:17–23.

La Scola, B., Audic, S., Robert, C., Jungang, L., de Lamballerie, X., Drancourt, M., Birtles, R., Claverie, J. M., and Raoult, D. (2003). A giant virus in amoebae. *Science* **299**:2033.

Langdon, J. S., Humphrey, J. D., Williams, L. M., Hyatt, A. D., and Westbury, H. (1986). First virus isolation from Australian fish: An iridovirus-like pathogen from redfin perch, *Perca fluviatilis* L. *J. Fish Dis.* **9**:263–268.

Langland, J. O., and Jacobs, B. L. (2002). The role of the PKR-inhibitory genes E3L and K3L, in determining vaccinia virus host range. *Virology* **299**:133–141.

LaPatra, S. E., Groff, J. M., Jones, G. R., Munn, B., Patterson, T. L., Holt, R. A., Hauck, A. K., and Hedrick, R. P. (1994). Occurrence of white sturgeon iridovirus infections among cultured white sturgeon in the Pacific Northwest. *Aquaculture* **126**:201–210.

LaPatra, S. E., Ireland, S., Groff, J. M., Clemens, K., and Siple, J. (1999). Adaptive disease management strategies for the endangered population of Kootenai River white sturgeon *Acipenser transmontanus*. *Fisheries* **24**:6–13.

Leibovitz, L., and Koulish, S. (1989). A viral disease of the ivory barnacle, *Balanus eburneus*, Gould (Crustacea, Cirripedia). *Biol. Bull. Mar. Biol. Lab. Woods Hole.* **176**:301–307.

Lembo, D., Donalisio, M., Hofer, A., Cornaglia, M., Brune, W., Koszinowski, U., Thelander, L., and Landolfo, S. (2004). The ribonucleotide reductase R1 homolog of murine cytomegalovirus is not a functional enzyme subunit but is required for pathogenesis. *J. Virol.* **78**:4278–4288.

Lightner, D. V., and Redman, R. M. (1993). A putative iridovirus from the penaeid shrimp *Protrachypene precipua* Burkenroad (Crustacea: Decapoda). *J. Invertebr. Pathol.* **62**:107–109.

Linley, J. R., and Nielsen, H. T. (1968a). Transmission of a mosquito iridescent virus in *Aedes taeniorhynchus*. I. Laboratory experiments. *J. Invertebr. Pathol.* **12**:7–16.

Linley, J. R., and Nielsen, H. T. (1968b). Transmission of a mosquito iridescent virus in *Aedes taeniorhynchus*. II. Experiments related to transmission in nature. *J. Invertebr. Pathol.* **12**:17–24.

Long, S. L., Wilson, M., Bengten, E., Bryan, L., Clem, L. W., Miller, N. W., and Chinchar, V. G. (2004a). Identification of a cDNA encoding channel catfish interferon. *Dev. Comp. Immunol.* **28**:97–111.

Long, S. L., Wilson, M., Bengten, E., Clem, L. W., Miller, N. W., and Chinchar, V. G. (2004b). Identification and characterization of a FasL-like protein and cDNAs encoding the channel catfish death-inducing signaling complex. *Immunogenetics* **56**:518–530.

López, M., Rojas, J. C., Vandame, R., and Williams, T. (2002). Parasitoid-mediated transmission of an iridescent virus. *J. Invertebr. Pathol.* **80**:160–170.

MacConnell, E., Hedrick, R. P., Hudson, C., and Speer, C. A. (2001). Identification of an iridovirus in cultured pallid (*Scaphirhynchus albus*) and shovelnose sturgeon (*S. platorynchus*). *Am. Fish. Soc. Fish Health Newsletter* **29**:1–3.

Madeley, C. R., Smail, D. A., and Egglestone, S. I. (1978). Observations on the fine structure of lymphocystis virus from European flounders and plaice. *J. Gen. Virol.* **40**:421–431.

Manyakov, V. F. (1977). Fine structure of the iridescent virus type 1 capsid. *J. Gen. Virol.* **36**:73–79.

Mao, J. H., Hedrick, R. P., and Chinchar, V. G. (1997). Molecular characterization, sequence analysis, and taxonomic position of newly isolated fish iridoviruses. *Virology* **229**:212–220.

Mao, J., Green, D. E., Fellers, G., and Chinchar, V. G. (1999a). Molecular characterization of iridoviruses isolated from sympatric amphibians and fish. *Virus Res.* **63**:45–62.

Mao, J., Wang, J., Chinchar, G. D., and Chinchar, V. G. (1999b). Molecular characterization of a ranavirus isolated from largemouth bass, *Micropterus salmoides*. *Dis. Aquat. Org.* **37**:107–114.

Marina, C. F., Arredondo-Jiménez, J., Castillo, A., and Williams, T. (1999). Sublethal effects of iridovirus disease in a mosquito. *Oecologia* **119**:383–388.

Marina, C. F., Feliciano, J. M., Valle, J., and Williams, T. (2000). Effect of temperature, pH, ion concentration and chloroform treatment on the stability of *Invertebrate iridescent virus 6*. *J. Invertebr. Pathol.* **75**:91–94.

Marina, C. F., Arredondo-Jiménez, J. I., Ibarra, J. E., Fernández-Salas, I., and Williams, T. (2003a). Effects of an optical brightener and an abrasive on iridescent virus infection and development of *Aedes aegypti*. *Entomol. Exp. Appl.* **109**:155–161.

Marina, C. F., Ibarra, J. E., Arredondo-Jiménez, J. I., Fernández-Salas, I., Liedo, P., and Williams, T. (2003b). Adverse effects of covert iridovirus infection on life history and demographic parameters of *Aedes aegypti*. *Entomol. Exp. Appl.* **106**:53–61.

Marina, C. F., Ibarra, J. E., Arredondo-Jiménez, J. I., Fernández-Salas, I., Valle, J., and Williams, T. (2003c). Sublethal iridovirus disease of the mosquito *Aedes aegypti* is due to viral replication not cytotoxicity. *Med. Vet. Entomol.* **17**:187–194.

Marina, C. F., Fernández-Salas, I., Ibarra, J. E., Arredondo-Jiménez, J. I., Valle, J., and Williams, T. (2005). Transmission dynamics of an iridescent virus in an experimental mosquito population: The role of host density. *Ecol. Entomol.* **30**:376–382.

Marschang, R. E., Becher, P., Posthaus, H., Wild, P., and Thiel, H. J. (1999). Isolation and characterization of an iridovirus from Hermans tortoises (*Testudo hermanni*). *Arch. Virol.* **144**:1909–1922.

Marsh, I. B., Whittington, R. J., O'Rouke, B., Hyatt, A. D., and Chisholm, O. (2002). Rapid identification of Australian, European, and American ranaviruses based on variation in major capsid protein gene sequences. *Mol. Cell. Probes* **16**:137–151.

Martínez, G., Christian, P., Marina, C. F., and Williams, T. (2003). Sensitivity of *Invertebrate iridescent virus 6* to organic solvents, detergents, enzymes and temperature treatment. *Virus Res.* **91**:249–254.

Matsuoka, S., Inouye, K., and Nakajima, K. (1996). Cultured fish species affected by red sea bream iridoviral disease from 1991 to 1995. *Fish. Pathol.* **31**:233–234.

McIntosh, A. H., and Kimura, M. (1974). Replication of the insect *Chilo* iridescent virus (CIV) in a poikilothermic vertebrate cell line. *Intervirology* **4**:257–267.

McLaughlin, R. E., Scott, H. A., and Bell, M. R. (1972). Infection of the boll weevil by *Chilo* iridescent virus. *J. Invertebr. Pathol.* **19**:285–290.

McMillan, N., and Kalmakoff, J. (1994). RNA transcript mapping of the *Wiseana* iridescent virus genome. *Virus Res.* **32**:343–352.

Means, J. C., Muro, I., and Clem, R. J. (2003). Silencing of the baculovirus *Op-iap3* gene by RNA interference reveals that it is required for prevention of apoptosis during *Orgyia pseudotsugata M* nucleopolyhedrovirus infection of Ld652Y cells. *J. Virol.* **77**:4481–4488.

Meininghaus, M., and Eick, D. (1999). Requirement of the C-terminal domain of RNA polymerase II for the transcriptional activation of the chromosomal c-fos and hsp70 genes. *FEBS Letts.* **446**:173–176.

Meister, G., and Tuschl, T. (2004). Mechanisms of gene silencing by double-stranded RNA. *Nature* **431**:343–349.

Mercer, E. H., and Day, M. F. (1965). The structure of *Sericesthis* iridescent virus and of its crystals. *Biochim. Biophys. Acta* **102**:590–599.

Miller, N. W., Wilson, M., Bengten, E., Stuge, T., Warr, G., and Clem, W. (1998). Functional and molecular characterization of teleost leukocytes. *Immunol. Rev.* **166**:187–197.

Mitsuhashi, J. (1967). Establishment of an insect cell strain persistently infected with an insect virus. *Nature* **215**:863–864.

Miyata, M., Matsuno, K., Jung, S. J., Danayadol, Y., and Miyazaki, T. (1997). Genetic similarity of iridoviruses from Japan and Thailand. *J. Fish Dis.* **20**:127–134.

Monini, M., and Ruggeri, F. M. (2002). Antigenic properties of the epizootic hematopoietic necrosis virus. *Virology* **297**:8–18.

Montanie, H., Bonami, J. R., and Comps, M. (1993). Irido-like virus infection in the crab *Macropipus depurator* L. (Crustacea, Decapoda). *J. Invertebr. Pathol.* **61**:320–322.

Moody, N. J. G., and Owens, L. (1994). Experimental demonstration of pathogenicity of a frog virus, bohle iridovirus, for a fish species, barramundi *Lates calcarifer*. *Dis. Aquat. Org.* **18**:95–102.

Mullens, B. A., Velten, R. K., and Federici, B. A. (1999). Iridescent virus infection in *Culicoides variipennis sonorensis* and interactions with the mermithid parasite *Heleidomermis magnapapula*. *J. Invertebr. Pathol.* **73**:231–233.

Muller, K., Tidona, C. A., and Darai, G. (1999). Identification of a gene cluster within the genome of *Chilo* iridescent virus encoding enzymes involved in viral DNA replication and processing. *Virus Genes* **18**:243–264.

Murali, S., Wu, M. F., Gou, I. C., Chen, S. C., Yang, H. W., and Chang, C. Y. (2002). Molecular characterization and pathogenicity of a grouper iridovirus (GIV) isolated from yellow grouper, *Epinephelus awoara* (Temminick and Schlegel). *J. Fish Dis.* **25**:91–100.

Murti, K. G., and Goorha, R. (1989). Synthesis of FV3 proteins occurs on intermediate filament-bound polyribosomes. *Biol. Cell* **65:**205–214.

Murti, K. G., and Goorha, R. (1990). Virus-cytoskeleton interaction during replication of frog virus 3. *In* "Molecular Biology of Iridoviruses" (G. Darai, ed.), pp. 137–162. Kluwer, Boston.

Murti, K. G., Goorha, R., and Klymkowsky, M. W. (1988). A functional role for intermediate filaments in the formation of FV3 assembly sites. *Virology* **162:**264–269.

Nakajima, K., and Maeno, Y. (1998). Pathogenicity of red sea bream iridovirus and other fish iridoviruses to red sea bream. *Fish Pathol.* **33:**143–144.

Nakajima, K., and Sorimachi, M. (1995). Production of monoclonal antibodies against red sea bream iridovirus. *Fish Pathol.* **30:**47–52.

Nakajima, K., Inouye, K., and Sorimachi, M. (1998). Viral diseases in cultured marine fish in Japan. *Fish Pathol.* **33:**181–188.

Nakajima, K., Maeno, Y., Honda, A., Yokoyama, K., Tooriyama, T., and Manabe, S. (1999). Effectiveness of a vaccine against red sea bream iridovirus disease in a field trial test. *Dis. Aquat. Org.* **36:**73–75.

Nalcacioglu, R., Marks, H., Vlak, J. M., Demirbag, Z., and van Oers, M. M. (2003). Promoter analysis of the *Chilo* iridescent virus DNA polymerase and major capsid protein genes. *Virology* **317:**321–329.

Nandhagopal, N., Simpson, A. A., Gurnon, J. R., Yan, X., Baker, T. S., Graves, M. V., Van Etten, J. L., and Rossmann, M. G. (2002). The structure and evolution of the major capsid protein of a large lipid-containing DNA virus. *Proc. Natl. Acad. Sci. USA* **99:**14758–14763.

Nash, P., Barrett, J., Cao, J. X., Hota-Mitchell, S., Lalani, A. S., Everett, H., Xu, X. M., Robichaud, J., Hnatiuk, S., Ainslie, C., Seet, B. T., and McFadden, G. (1999). Immunomodulation by viruses: The myxoma virus story. *Immunol. Rev.* **168:**103–120.

Neuman, B. W., Stein, D. A., Kroeker, A. D., Paulino, A. D., Moulton, H. M., Iversen, P. L., and Buchmeier, M. J. (2004). Antisense morpholino-oligomers directed against the 5′ end of the genome inhibit coronavirus proliferation and growth. *J. Virol.* **78:**5891–5899.

Office International des Epizooties(2000). Epizootic haematopoietic necrosis virus . *In* "Diagnostic Manual for Aquatic Animal Diseases," pp. 17–25. Office International des Epizooties, Paris, France.

Ohba, M. (1975). Studies on the parthogenesis of *Chilo* iridescent virus. 3. Multiplication of CIV in the silkworm *Bombyx mori* L. and field insects. *Sci. Bull. Fac. Agr. Kyushu Univ.* **30:**71–81.

Ohba, M., and Aizawa, K. (1979). Multiplication of *Chilo* iridescent virus in noninsect arthropods. *J. Invertebr. Pathol.* **33:**278–283.

Ohba, M., and Aizawa, K. (1982). Failure of *Chilo* iridescent virus to replicate in the frog *Rana limnocharis. Proc. Assoc. Plant Protec. Kyushu* **28:**164–166 (in Japanese).

Ohba, M., Kanda, K., and Aizawa, K. (1990). Cytotoxicity of *Chilo* iridescent virus to *Antheraea eucalypti* cultured cells. *Appl. Entomol. Zool.* **25:**528–531.

Oshima, S., Hata, J., Segawa, C., Hirasawa, N., and Yamashita, S. (1996). A method for direct DNA amplification of uncharacterized DNA viruses and for development of a viral polymerase chain reaction assay: Application to the red sea bream iridovirus. *Analyt. Biochem.* **242:**15–19.

Oshima, S. K., Hata, J. I., Hirasawa, N., Ohtaka, T., Hirona, I., Aoki, T., and Yamashita, S. (1998). Rapid diagnosis of red sea bream iridovirus infection using the polymerase chain reaction. *Dis. Aquat. Org.* **32:**87–90.

Paperna, I., Vilenkin, M., and Alves de Matos, A. P. (2001). Iridovirus infections in farm-reared tropical ornamental fish. *Dis. Aquat. Org.* **48:**17–25.

Pietrokovski, S. (1998). Identification of a virus intein and a possible variation in the protein-splicing reaction. *Curr. Biol.* **8:**R634–R635.

Plumb, J. A., Grizzle, J. M., Young, H. E., and Noyes, A. D. (1996). An iridovirus isolated from wild largemouth bass. *J. Aquat. Anim. Health* **8:**265–270.

Poinar, G. O., Jr., Hess, R., and Cole, A. (1980). Replication of an iridovirus in a nematode (Mermithidae), *Intervirology* **14:**316–320.

Poinar, G. O., Jr., Hess, R. T., and Stock, J. H. (1985). Occurrence of the isopod iridovirus in European *Armadillidium* and *Porcellio* (Crustacea, Isopoda). *Bijdragen Dierkunde* **55:**280–282.

Popelkova, Y. (1982). Coelomomyces from *Aedes cinereus* and a mosquito iridescent virus of *Aedes cantans* in Sweeden. *J. Invertebr. Pathol.* **40:**148–149.

Pozet, F., Moussa, A., Torhy, C., and de Kinkelin, P. (1992). Isolation and preliminary characterization of a pathogenic icosahedral deoxyribovirus from the catfish *Ictalurus melas*. *Dis Aquat. Org.* **14:**35–42.

Qin, Q. W., Shi, C., Gin, K. Y. H., and Lam, T. J. (2002). Antigenic characterization of a marine fish iridovirus from grouper *Epinephelus* spp. *J. Virol. Meth.* **106:**89–96.

Quiniou, S., Bigler, S., Clem, L. W., and Bly, J. E. (1998). Effects of water temperature on mucous cell distribution in channel catfish epidermis: A factor in winter saprolegniasis. *Fish Shellfish Immunol.* **8:**1–11.

Raghow, R., and Granoff, A. (1979). Macromolecular synthesis in cells infected with frog virus 3. X. Inhibition of cell protein synthesis by heat-inactivated frog virus 3. *Virology* **98:**319–327.

Raghow, R., and Granoff, A. (1980). Macromolecular synthesis in cells infected by frog virus 3: XIV. Characterization of the methylated nucleotide sequences in viral messenger RNAs. *Virology* **107:**283–294.

Raghow, R., and Granoff, A. (1983). Cell-free translation of FV3 mRNA: Initiation factors from infected cells discriminate between early and late viral mRNAs. *J. Biol. Chem.* **258:**571–578.

Reyes, A., Christian, P., Valle, J., and Williams, T. (2004). Persistence of *Invertebrate iridescent virus 6* in soil. *BioContr.* **49:**433–440.

Risco, C., Rodriguez, J. R., Lopez-Iglesias, C., Carrascosa, J. L., Esteban, M., and Rodriguez, D. (2002). Endoplasmic reticulum-Golgi intermediate compartment membranes and vimentin filaments participate in vaccinia virus assembly. *J. Virol.* **76:**1839–1855.

Rohozinski, J., and Goorha, R. (1992). A frog virus 3 protein codes for a protein containing the motif characteristic of the INT family of integrases. *Virology* **186:**693–700.

Rokas, A., Williams, B. L., King, N., and Carroll, S. B. (2003). Genome-scale approaches to resolving incongruence in molecular phylogenies. *Nature* **425:**798–804.

Rondelaud, D., and Barthe, D. (1992). Epidemiological observations on iridovirosis of *Lymnaea truncatula*, host mollusca of *Fasciola hepatica*. *C. R. Acad. Sci. Paris Ser. D.* **314:**609–612.

Rothman, L. D., and Myers, J. H. (1996). Debilitating effects of viral diseases on host Lepidoptera. *J. Invertebr. Pathol.* **67:**1–10.

Rouiller, I., Brookes, S. M., Hyatt, A. D., Windsor, M., and Wileman, T. (1998). African swine fever virus is wrapped by the endoplasmic reticulum. *J. Virol.* **72:**2373–2387.

Ruellan, L. (1992). Contribution à l'étude du tissu hémocytaire et des cellules circulantes chez *Lymnaea truncatula* Müller (Mollusque Gastéropode pulmoné). Impact du parasitisme et d'une iridovirose. *Bull. Soc. Zool. Fr.* **117:**116–117.

Rungger, D., Rastelli, M., Braendle, E., and Malsberger, R. G. (1971). A virus-like particle associated with lesions in the muscles of *Octopus vulgaris*. *J. Invertebr. Pathol.* **117:**72–80.

Russell, P. H. (1974). Lymphocystis in wild plaice (*Pleuronectes platessa* L.) and flounder (*Platichtys flesus* L.) in British coastal waters. A histopathological and serological study. *J. Fish Biol.* **6:**771–778.

Russell, J. D., Walker, G., and Woollen, R. (2000). Observations on two infectious agents found within rootlets of the parasitic barnacle, *Sacculina carcini*. *J. Mar. Biol. Assoc. UK* **80:**373–374.

Schmelz, M., Sodeik, B., Ericsson, M., Wolffe, E. J., Shida, H., Hiller, G., and Griffiths, G. (1994). Assembly of vaccinia virus: The second wrapping cisterna is derived from the trans Golgi network. *J. Virol.* **68:**130–147.

Schultz, G. A., Garthwaite, R. L., and Sassaman, C. (1982). A new family placement for *Mauritaniscus littorinus* (Miller) N. Comb. from the west coast of North America with ecological notes (Crustacea: Isopoda: Oniscoidea: Bathytropidae). *Wasmann J. Biol.* **40:**77–89.

Schwartz, A., and Koella, J. C. (2004). The cost of immunity in the yellow fever mosquito *Aedes aegypti* depends on immune activation. *J. Evol. Biol.* **17:**834–840.

Seshagiri, S., and Miller, L. K. (1997). Baculovirus inhibitors of apoptosis (IAPs) block activation of Sf-caspase-1. *Proc. Natl. Acad. Sci. USA* **94:**13606–13611.

Shen, L., Stuge, T. B., Zhou, H., Khayat, M., Barker, K. S., Quiniou, S. M. A., Wilson, M., Bengten, E., Chinchar, V. G., Clem, L. W., and Miller, N. W. (2002). Channel catfish cytotoxic cells: A mini-review. *Dev. Comp. Immunol.* **26:**141–149.

Shi, C., Wei, Q., Gin, K. Y. H., and Lam, T. J. (2003). Production and characterization of monoclonal antibodies to a grouper iridovirus. *J. Virol. Meth.* **107:**147–154.

Shi, C. Y., Wang, Y. G., Yang, S. L., Huang, J., and Wang, Q. Y. (2004). The first report of an iridovirus like agent infection in farmed turbot *Scophthalmus maximus* in China. *Aquaculture* **236:**11–25.

Siwicki, A. K., Pozet, F., Morand, M., Volatier, C., and Terech-Majewska, E. (1999). Effects of iridovirus like agent on the cell mediated immunity in sheatfish (*Silurus glanis*) - an *in vitro* study. *Virus Res.* **63:**115–119.

Smith, K. M., Hill, G. J., and Rivers, C. F. (1961). Studies on the cross-inoculation of the *Tipula* iridescent virus. *Virology* **13:**233–241.

Sodeik, B., Griffiths, G., Ericsson, M., Moss, B., and Doms, R. W. (1994). Assembly of vaccinia virus: Effects of rifampin on the intracellular distribution of viral protein p65. *J. Virol.* **68:**1103–1114.

Somamoto, T., Nakanishi, T., and Okamoto, N. (2002). Role of cell-mediated cytotoxicity in protecting fish from virus infections. *Virology* **297:**120–127.

Song, W. J., Qin, Q. W., Qiu, J., Huang, C. H., Wang, F., and Hew, C. L. (2004). Functional genomics analysis of Singapore grouper iridovirus: Complete sequence determination and proteomic analysis. *J. Virol.* **78:**12576–12590.

Stadelbacher, E. A., Adams, J. R., Faust, R. M., and Tompkins, G. J. (1978). An iridescent virus of the bollworm *Heliothis zea* (Lepidoptera: Noctuidae). *J. Invertebr. Pathol.* **32:**71–76.

Stasiak, K., Demattei, M. V., Federici, B. A., and Bigot, Y. (2000). Phylogenetic position of the *Diadromus pulchellus* ascovirus DNA polymerase among viruses with large double-stranded DNA genomes. *J. Gen. Virol.* **81:**3059–3072.

Stasiak, K., Renault, S., Demattei, M. V., Bigot, Y., and Federici, B. A. (2003). Evidence for the evolution of ascoviruses from iridoviruses. *J. Gen. Virol.* **84:**2999–3009.

Stohwasser, R., Raab, K., Schnitzler, P., Janssen, W., and Darai, G. (1993). Identification of the gene encoding the major capsid protein of insect iridescent virus type 6 by polymerase chain reaction. *J. Gen. Virol.* **74:**873–879.

Stoltz, D. B. (1971). The structure of icosahedral cytoplasmic deoxyriboviruses. *J. Ultrastruc. Res.* **37:**219–239.

Stoltz, D. B. (1973). The structure of icosahedral cytoplasmic deoxyriboviruses II. An alternative model. *J. Ultrastruc. Res.* **43:**58–74.

Stoltz, D. B., and Summers, M. D. (1971). Pathway of infection of mosquito iridescent virus. I. Preliminary observations on the fate of ingested virus. *J. Virol.* **8:**900–909.

Sudthongkong, C., Miyata, M., and Miyazaki, T. (2002). Viral DNA sequences of genes encoding the ATPase and the major capsid protein of tropical iridovirus isolates which are pathogenic to fishes in Japan, South China Sea, and Southeast Asian countries. *Arch. Virol.* **147:**2089–2109.

Tajbakhsh, S., Kiss, G., Lee, P. E., and Seligy, V. L. (1990). Semipermissive replication of *Tipula* iridescent virus in *Aedes albopictus* C6/36 cells. *Virology* **174:**264–275.

Tamai, T., Tsujimura, K., Shirahata, S., Oda, H., Noguchi, T., Kusuda, R., Sato, N., Kimura, S., Katakura, Y., and Murakami, H. (1997). Development of DNA diagnostic methods for the detection of new fish iridoviral diseases. *Cytotechnol.* **23:**211–220.

Tan, W. G. H., Barkman, T. J., Chinchar, V. G., and Essani, K. (2004). Comparative genomic analysis of frog virus 3, type species of the genus *Ranavirus* (family Iridoviridae). *Virology* **323:**70–84.

Tanaka, H., Sato, K., Saito, Y., Yamashita, T., Agoh, M., Okunishi, J., Tachikawa, E., and Suzuki, K. (2003). Insect diapause-specific peptide from the leaf beetle has consensus with a putative iridovirus peptide. *Peptides* **24:**1327–1333.

Taylor, S. K., Williams, E. S., and Mills, K. W. (1999). Effects of malathion on disease susceptibility in Woodhouses's toads. *J. Wildlife Res.* **35:**536–541.

Telford, S. R., Jr, and Jacobson, E. R. (1993). Lizard erythrocytic virus in east African chameleons. *J. Wildlife Dis.* **29:**57–63.

Tesh, R. B., and Andreadis, T. G. (1992). Infectivity and pathogenesis of iridescent virus type 22 in various insect hosts. *Arch. Virol.* **126:**57–65.

Thompson, J. P., Granoff, A., and Willis, D. B. (1986). Trans-activation of a methylated adenovirus promoter by a frog virus 3 protein. *Proc. Natl. Acad. Sci. USA* **83:**7688–7692.

Thompson, J. P., Granoff, A., and Willis, D. B. (1988). Methylation of the promoter for an immediate-early frog virus 3 gene does not inhibit transcription. *J. Virol.* **62:**4680–4685.

Tidona, C. A., and Darai, G. (1997). The complete DNA sequence of lymphocystis disease virus. *Virology* **230:**207–216.

Tidona, C. A., and Darai, G. (2000). Iridovirus homologues of cellular genes - implications for the molecular evolution of large DNA viruses. *Virus Genes* **21:**77–81.

Tidona, C. A., Schnitzler, P., Kehm, R., and Darai, G. (1998). Is the major capsid protein of iridoviruses a suitable target for the study of viral evolution? *Virus Genes* **16:**59–66.

Ting, J. W., Wu, M. F., Tsai, C. T., Lin, C. C., Guo, I. C., and Chang, C. Y. (2004). Identification and characterization of a novel gene of grouper iridovirus encoding a purine nucleoside phosphorylase. *J. Gen. Virol.* **85:**2883–2892.

Tonka, T., and Weiser, J. (2000). Iridovirus infection in mayfly larvae. *J. Invertebr. Pathol.* **76:**229–231.

Tortorella, D., Gewurz, B. E., Furman, M. H., Schust, D. J., and Ploegh, H. L. (2000). Viral subversion of the immune system. *Annu. Rev. Immunol.* **18:**861–926.

Tweedel, K., and Granoff, A. (1968). Viruses and renal carcinoma of *Rana pipiens*. V. Effect of frog virus 3 on developing frog embryos and larvae. *J. Natl. Cancer Inst.* **40**:407–409.

Undeen, A. H., and Fukuda, T. (1994). Effects of host resistance and injury on the susceptibility of *Aedes taeniorhynchus* to mosquito iridescent virus. *J. Amer. Mosq. Contr. Assoc.* **10**:64–66.

Van Etten, J. L., Graves, M. V., Muller, D. G., Boland, W., and Delaroque, N. (2002). *Phycodnaviridae* - large DNA algal viruses. *Arch. Virol.* **147**:1479–1516.

Van Regenmortel, M. H. V. (2000). Introduction to the species concept in virus taxonomy. *In* "Virus Taxonomy, 7th Report of the International Committee on Virus Taxonomy" (M. H. V. van Regenmortel, C. M. Fauquet, D. H. L. Bishop, E. B. Carstens, M. K. Estes, S. M. Lemon, J. Maniloff, M. A. Mayo, D. J. McGeoch, C. R. Pringle, and R. B. Wickner, eds.), pp. 3–16. Academic Press, New York.

Wagner, G. W., and Paschke, J. D. (1977). A comparison of the DNA of R and T strains of mosquito iridescent virus. *Virology* **81**:298–308.

Wang, C. S., Shih, H. H., Ku, C. C., and Chen, S. N. (2003a). Studies on epizootic iridovirus infection among red sea bream, *Pagrus major* (Temminck and Schlegel), cultured in Taiwan. *J. Fish Dis.* **26**:127–133.

Wang, J. W., Deng, R. Q., Wang, X. Z., Huang, Y. S., Xing, K., Feng, J. H., He, J. G., and Long, Q. X. (2003b). Cladistic analysis of iridoviruses based on protein and DNA sequences. *Arch. Virol.* **148**:2181–2194.

Ward, V. K., and Kalmakoff, J. (1991). Invertebrate Iridoviridae. *In* "Viruses of Invertebrates" (E. Kurstak, ed.), pp. 197–226. Marcel Dekker, New York.

Watson, D., and Seligy, V. L. (1997). Characterization of iridovirus IV1 polypeptides: Mapping by surface labelling. *Res. Virol.* **148**:239–250.

Watson, L. R., Groff, J. M., and Hedrick, R. P. (1998a). Replication and pathogenesis of white sturgeon iridovirus (WSIV) in experimentally infected white sturgeon *Acipenser transmontanus* juveniles and sturgeon cell lines. *Dis. Aquat. Org.* **32**:173–184.

Watson, L. R., Milani, A., and Hedrick, R. P. (1998b). Effects of water temperature on experimentally-induced infections of juvenile white sturgeon (*Acipenser transmontanus*) with thc white sturgeon iridovirus (WSIV). *Aquaculture* **166**:213–228.

Webb, S. R., Paschke, J. D., Wagner, G. W., and Campbell, W. R. (1975). Bioassay of mosquito iridescent virus of *Aedes taeniorhynchus* in cell cultures of *Aedes aegypti*. *J. Invertebr. Pathol.* **26**:205–212.

Webby, R., and Kalmakoff, J. (1998). Sequence comparison of the major capsid protein gene from 18 diverse iridoviruses. *Arch. Virol.* **143**:1949–1966.

Webby, R. J., and Kalmakoff, J. (1999). Comparison of the major capsid protein genes, terminal redundancies, and DNA-DNA homologies of two New Zealand iridoviruses. *Virus Res.* **59**:179–189.

Weissenberg, R. (1965). Fifty years of research on the lymphocystis virus disease of fishes (1914–1964). *Ann. N. Y. Acad. Sci.* **126**:362–374.

Weng, S. P., Wang, Y. Q., He, J. G., Deng, M., Lu, L., Guan, H. J., Liu, Y. J., and Chan, S. M. (2002). Outbreaks of an iridovirus in red drum, *Sciaenops ocellata* (L.), cultured in southern China. *J. Fish Dis.* **25**:681–685.

Whittington, R. J., Kearns, C., and Speare, R. (1997). Detection of antibodies against iridoviruses in the serum of the amphibian *Bufo marinus*. *J. Virol. Meth.* **68**:105–108.

Whittington, R. J., Reddacliff, L. A., Marsh, I., Kearns, C., Zupanovic, Z., and Callinan, R. B. (1999). Further observations on the epidemiology and spread of epizootic haematopoietic necrosis virus (EHNV) in farmed rainbow trout *Oncorhynchus mykiss* in southeastern Australia and a recommended sampling strategy for surveillance. *Dis. Aquat. Org.* **35:**125–130.

Wijnhoven, H., and Berg, M. P. (1999). Some notes on the distribution and ecology of iridovirus (Iridovirus, Iridoviridae) in terrestrial isopods (Isopoda, Oniscidae). *Crustaceana* **72:**145–156.

Williams, T. (1993). Covert iridovirus infection of blackflies. *Proc. R. Soc. Lond. B* **251:**225–230.

Williams, T. (1994). Comparative studies of iridoviruses: Further support for a new classification. *Virus Res.* **33:**99–121.

Williams, T. (1995). Patterns of covert infection by invertebrate pathogens: Iridescent viruses of blackflies. *Mol. Ecol.* **4:**447–457.

Williams, T. (1996). The iridoviruses. *Adv. Virus Res.* **46:**347–412.

Williams, T. (1998). Invertebrate iridescent viruses. *In* "The Insect Viruses" (L. K. Miller and L. A. Ball, eds.), pp. 31–68. Plenum Press, New York.

Williams, T., and Cory, J. S. (1993). DNA restriction fragment polymorphism in iridovirus isolates from individual blackflies (Diptera: Simuliidae). *Med. Vet. Entomol.* **7:**199–201.

Williams, T., and Cory, J. S. (1994). Proposals for a new classification of iridescent viruses. *J. Gen. Virol.* **75:**1291–1301.

Willis, D. B., and Granoff, A. (1978). Macromolecular synthesis in cells infected by frog virus 3. IX. Two temporal classes of early viral RNA. *Virology* **86:**443–453.

Willis, D. B., and Granoff, A. (1980). Frog virus 3 DNA is heavily methylated at CpG sequences. *Virology* **107:**250–257.

Willis, D. B., and Granoff, A. (1985). Trans-activation of an immediate-early frog virus 3 promoter by a virion protein. *J. Virol.* **56:**495–501.

Willis, D. B., Goorha, R., Miles, M., and Granoff, A. (1977). Macromolecular synthesis in cells infected by frog virus 3: VII Transcriptional and post-transcriptional regulation of virus gene expression. *J. Virol.* **24:**326–342.

Willis, D. B., Goorha, R., and Granoff, A. (1979). Nongenetic reactivation of FV3 DNA. *Virology* **98:**476–479.

Willis, D. B., Goorha, R., and Granoff, A. (1984). DNA methyltransferase induced by frog virus 3. *J. Virol.* **49:**86–91.

Willis, D. B., Goorha, R., and Chinchar, V. G. (1985). Macromolecular synthesis in cells infected by frog virus 3. *Curr. Topics Microbiol. Immunol.* **116:**77–106.

Willis, D. B., Thompson, J. P., Essani, K., and Goorha, R. (1989). Transcripiton of methylated viral DNA by eukaryotic RNA polymerase II. *Cell Biophys.* **15:**97–111.

Willis, D. B., Essani, K., Goorha, R., Thompson, J. P., and Granoff, A. (1990a). Transcription of a methylated DNA virus. Nucleic Acid Methylation. *UCLA Symp. Mol. Cell. Biol.* **128:**139–151.

Willis, D. B., Thompson, J. P., and Beckman, W. (1990b). Transcription of frog virus 3. *In* "Molecular Biology of Iridoviruses" (G. Darai, ed.), pp. 173–186. Kluwer, Boston.

Wilson, M. R., Zhou, H., Bengten, E., Clem, L. W., Stuge, T. B., Warr, G. W., and Miller, N. W. (1998). T-cell receptors in channel catfish: Structure and expression of TCR α and β genes. *Mol. Immunol.* **35:**545–557.

Wolf, K., Bullock, G., Dunbar, C., and Quimby, M. (1968). Tadpole edema virus: Viscerotropic pathogen for anuran amphibians. *J. Infect. Dis.* **118:**253–262.

Woodard, D. B., and Chapman, H. C. (1968). Laboratory studies with the mosquito iridescent virus (MIV). *J. Invertebr. Pathol.* **11:**296–301.

Woodland, J. E., Noyes, A. D., and Grizzle, J. M. (2002). A survey to detect largemouth bass virus among fish from hatcheries in the southeastern USA. *Trans. Amer. Fish. Soc.* **131:**308–311.

Wrigley, N. G. (1970). An electron microscope study of the structure of *Tipula* iridescent virus. *J. Gen. Virol.* **6:**169–173.

Yan, X., Olson, N. H., Van Etten, J. L., Bergoin, M., Rossmann, M. G., and Baker, T. S. (2000). Structure and assembly of large lipid-containing dsDNA viruses. *Nature Struct. Mol. Biol.* **7:**101–103.

Yoder, J. A. (2004). Investigating the morphology, function, and genetics of cytotoxic cells in bony fish. *Comp. Biochem. Physiol. C* **138:**271–280.

Yoder, J. A., Litman, R. T., Mueller, M. G., Desai, S., Dobrinski, K. P., Montgomery, J. S., Buzzeo, M. P., Ota, T., Amemiya, C. T., Trede, N. K., Wei, S., Djeu, J. Y., Humphray, S., Jekosch, K., Hernandez-Prada, J. A., Ostrov, D. A., and Litman, G. W. (2004). Resolution of the novel immune-type receptor gene cluster in zebrafish. *Proc. Natl. Acad. Sci. USA* **101:**15706–15711.

Yu, Y. X., Bearzotti, M., Vende, P., Ahne, W., and Bremont, M. (1999). Partial mapping and sequencing of a fish iridovirus genome reveals genes homologous to the frog virus 3 p31, p40 and human eIF2 alpha. *Virus Res.* **63:**53–63.

Zemskov, E. A., Kang, W., and Maeda, S. (2000). Evidence for nucleic acid binding ability and nucleosome association of *Bombyx mori* nucleopolyhedrovirus BRO proteins. *J. Virol.* **74:**6784–6789.

Zhang, Q. Y., Xiao, F., Li, Z. Q., Gui, J. F., Mao, J., and Chinchar, V. G. (2001). Characterization of an iridovirus from the cultured pig frog (*Rana grylio*) with lethal syndrome. *Dis. Aquat. Org.* **48:**27–36.

Zhang, Q. Y., Xiao, F., Xie, J., Li, Z. Q., and Gui, J. F. (2004a). Complete genome sequence of lymphocystis disease virus isolated from China. *J. Virol.* **78:**6982–6994.

Zhang, X., Huang, C., and Qin, Q. (2004b). Antiviral properties of hemocyanin isolated from shrimp *Penaeus monodon*. *Antiviral Res.* **61:**93–99.

Zhao, K., and Cui, L. (2003). Molecular characterization of the major virion protein gene from the *Trichoplusia ni* ascovirus. *Virus Genes* **27:**93–102.

Zou, J., Secombes, C. J., Long, S., Miller, N., Clem, L. W., and Chinchar, V. G. (2003). Molecular identification and expression analysis of tumor necrosis factor in channel catfish (*Ictalurus punctatus*). *Dev. Comp. Immunol.* **27:**845–858.

Zupanovic, Z., Lopez, G., Hyatt, A., Shiell, B. J., and Robinson, A. J. (1998a). An improved enzyme linked immunosorbent assay for detection of anti-ranavirus antibodies in the serum of the giant toad (*Bufo marinus*). *Dev. Comp. Immunol.* **22:**573–585.

Zupanovic, Z., Musso, C., Lopez, C., Louriero, C. L., Hyatt, A. D., Hengstberger, S., and Robinson, A. J. (1998b). Isolation and characterization of iridoviruses from the giant toad *Bufo marinus* in Venezuela. *Dis. Aquat. Org.* **33:**1–9.

INDEX

FUJIYUKI *ET AL.*, FIG 1. Attacking behavior of the honeybees. The guard bees engulf a forager giant hornet, *Vespa mandarinia japonica*.

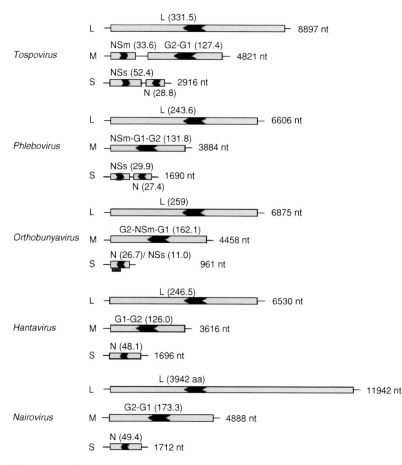

SNIPPE *ET AL.*, FIG 3. Comparison of viral genomes from representatives of all *Bunyaviridae* genera. Sizes of the RNA segments (in nucleotides) and encoded gene products (in kilodaltons, except for Nairovirus L) are indicated. Arrows reflect the orientation of the ORF relative to the viral genomic strand. For the tospoviruses and phleboviruses, the NSs gene is located in an ambisense arrangement with the N protein; whereas, for the orthobunyaviruses, the NSs gene is oriented in the same polarity, overlapping the N ORF.

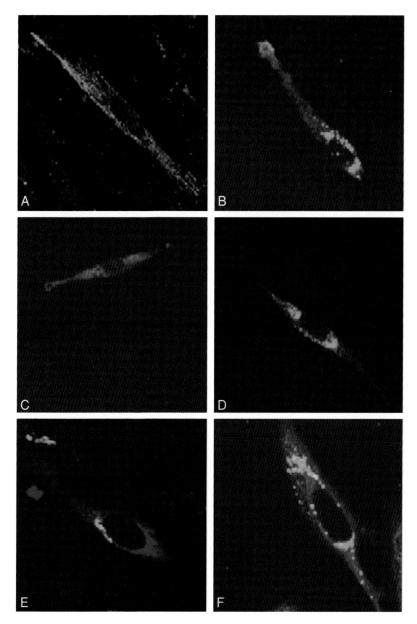

SNIPPE *ET AL.*, FIG 8. Time course analysis of wild-type TSWV N protein (wt N) and N-YFP expressed in BHK21 cells. Images were taken at 4 (A, C) and 24 (B, D, E, F) hours post transfection. Both N (A, B) and N-YFP (C, D) are observed dispersed thoughout the cytoplasm early in transfection. Upon longer incubation, both proteins are gradually observed to form clusters that are ultimately most abundantly localized near the nucleus. Panels E and F show N (E) and N-YFP (F) in combination with a marker for the trans-Golgi system (E: marker in green, N in red) or for the ERGIC (F: ERGIC in red, N-YFP in green).

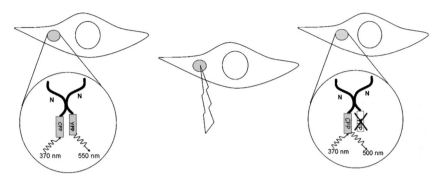

SNIPPE *ET AL.*, FIG 10. Principles of FRET and bleaching.

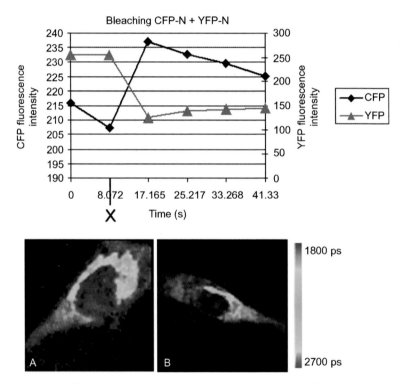

SNIPPE *ET AL.*, FIG 12. Acceptor photobleaching (top panel) and FLIM (lower panel). Top Panel: CFP and YFP fluorescence intensities (in arbitrary units) of a region in a cell, cotransfected with CFP-N and YFP-N, measured before and after YFP bleaching. The "X" indicates the time point at which the bleaching pulse was active. A decrease in YFP fluorescence caused by bleaching coincides with a significant CFP fluorescence increase, indicating the occurrence of FRET before photobleaching. Lower panel: Fluorescence lifetime of CFP in a cell transfected with (A): CFP-N, and (B): CFP-N and YFP-N. A shorter fluorescence lifetime of CFP(-N) is clearly observed in the presence of YFP-N (B), as indicated by the pseudocolor change into yellow. The legend for the pseudocolours representing CFP fluorescence lifetime is provided in the color scale on the right. The decrease in lifetime is a consequence of fluorescence energy transfer and thus, of interaction between the fusion proteins.

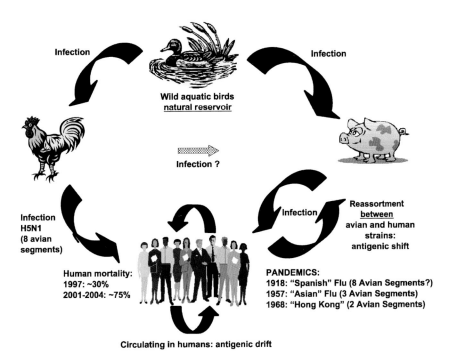

Infection

Infection

**Wild aquatic birds
natural reservoir**

Infection ?

**Infection
H5N1
(8 avian
segments)**

Infection

**Reassortment
between
avian and human
strains:
antigenic shift**

**Human mortality:
1997: ~30%
2001-2004: ~75%**

**PANDEMICS:
1918: "Spanish" Flu (8 Avian Segments?)
1957: "Asian" Flu (3 Avian Segments)
1968: "Hong Kong" (2 Avian Segments)**

Circulating in humans: antigenic drift

NOAH AND KRUG, FIG 2. The reservoir of influenza A viruses in wild aquatic birds and the routes of transmission of these viruses to humans. The infection of wild aquatic birds with most strains of influenza A virus is asymptomatic, so these viruses can be maintained and propagated in aquatic birds without significant mortality. The routes of transmission are described in the text.

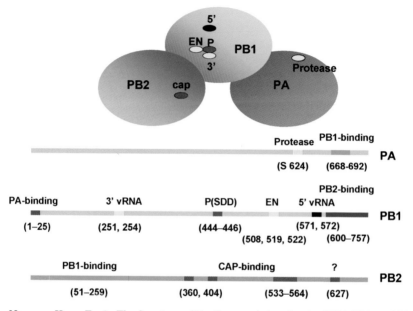

NOAH AND KRUG, FIG 3. The functions of the three protein subunits (PB1, PB2, and PA) of the influenza A virus polymerase have been mapped to specific regions of these proteins. The amino acid at position 627 in the PB2 protein has been implicated in the virulence of H5N1 avian viruses, but it has not been associated with known functions of the viral RNA polymerase. Two functions of the PB2 protein have been reported: the region (amino acids 51–259) that binds to PB1 and the cap-binding site (either amino acids 360 and 404 or the region from amino acids 533–564). EN, the active site for endonuclease cleavage. P(SDD), the polymerization active site which has the sequence SDD.

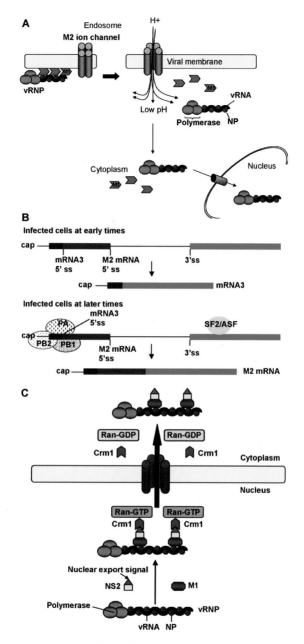

NOAH AND KRUG, FIG 4. (A) Role of the M2 ion channel protein in the uncoating of influenza A virus. By promoting the influx of H$^+$ ions into the interior of the virion, the M2 protein causes the disruption of protein–protein interactions, including that between the M1 protein and vRNPs. The dissociation of M1 from the vRNPs enables the vRNPs to be transported into the nucleus. (B) The viral polymerase and the cellular SF2/ASF splicing enhancer regulate the alternative splicing of M1 mRNA, thereby regulating M2 protein synthesis. See text for the description of the mechanism. Redrawn with permission from Shih and Krug (1996). (C) A M1-NS2 protein complex mediates the nuclear export of influenza vRNPs. The nuclear export signal (NES) interacts with Crm1, the protein that mediates RanGTP-dependent nuclear export.

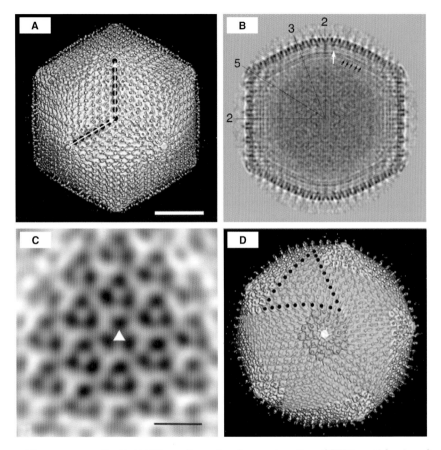

WILLIAMS *ET AL.*, FIG 5. (A) Three dimensional reconstruction of IIV-6 particle viewed down a threefold axis of symmetry by cryo-electron microscopy. Circles indicate the position of capsomers along the h and k lattice used to calculate the triangulation number (T). The proximal ends of surface fibers are visible, whereas the distal ends have become lost during the reconstruction process because of their flexibility (white bar = 50 nm applies to images A, B, and D). (B) Central section of the reconstruction density map viewed along twofold axis indicating twofold, threefold, and fivefold axes of symmetry. A lipid bilayer is seen beneath the capsid shell (black arrows) with numerous connections to an additional shell (white arrow) beneath the outer shell. (C) Close-up view of the trimers comprising each capsomer shown as a planar section through a reconstruction density map viewed along threefold axis (triangle) (black bar = 10 nm). (D) Facets of capsid with five trisymmetrons (highlighted in green) arranged around one pentasymmetron (in pink). The central capsomer of the pentasymmetron (uncolored) is pentavalent. The edge of each trisymmetron comprises ten capsomers (black dots). Images reproduced with permission from Yan *et al.* (2000).